Complex Systems and Their Applications

Eric Campos-Cantón • Guillermo Huerta-Cuellar •
Ernesto Zambrano-Serrano • Esteban Tlelo-Cuautle
Editors

Complex Systems and Their Applications

Fourth International Conference (EDIESCA 2023)

Springer

Editors
Eric Campos-Cantón
Instituto Potosino de Investigación Científica y Tecnológica (IPICYT)
San Luis Potosi, San Luis Potosí, Mexico

Guillermo Huerta-Cuellar
Centro Universitario de los Lagos, Universidad de Guadalajara
Lagos De Moreno, Jalisco, Mexico

Ernesto Zambrano-Serrano
Universidad Autónoma de Nuevo León
San Nicolas De Los Garza, Nuevo León, Mexico

Esteban Tlelo-Cuautle
Instituto Nacional de Astrofísica, Optica y Electrónica
San Andres Cholula, Puebla, Mexico

ISBN 978-3-031-51223-0 ISBN 978-3-031-51224-7 (eBook)
https://doi.org/10.1007/978-3-031-51224-7

This Springer imprint is published by the registered company Springer Nature Switzerland AG
The registered company address is: Gewerbestrasse 11, 6330 Cham, Switzerland

Paper in this product is recyclable.

Preface

This volume is intended for use by students, engineers, and researchers interested in complex systems and their applications. The reader is assumed to have a basic knowledge of complex systems and dynamical nonlinear systems. EDIESCA is the Spanish acronym of Dissemination and Research in the Study of Complex Systems and their Applications. The first edition EDIESCA-2019 was held at Instituto Potosino de Investigación Científica y Tecnológica (IPICyT, Mexico) in September 25–27, 2019, and the book of abstracts is available at https://sites.google.com/view/mesdia/sociedad-mexicana-de-sistemas-din%C3%A1micos-y-sus-aplicaciones/eventos/en-EDIESCA2021.

The second edition of EDIESCA-2021 was held virtually at Universidad de Guadalajara, México, Lagos campus, in November 17–19, 2021, and the book of chapters is available at https://books.google.com.mx/books?id=yaV0EAAAQBAJ&dq=info:6eY1HMa6BkwJ:scholar.google.com&lr=&source=gbs_navlinks_s.

The third edition of EDIESCA-2022 was held at Universidad Autónoma de Baja California (UABC), Ensenada campus, during September 27–30, 2022, and the presented works were published in *Special Issues of Journals*.

The fourth edition of EDIESCA-2023 was held during September 20–22, 2023, having as headquarters the facilities of the Facultad de Ingeniería Mecánica y Eléctrica, of the Universidad Autónoma de Nuevo León (UANL), San Nicolás de los Garza, México. The meeting encouraged the participation of researchers and students whose interests range from mathematical modeling to application of results in the area of complex systems, with an emphasis on chaos and dynamic systems. This book is the result of the presentations at EDIESCA-2023, integrating the work of students, research groups, and academic groups from related complex systems, as well as technological developments at national and international level. It is divided into four parts, covering the following topics:

Part I, Biomedical Advancements in Rehabilitation Through Complex Systems includes three chapters: Classification of Chaotic Dynamics Through Time-Frequency Representations and Machine Learning, Optimization of Echo State Neural Networks to Solve Classification Problems, and Deep Learning in the Expansion of the Urban Spot.

Part II, Biomedical Advancements in Rehabilitation Through Complex Systems, includes four chapters: Solving Inverse Kinematics Problem for Manipulator Robots Using Artificial Neural Network with Varied Dataset Formats, CRNN-Based Classification of EMG Signals for the Rehabilitation of the Human Arm, LMI-Based Design of a Robust Affine Control law for the Position Control of a Knee Exoskeleton Robot: Comparative Analysis of Stability Conditions, and Position Control of Robotic Systems via Linear Controllers with Application to a Lower Limb Rehabilitation Exoskeleton Robot: Design and Comparative Analysis.

Part III, Controlling Dynamical and Complex Systems, includes five chapters: Synchronization of memristive Hindmarsh-Rose neurons connected by memristive synapses, A Systematic Approach for Multi-switching Compound Synchronization of Non-identical Chaotic Systems Using Optimal Control, Limit Cycle Generation by Inducing the Controllable Hopf Bifurcation, Cascading Timers the New Method to Solve the Sequential Problems Including Timers in Automation Systems, and Mixed Sensitivity Control of Euler-Lagrange Models.

Part IV, Applications of Chaotic and Complex Systems, includes five chapters: A New 4-D Four-Scroll Hyperchaotic System with Multistability, Coexisting Attractors and Its Circuit Realization, Secure Communication System Based on Multi-stability: Evaluation Using the SCAMPER Method for Innovation Projects, An Image Compression and Encryption Approach with Convolutional Layers, Two-Dimensional Sparse Recovery and Chaotic Dynamics, Comparative Study of the Viscoelastic Behavior on PLA Filaments, a Fractional Calculus Approach, and A New 4-D Highly Chaotic Two-Scroll System with a Hyperbola of Equilibrium Points and Its Circuit Simulation.

You are cordially invited to attend the following editions of EDIESCA. The fifth one will be held in 2024 at Universidad Panamericana, Aguascalientes campus, México. Enjoy the book collection of Chapters of EDIESCA2023.

México
January 2024

Cornelio Posadas-Castillo[1]
Eric Campos-Cantón
Guillermo Huerta-Cuellar
Ernesto Zambrano-Serrano[2]
Esteban Tlelo-Cuautle[3]

[1] Passed on July 2023.

[2] Ernesto Zambrano-Serrano acknowledges the research grant CF-2023-I-1110 from the Ciencia de Frontera 2023 program of CONAHCYT (Mexico) and the support provided by the Departamento de Electrónica y Automatización at the Facultad de Ingeniería Mecánica y Eléctrica, UANL.

[3] Esteban Tlelo-Cuautle is on sabbatical leave from INAOE collaborating at CINVESTAV-CDMX under the funding support of CONAHCyT-México for the program: Apoyos Complementarios Para Estancias Sabáticas Vinculadas a la Consolidación de Grupos de Investigación 2023.

In Memorian of Dr. Cornelio Posadas Castillo 1973–2023

Dr. Posadas-Castillo was a full-time professor at Facultad de Ingeniería Mecánica y Eléctrica (FIME), Universidad Autónoma de Nuevo León (UANL), México, during 1997–2023. He participated as reviewer in *IEEE Transactions on Circuits and Systems II*, *Journal of Computational Methods in Sciences and Engineering*, *Communication in nonlinear Science and Numerical Simulation*, *Chaos Solitons & Fractals*, *Applied Mathematics and Computation*, and *Mathematic and Computers in Simulations*. He published more than 25 scientific works, 8 book chapters, and more than 50 participations in national and international conferences. He was Academic Coordinator of the graduate program in Electrical Engineering (2012–2014) for Master of Science and Doctorate in Electrical Engineering. Lastly, he was appointed as Administrative Secretary of Postgraduate Studies at FIME-UANL. His research interest included synchronization, complex networks, data encryption, and multi-agent systems.

Contents

Editors and Contributors

About the Editors

Eric Campos-Cantón received his B.Sc., M.Sc. and Ph.D. degrees from Universidad Autónoma de San Luis Potosí (UASLP) México, in 1997, 1999 and 2003, respectively. In 2010 he was appointed as Professor-Researcher at IPICYT. He has been Visiting Researcher in the Department of Mathematics at University of Houston, USA. He serves as Associate Editor in *Journal of Applied Nonlinear Dynamics*, and *Complexity and Mathematical Problems in Engineering*. He has coedited a conference book *Complex Systems and Their Applications*, Springer. He has 13 international patents and his research interests include dynamical systems with chaotic behavior, in which he has the following lines of research: Analysis and modeling of dynamical systems in continuous and discrete time with chaotic behavior; Application of dynamical systems in cryptology; Experimental implementations of mathematical models with nonlinear and chaotic dynamics. https://urldefense.com/v3/__https://publons.com/researcher/697691/eric-campos-canton/__;!!NLFGqXoFfo8MMQ!tPXlNHJlkyG3AZQ2SEMtOrBaZqPwMRxyaWEHfJCab0BtYaExnSjWnt4QWt2-7DbosOIvN1nwT_qhR6cJegmwUYR1flBN$

Guillermo Huerta-Cuellar received a B.Sc. degree from Instituto de Investigación en Comunicaciones Ópticas, Universidad Autónoma de San Luis Potosí, México, in 2004. He received a Ph.D. degree from Centro de Investigaciones en Óptica, León Guanajuato, Mexico, in 2009. From 2010 to the present, he has been working at the Exact Sciences and Technology Department in Centro Universitario de los Lagos, Universidad de Guadalajara. He has been Visiting Researcher in the Department of Theory of Oscillations and Automatic Control, Faculty of Radiophysics, Lobachevsky State University of Nizhny Novgorod, Russia, and Department of Physics and Environmental Science at St. Mary's University, San Antonio, TX, USA. He is editor of 4 books and more than 70 high-impact publications. His research interests include study, characterization, dynamical behavior and design in nonlinear dynamical systems such as lasers, electronics and numerical models. https://urldefense.com/v3/__https://publons.com/researcher/2237811/guillermo-huerta-cuella__;!!NLFGqXoFfo8MMQ!tPXlNHJlkyG3AZQ2SEMtOrBaZqPwMRxyaWEHfJCab0BtYaExnSjWnt4QWt2-7DbosOIvN1nwT_qhR6cJegmwUZfexCPG$

Ernesto Zambrano-Serrano received both B.Sc. and M.Sc. degrees from Benemérita Universidad Autónoma de Puebla (BUAP), Mexico, in 2009 and 2012, respectively. He then received a Ph.D. degree from Instituto Potosino de Investigación Científica y Tecnológica (IPICYT), Mexico, in 2017. He is with Facultad de Ingeniería Mecánica y Eléctrica de la Universidad Autónoma de Nuevo León. He has edited special issues and participated as editor in several international journals. He is reviewer in *Nonlinear Dynamics, Entropy, Applied Mathematics and Computation, Entropy, Complexity, International Journal of Dynamics and Control*, and *Mathematics Problems in Engineering, Mathematics*. His research interests include chaotic behavior, nonlinear circuits, complex networks, fractional-order dynamical systems, stability, control and synchronization. https://orcid.org/0000-0002-2115-0097

Esteban Tlelo-Cuautle received his PhD degree from INAOE in 2000. He is appointed at INAOE from 2001. He has authored 5 books, edited 12 books and published more than 300 works in journals, book chapters and conference proceedings. He serves as Associate Editor in: *Engineering Applications of Artificial intelligence, Fractal and fractional, International Journal of Circuit Theory and Applications, IEEE Transactions on Circuits and Systems I: Regular Papers, Integration—the VLSI Journal, Frontiers of Information Technology & Electronic Engineering, MDPI Electronics, IEEE Transactions on Circuits and Systems II*, and *Journal of Engineering and Applied Research*. His research focuses on: integrated circuit design and synthesis, artificial intelligence, design and applications of (fractional-order) chaotic systems, symbolic circuit analysis, modeling and simulation of circuits and systems, optimization by metaheuristics, and analog/RF and mixed-signal design automation tools. https://urldefense.com/v3/__https://publons.com/wos-op/researcher/1487250/esteban-tlelo-cuautle/__;!!NLFGqXoFfo8MMQ!tPXlNHJlkyG3AZQ2SEMtOrBaZqPwMRxyaWEHfJCab0BtYaExnSjWnt4QWt2-7DbosOIvN1nwT_qhR6cJegmwUQnk2ILG$

Contributors

A. Anzo-Hernandez Investigadoras e Investigadores por Mexico, CONACYT, CEMMAC-FCFM-BUAP, Puebla, Mexico

J. G. Barajas-Ramirez Instituto Potosino de Investigación Cientifica y Tecnológica, División de Control y Sistemas Dinámicos, Camino a la Presa San José, Mexico

Safya Belghith Laboratory of Robotics, Informatics and Complex Systems (RISC Lab LR16ES07), National Engineering School of Tunis, University of Tunis El Manar, Tunis, Tunisia

B. Bonilla-Capilla Investigadoras e Investigadores por Mexico, CONACYT, CEMMAC-FCFM-BUAP, Puebla, Mexico

Rania Bouzid Higher Institute of Information and Communication Technologies, University of Carthage, Tunis, Tunisia

Sami Briouza Laboratory of Robotics, Informatics and Complex Systems (RISC Lab LR16ES07), National Engineering School of Tunis, University of Tunis El Manar, Tunis, Tunisia

Selene L. Cardenas-Maciel Tecnologico Nacional de Mexico, Instituto Tecnologico de Tijuana, Calzada Tecnologico S/N, Fraccionamiento Tomas Aquino, Tijuana, Mexico

I. Carro-Perez Instituto Potosino de Investigacion Cientifica y Tecnologica, Division de Control y Sistemas Dinamicos, Camino a la Presa San Jose, Mexico

Nohe R. Cazarez-Castro Tecnologico Nacional de Mexico, Instituto Tecnologico de Tijuana, Calzada Tecnologico S/N, Fraccionamiento Tomas Aquino, Tijuana, Mexico

Luis Gerardo de la Fraga Cinvestav, Computer Science Department, Mexico City, Mexico

R. Galindo Faculty of Mechanical Engineering, Autonomous University of Nuevo Leon, San Nicolas de los Garza, Mexico

Aboozar Ghaffari Department of Electrical Engineering, Iran University of Science and Technology, Tehran, Iran

G. Rodriguez Gomez INAOE, Puebla, Mexico

N. A. Gomez-Torres Departmento de Ciencias Basicas, Centro Universitario de la Cienega, Universidad de Guadalajara, Ocotlan, Mexico

Hassene Gritli Higher Institute of Information and Communication Technologies, University of Carthage, Tunis, Tunisia
Laboratory of Robotics, Informatics and Complex Systems (RISC Lab LR16ES07), National Engineering School of Tunis, University of Tunis El Manar, Tunis, Tunisia

Fareh Hannachi Larbi Tebessi University - Tebessi, Tebessa, Algeria

Sajad Jafari Department of Biomedical Engineering, Amirkabir University of Technology (Tehran polytechnic), Tehran, Iran
Health Technology Research Institute, Amirkabir University of Technology (Tehran polytechnic), Tehran, Iran

R. Jaimes-Reátegui Optics, Complex Systems and Innovation Laboratory, Centro Universitario de los Lagos, Universidad de Guadalajara, Lagos de Moreno, Mexico

Sahar Jenhani Laboratory of Robotics, Informatics and Complex Systems (RISC Lab LR16ES07), National Engineering School of Tunis, University of Tunis El Manar, Tunis, Tunisia

Nahla Khraief Laboratory of Robotics, Informatics and Complex Systems (RISC Lab LR16ES07), National Engineering School of Tunis, University of Tunis El Manar, Tunis, Tunisia

Pooyan Rezaeipour Lasaki Department of Electrical Engineering, Iran University of Science and Technology, Tehran, Iran

Eduardo Jimenez Lopez El Colegio Mexiquense A.C., Cerro del Murcielago, Mexico

Jyotindra Narayan Mechatronics and Robotics Laboratory, Department of Mechanical Engineering, Indian Institute of Technology Guwahati, Guwahati, India

Fahimeh Nazarimehr Department of Biomedical Engineering, Amirkabir University of Technology (Tehran Polytechnic), Tehran, Iran

F. Orihuela-Espina University of Birmingham (UoB), Birmingham, UK

Miguel Angel Platas-Garza Universidad Autónoma de Nuevo León, Ciudad Universitaria, San Nicolás de los Garza, México

Jesus Gabino Puente-Cordova Facultad de Ingenieria Mecanica y Electrica, Universidad Autonoma de Nuevo Leon, San Nicolas de los Garza, Mexico

K. Pulido-Hernandez Departmento de Ciencias Basicas, Centro Universitario de la Cienega, Universidad de Guadalajara, Ocotlan, Mexico

Jesus R. Pulido Luna Tecnologico Nacional de Mexico, Instituto Tecnologico de Tijuana, Fraccionamiento Tomas Aquino, Tijuana, Mexico

Andres Cureño Ramirez Cinvestav, Computer Science Department, Mexico City, Mexico

M.R. Ramirez-Jimenez Optics, Complex Systems and Innovation Laboratory, Centro Universitario de los Lagos, Universidad de Guadalajara, Lagos de Moreno, Mexico

Jorge A. Lopez Renteria CONAHCYT/Tecnologico Nacional de Mexico, Instituto Tecnologico de Tijuana, Calzada Tecnologico S/N, Fraccionamiento Tomas Aquino, Tijuana, Mexico

C.E. Rivera-Orozco Optics, Complex Systems and Innovation Laboratory, Centro Universitario de los Lagos, Universidad de Guadalajara, Lagos de Moreno, Mexico

Aceng Sambas Faculty of Informatics and Computing, University Sultan Zainal Abidin, Gong Badak, Malaysia
Department of Mechanical Engineering, Universitas Muhammadiyah Tasikmalaya, Jawa Barat, Indonesia

Karla Louisse Segura-Mendez Facultad de Ingenieria Mecanica y Electrica, Universidad Autonoma de Nuevo Leon, San Nicolas de los Garza, Mexico

Ernesto Zambrano Serrano Universidad Autónoma de Nuevo León, Ciudad Universitaria, San Nicolás de los Garza, Mexico

L. Serrano-Zuniga Departmento de Ciencias Basicas, Centro Universitario de la Cienega, Universidad de Guadalajara, Ocotlan, Mexico

M.T. Solorio-Nunez Optics, Complex Systems and Innovation Laboratory, Centro Universitario de los Lagos, Universidad de Guadalajara, Lagos de Moreno, Mexico

Saturnino Soria-Tello FIME of the UANL, Pedro de Alba S/N CD Universitaria, San Nicolas de los Garza, Mexico

Sundarapandian Vaidyanathan Centre for Control Systems, Vel Tech University, Chennai, India
Centre of Excellence for Research, Value Innovation & Entrepreneurship (CERVIE), UCSI University, Kuala Lumpur, Malaysia

J. Zaqueros-Martinez Computer Sciences, Instituto Nacional de Astrofísica, Óptica y Electrónica (INAOE), Puebla, Mexico

Part I
Artificial Intelligence Applied to Dynamics and Complexity

Chapter 1
Classification of Chaotic Dynamics Through Time–Frequency Representations and Machine Learning

Miguel Angel Platas-Garza and Ernesto Zambrano-Serrano

1.1 Introduction

Chaos in dynamics emerges as the cloud of representative points in the phase space repeatedly undergoes deformations, characterized by stretching, folding, and transversal compression [1]. Chaotic phenomena, characterized by their complex beauty and apparent disorder, challenge intuitive understanding while simultaneously unveiling an underlying layer of order. These systems, renowned for their sensitivity to initial conditions and unpredictable long-term evolution, manifest across diverse scientific fields, spanning from meteorology to data security and beyond [2–5].

In the field of secure communication schemes, chaotic dynamics have emerged as a promising approach to enhance data encryption and transmission security since they introduce an additional layer of complexity and security to fortify the confidentiality and integrity of the transmission of information [6, 7]. They often involve the utilization of one state variable from a particular dynamical system for the encryption of data, while another state variable serves as the cryptographic key [8, 9]. The synchronization of these states is a fundamental prerequisite for successful communication, ensuring that the intended recipient can accurately decipher the transmitted information [10–12].

The task of classifying such chaotic systems based on time-domain signals is not trivial [13], as the chaotic nature inherently implies high complexity and unpredictability. However, the ability to accurately classify and predict chaotic systems has profound implications, especially in scenarios where understanding the dynamic behavior can lead to significant advancements. In [14], the authors explore the performance of various machine learning techniques for forecasting chaotic

M. A. Platas-Garza (✉) · E. Zambrano-Serrano
Universidad Autónoma de Nuevo León, San Nicolás de los Garza, México
e-mail: miguel.platasgrz@uanl.edu.mx; ernesto.zambranos@uanl.edu.mx

E. Campos-Cantón et al. (eds.), *Complex Systems and Their Applications*,
https://doi.org/10.1007/978-3-031-51224-7_1

time series data. The techniques include gated recurrent neural networks, reservoir computing such as echo state networks and hybrid physics-informed, and the nonlinear vector autoregression approach. In [15], the use of hidden Markov model–based classifiers for classifying chaotic electroencephalogram signals in mental task classification is investigated. In [16], also an artificial neural network is developed to distinguish chaotic from regular dynamics in the two-dimensional Chirikov standard map. An integrated architecture for predicting cardiac arrhythmias, addressing the chaotic nature of cardiac voltage time series, where chaos is a common feature in the data is given in [17]. The article in [18] introduces an approach for echo state networks, designed to overcome the computational challenges and irreversibility of traditional methods, particularly in handling chaotic data in the era of big data. The method employs a recursive inverse-free algorithm and an adaptive activation function based on prediction errors to enhance nonlinear mapping capabilities. In this case, the identification of the used system may highlight potential vulnerabilities as are described in [19]. This identification is not only essential for preserving data integrity but also for thwarting potential security breaches.

In response to these challenges, this chapter presents an approach to the classification of chaotic dynamics by generating data from a set of popular chaotic systems and focusing on identifying both the chaotic system and its state from measurements of a single state. Specifically, the proposed method integrates Support Vector Machine (SVM) classifiers with Time–Frequency Representations (TFR) [20, 21].

SVM classifiers, operating on the principle of identifying optimal hyperplanes in high-dimensional spaces, provide a powerful tool for classifying complex systems [22]. This approach extends beyond the binary domain, encompassing multiclass classification. To enhance the scalability and adaptability of the classification process.

Whereas the application of TFRs is prominent in various real-world scenarios, for instance, the Shazam music identification service, which utilizes spectrogram-like representations or "fingerprints" for precise audio recognition [23]. This exemplifies the critical role of TFRs in enhancing discriminative capabilities in signal processing systems. The proposed approach considers the inherent adaptability of wavelet transforms in TFRs [24], providing a variable resolution that finely tunes to the dynamics of the signal. This feature is particularly beneficial when dealing with systems where changes can be abrupt or subtle over time. Unlike traditional spectrograms, which maintain a fixed resolution, wavelet-based TFRs, or scalograms, offer an adaptive resolution essential for capturing transient events and nonlinearities characteristic of chaotic systems.

Also, it is important to highlight that the proposed methodology uses the analysis of time-domain signals corresponding to only one state, in contrast to other classification methods. The implications of accurate chaotic system classification are vast, with particular relevance in secure communications utilizing chaotic synchronization. Herein, different system states serve multifaceted roles, such as encryption tools or cryptographic keys, necessitating precise system and state identification for effective communication and security.

This study unfolds as follows: Sect. 1.2 presents a detailed methodology, encompassing the stages of chaotic dynamics generation, data-set creation, data preprocessing, feature extraction, and classification. In Sect. 1.3, the results are shown. We evaluate the performance of the classifier via accuracy, precision, recall, and F1-score. Finally, in Sect. 1.4, the conclusions and future work close the chapter.

1.2 Methodology

Outlined below is the structured progression of phases that form the proposed methodology for the classification and state identification of chaotic systems.

1. **Chaotic Dynamics Generation**: This primary stage involves the selection of specific chaotic systems for examination. Four different chaotic systems were selected, each one of them displaying three distinct states. Time-domain data for each state of every system are generated by integrating their respective system equations via the Runge–Kutta fourth-order algorithm. Initial conditions were randomly selected.
2. **Data-set Creation**: We select randomly time signals, which are derived from a state chosen randomly from a chaotic system selected in a similar random manner. For each data in the dataset, a label is generated to clarify the originating chaotic system and the particular state responsible for producing the data.
3. **Data Preprocessing**: During this phase, a dual preprocessing is conducted on the data. A TFR and certain statistical measures in the time-domain are deduced. This processed information is structured into an array, which functions as the classifier input. The processed dataset, coupled with its labels, gets partitioned into the training and validation sets, respectively.
4. **Feature Extraction and Classification**: The training of the classifier is done using the training dataset. Classification is orchestrated employing a multiclass SVM.
5. **Validation**: Finally, the robustness and accuracy of the model are evaluated using the validation dataset employing diverse metrics. The results are presented into a confusion matrix. Accuracy, precision, recall, and F1-score are computed. These values for these metrics are shown in Sect. 1.3.

Next, a comprehensive description detailing the procedures behind each step is shown.

1.2.1 *Chaotic Dynamics Generation*

Central to this study is the investigation of various dynamic systems, each characterized by its distinct chaotic behavior. The selection of the chaotic systems to be used in this work was based on their popularity. The selection of popular systems is important because these are the systems that are used in the majority of applications. We focus on four different systems, each one of them with three distinct states. The states of the systems were labeled as $x_i(t) \in \mathbb{R}$ for $\{i \in \mathbb{Z} \mid 1 \leq i \leq 12\}$. Below, we outline the specific systems employed for this study.

- **Lorenz System**. Characterized by:

$$\frac{dx_1(t)}{dt} = 10(x_2(t) - x_1(t)), \tag{1.1}$$

$$\frac{dx_2(t)}{dt} = x_1(t)(28 - x_3(t)) - x_2(t), \tag{1.2}$$

$$\frac{dx_3(t)}{dt} = x_1(t)x_2(t) - \frac{8}{3}x_3(t). \tag{1.3}$$

- **Rossler System**. The dynamics are represented by:

$$\frac{dx_4(t)}{dt} = -x_5(t) - x_6(t), \tag{1.4}$$

$$\frac{dx_5(t)}{dt} = x_4(t) + \frac{1}{5}x_5(t), \tag{1.5}$$

$$\frac{dx_6(t)}{dt} = \frac{1}{5} + x_6(t)\left(x_4(t) - \frac{57}{10}\right). \tag{1.6}$$

- **Chen System**. The system dynamics are described by:

$$\frac{dx_7(t)}{dt} = 40x_8(t) - 3x_7(t), \tag{1.7}$$

$$\frac{dx_8(t)}{dt} = 28x_8(t) - x_7(t)x_9(t) + x_8(t), \tag{1.8}$$

$$\frac{dx_9(t)}{dt} = x_7(t)x_8(t) - 12x_9(t). \tag{1.9}$$

- **Chua System**. Described by:

$$\frac{dx_{10}(t)}{dt} = 10(x_{11}(t) - x_{10}(t) - f(x_{10}(t)), \tag{1.10}$$

$$\frac{dx_{11}(t)}{dt} = x_{10}(t) - x_{11}(t) + x_{12}(t), \tag{1.11}$$

$$\frac{dx_{12}(t)}{dt} = -\frac{392}{25}x_{11}(t), \tag{1.12}$$

with $f(x) = -\frac{861}{1250}x + \frac{147}{500}\left(|x+1| - |x-1|\right)$.

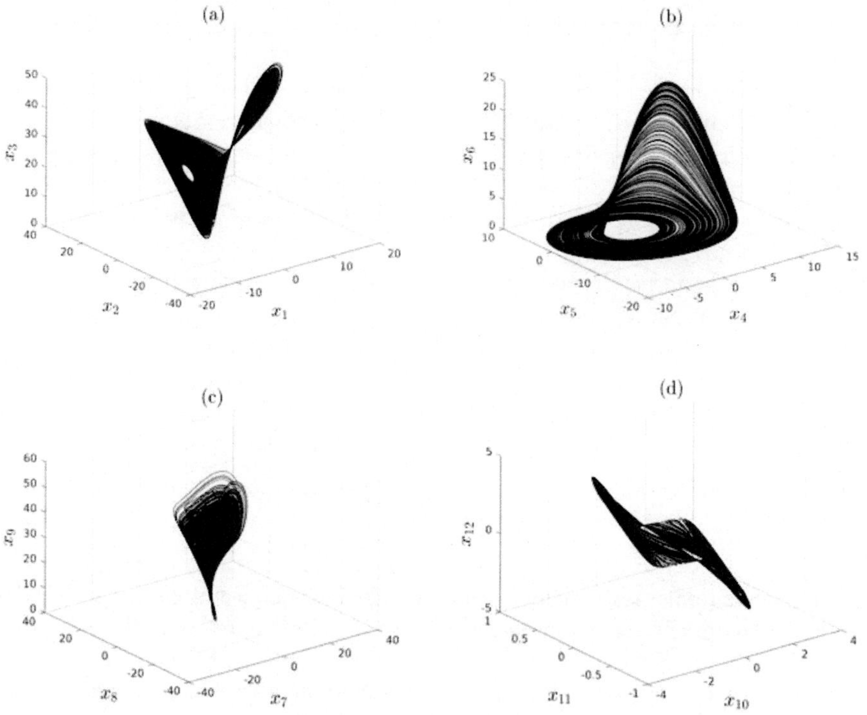

Fig. 1.1 Phase portraits for the chaotic systems analyzed in this study: (**a**) Lorenz, (**b**) Rossler, (**c**) Chen, and (**d**) Chua systems, respectively

Time-domain data from these systems were generated using the fourth-order Runge–Kutta method with a uniform step size of $h = 0.01$ seconds. Random initial conditions were chosen for the numerical integration.

The phase portraits of the used systems can be found in Fig. 1.1, where the related systems to each phase portrait in Fig. 1.1 are (a) Lorenz, (b) Rossler, (c) Chen, and (d) Chua systems, respectively.

In this application, for each scenario, $N \in \mathbb{Z}$ samples of each system were collected, with N defined by the equation $N = \kappa\eta + \delta$. Here, $\kappa \in \mathbb{Z}$ indicates the total number of elements in the data-set, $\eta \in \mathbb{Z}$ refers to the number of samples for each element, and $\delta \in \mathbb{Z}$ constitutes an initial offset in the numerical integration. This offset is subsequently discarded to account for the transient response of the systems.

All the generated information was consolidated into an array $\mathbf{R} \in \mathbb{R}^{12\times N}$. In this array, individual rows represent distinct states, whereas each column corresponds to a specific time step. That is, the c-th element in the r-th row of $\mathbf{R}$ denotes the c-th sample of the r-th state as follows

$$\mathbf{R}(r, c) = x_r(t)|_{t=ch} \,. \tag{1.13}$$

1.2.2 Data Segmentation

At this stage, the data obtained from the numerical simulation were segmented to create individual elements of data. Initially, the first δ columns in **R** were eliminated due for the transient response arising from the initial conditions. Subsequently, the remaining $\eta - \delta$ columns in **R** were partitioned into κ subsets, each one with $\eta = \frac{N-\delta}{\kappa}$ samples, resulting in individual segments denoted as $\mathbf{R}_i \in \mathbb{R}^{12\times\eta}$, where $i = 1, 2, \ldots, \kappa$.

Next, we performed a ramdom selection of a state. A random sequence $\xi[j]$ representing an integer such that $1 \leq \xi[j] \leq 12$ for $j = 1, 2, \ldots, \kappa$ was used to select an arbitrary row of each $\mathbf{R}_i$. The distribution of ξ is uniform, meaning each integer in the range [1, 12] has an equal probability of appearing. The array $\mathbf{D} \in \mathbb{R}^{\kappa\times\eta}$ was constructed by rows, selecting for each $\mathbf{R}_i$ an arbitrary row (state) as:

$$\mathbf{D}(r, c) = \mathbf{R}_r(\xi[r], c), \quad \begin{cases} r = 1, 2, \ldots, \kappa, \\ c = 1, 2, \ldots, \eta. \end{cases} \tag{1.14}$$

Using the former segmentation procedure, the c-th element of the r-th row of the dataset **D** represents a random state $x_{\xi[r]}(t)$ at $t = (r-1)\eta h + (c-1)h$.

For each element in the dataset, represented as a row in **D**, the corresponding system and state responsible for that time-domain data were denoted with a tag. This information is used as desired output for our classificator and was stored in an array **Y** such that $\mathbf{Y} \in \mathbb{Z}^{\kappa}$. Each class, representing a unique combination of a system and a state, was mapped to an integer between 1 and 12. Hence, the elements y_i of **Y** satisfy:

$$y_i \in \{1, 2, \ldots, 12\},$$

for all i with $1 \leq i \leq \kappa$. The value of y_i provides the system and state information. For instance, if $y_{22} = 5$, it signifies that the 22nd element in the dataset (row number 22 in **D**) corresponds to the second state of the Rossler system, denoted as $x_5(t)$.

The generated dataset **D** and its corresponding output tags **Y** were randomly partitioned into training and validation sets, using a predefined split ratio of 75%.

1.2.3 Data Preprocessing

The time-domain array **D** generated in the previous section was subjected to wavelet transform for time–frequency feature extraction, as well as thorough statistical analysis to discern inherent patterns and characteristics.

1.2.3.1 Statistical Properties of Time-Domain Signal

The following statistical moments were computed for each entry in the dataset, represented by each row in $\mathbf{D}$. All these metrics were calculated and stored. Let $\mathbf{d}$ be an arbitrary row in $\mathbf{D}$, the following metrics were computed and related to these data.

- **Mean.** The mean (or average) of $\mathbf{d}$ is given by:

$$\text{mean}(\mathbf{d}) = \frac{1}{\eta} \sum_{i=1}^{\eta} d[i], \tag{1.15}$$

  where η represents the total number of samples in $\mathbf{d}$ and $d[i]$ is the ith sample.
- **Standard deviation (std).** The standard deviation quantifies the amount of variation or dispersion of a set of values. It is calculated as:

$$\text{std}(\mathbf{d}) = \sqrt{\frac{1}{\eta - 1} \sum_{i=1}^{\eta} (d[i] - \text{mean}(\mathbf{d}))^2}. \tag{1.16}$$

- **Maximum (max) and Minimum (min).** The maximum and minimum values in signal $\mathbf{d}$ are denoted as $\max(\mathbf{d})$ and $\min(\mathbf{d})$, respectively.
- **Skewness (skew).** Skewness measures the asymmetry of the probability distribution of a real-valued random variable about its mean. For a signal $\mathbf{d}$, skewness is:

$$\text{skew}(\mathbf{d}) = \frac{\sum_{i=1}^{\eta} (d[i] - \text{mean}(\mathbf{d}))^3/\eta}{(\text{std}(\mathbf{d}))^3}. \tag{1.17}$$

- **Kurtosis (kurt).** Kurtosis measures the "tailednes" of the probability distribution of a real-valued random variable. For a signal $\mathbf{d}$, kurtosis is defined as:

$$\text{kurt}(\mathbf{d}) = \frac{\sum_{i=1}^{\eta} (d[i] - \text{mean}(\mathbf{d}))^4/\eta}{(\text{std}(\mathbf{d}))^4} - 3. \tag{1.18}$$

1.2.3.2 Scalogram

Time–frequency representations provide a comprehensive view of signals by offering insights into how their spectral content changes over time. This dual perspective captures transient and non-stationary features in signals that might be elusive in purely time-based or frequency-based analyses.

While both scalograms and spectrograms offer TFRs of signals, scalograms provide multiresolution analysis, allowing for better resolution in both time and frequency domains. The magnitude of the scalogram indicates the amplitude of wavelet coefficients across various scales and temporal instances. Next, we show how the scalogram was computed for an arbitrary row **d** in **D**. This process was repeated for all the rows in **D**.

1. **Data Segmentation and Normalization.**
   First, the η samples in the data **d** were normalized as:

$$\mathbf{z} = \frac{\mathbf{d} - \text{mean}(\mathbf{d})}{\max(|\mathbf{d}|)}, \tag{1.19}$$

   where $\mathbf{z} \in \mathbb{R}^{\eta}$ and $|\cdot|$ represents absolute value.
2. **Wavelet Decomposition.**
   The "db4" wavelet, symbolized as $\psi(t)$, was employed to decompose the data segment across different scales. The resulting array of coefficients was denoted by $W(\alpha, \beta) \in \mathbb{C}^{a\times\eta}$, with α representing the scale and β the time shift. The wavelet coefficients were computed at $a \in \mathbb{Z}$ scale levels.
3. **Magnitude Calculation.**
   The magnitude of the complex-valued wavelet coefficients was then evaluated as:

$$M(\alpha, \beta) = |W(\alpha, \beta)|, \tag{1.20}$$

   producing the scalogram values.
4. Finally, the scalogram is plotted with scales (or levels of decomposition) on one axis and time on the other, where the intensity of the color indicates the energy magnitude at that scale and time.

Illustrative examples of scalograms for various chaotic time series are presented in Figs. 1.2 and 1.3. In these representations, the horizontal axis represents time, whereas the vertical axis represents scale. Specifically, Fig. 1.2 depicts the scalogram corresponding to all the states of the Lorenz and Rossler systems, providing a unique visual representation of their dynamic behavior over frequency and time. In addition, Fig. 1.3 illustrates the scalogram for all the states of the Chen and Chua systems. The scalogram of each state within each system bears a specific form or footprint, indicative of the inherent characteristics and the dynamical complexities of the respective systems.

1.2.3.3 Integration of the Preprocessing Data

Prior to classification, all the data analyzed (statistical information and scalogram) were merged into a single 1D array of elements. This array represents the input to the classifier.

For each row $\mathbf{d}_r$, $r = 1, 2, \ldots, \kappa$, of the matrix **D**, a new array $\mathbf{x}_r \in \mathbb{R}^{\gamma}$, with $\gamma = 6 + \alpha\eta$ was generated. The vector $\mathbf{x}_r$ encapsulates the statistical measures of $\mathbf{d}_r$ including its mean, standard deviation (std), maximum (max), minimum (min),

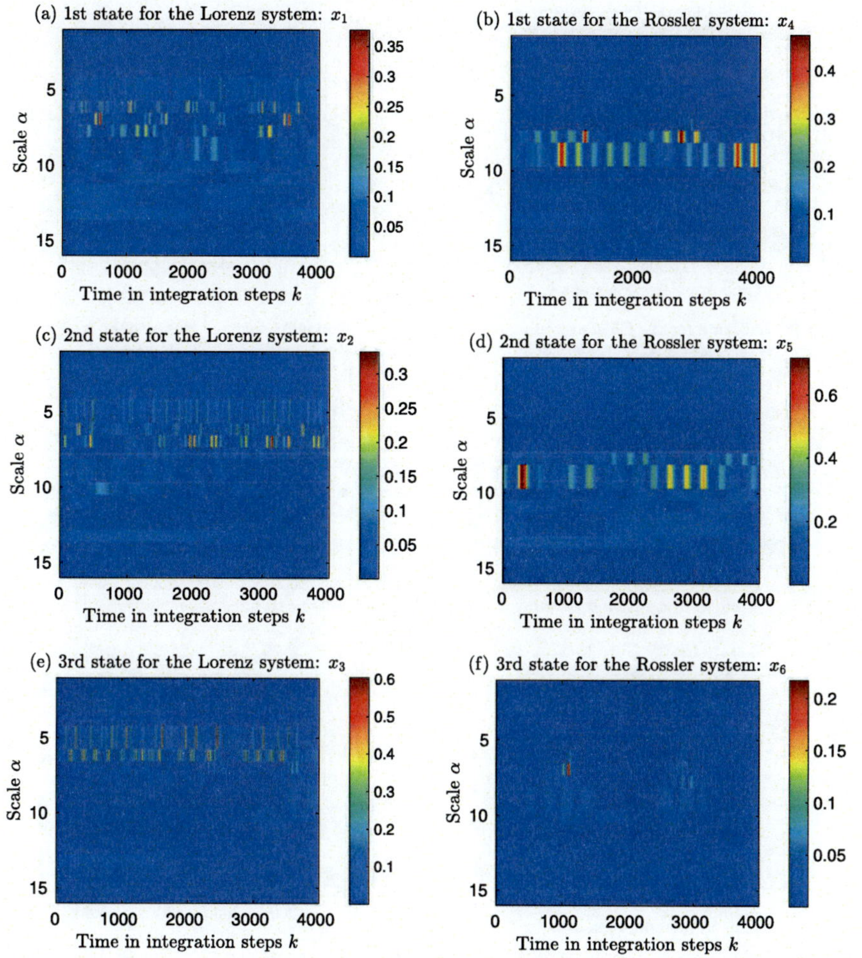

Fig. 1.2 Scalogram for chaotic time series in the dataset. The TFR acts like a footprint. Examples are shown for **(a)** $x_1(t)$, **(b)** $x_4(t)$, **(c)** $x_2(t)$, **(d)** $x_5(t)$, **(e)** $x_3(t)$ and **(f)** $x_6(t)$

skewness (skew) and Kurtosis (kurt). Additionally, $\mathbf{x}_r$ also contains all the scalogram points represented as a 1D array. The set of all the κ elements $\mathbf{x}_r$ is grouped in rows into an array $\mathbf{X} \in \mathbb{R}^{\kappa \times \gamma}$.

Finally, the data $\mathbf{X}$ obtained in this section and its tags $\mathbf{Y}$ will be used as inputs to the classifier. In this context, the κ rows of $\mathbf{X}$ represent the data to be classified, and the κ elements of $\mathbf{Y}$ correspond to the true tags or labels for each data point.

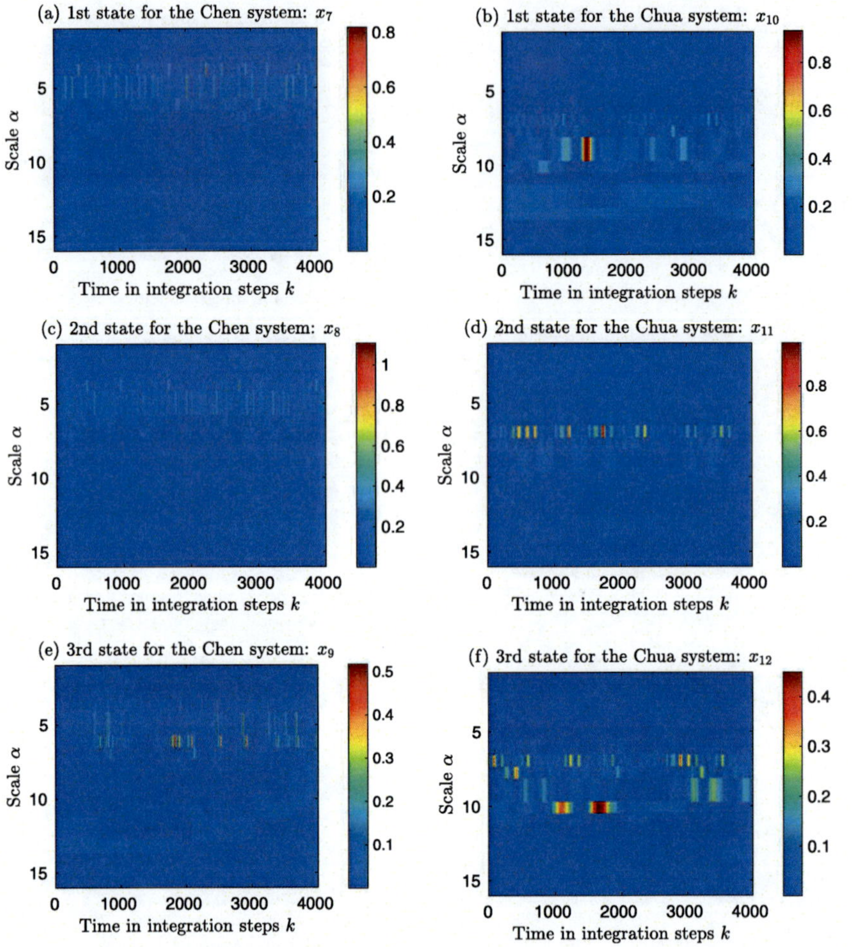

Fig. 1.3 Scalogram for chaotic time series in the dataset. The TFR acts like a footprint. Examples are shown for **(a)** $x_7(t)$, **(b)** $x_{10}(t)$, **(c)** $x_8(t)$, **(d)** $x_{11}(t)$, **(e)** $x_9(t)$ and **(f)** $x_{12}(t)$

1.2.4 Feature Extraction and Classification

The task of classification was accomplished using a multiclass SVM, a supervised learning algorithm renowned for its efficacy in complex categorization scenarios.

We divided the data, denoted as the rows of $\mathbf{X}$, and its corresponding class tags, denoted as the elements of $\mathbf{Y}$, into training and validation sets. A split ratio of 0.75 was employed, resulting in the random selection of 75% of the rows in $\mathbf{X}$ and their associated class tags in $\mathbf{Y}$ to compose the training set. This training set is represented as $\mathbf{X}_T$ and $\mathbf{Y}_T$. Similarly, the remaining rows in $\mathbf{X}$ and their corresponding class tags in $\mathbf{Y}$ were designated as the validation set, denoted as $\mathbf{X}_V$ and $\mathbf{Y}_V$.

An SVM with Error-Correcting Output Codes (ECOC) was used for multiclass classification. ECOC is a technique used to extend binary classifiers, such as SVMs, to handle multiclass classification problems. The idea is to represent each unique class as a binary code, and then train multiple binary classifiers, one for each code. In our case, we have 12 distinct classes, and we will represent each class as a binary code of length 12.

To apply SVM with ECOC for multiclass classification, we follow these steps:

1. **Binary Classification.** For each binary code digit j in the range from 1 to 12, create a binary classification problem.
2. **Train Binary Classifiers.** Train a binary SVM classifier for each binary problem using the training data $\mathbf{X}$ and the corresponding binary code digit.
3. **Classification.** To classify a new data point $\mathbf{x}$ into one of the 12 classes, we use each of the trained binary classifiers to produce a binary decision for each binary code digit j. The final class label is determined by decoding the binary decisions based on a binary code matrix.

1.3 Results and Discussion

In this section, we present the results of our proposal. We will discuss the performance metrics and any notable observations. For the results shown, the parameters shown in Table 1.1 were used. We remark that the input to the algorithm is the time-domain data for only one state of a given set of chaotic systems. These data were processed and then classified.

1.3.1 Performance Metrics

We evaluated the performance of the proposed method using various performance metrics, including accuracy, confusion matrix, precision, recall, and F1-score.

Precision and recall are computed as follows, let $c_{i,j}$ be the elements of the confusion matrix $\mathbf{C} \in \mathbb{R}^{\kappa \times \kappa}$, then accuracy, precision, and recall for each class

Table 1.1 Summary of parameters and configuration settings used

Description	Symbol	Value
Number of data	κ	4096 elements
Time samples per data	η	4096 samples
Number of classes	–	12 classes
Scale	a	32 levels
Wavelet	–	DWT db4
Offset	δ	1000 samples
Step size	h	0.01 seconds

can be computed as follows: The overall accuracy is the ratio between the cases classified corrected and all the cases:

$$A = \sum_i \frac{c_{i,i}}{\sum_i \sum_j c_{i,j}}, \quad i, j = 1, 2, \ldots, \kappa. \tag{1.21}$$

Precision and recall for each class can be defined as

$$P_i = \frac{c_{i,i}}{\sum_i c_{i,j}}, \quad i, j = 1, 2, \ldots, \kappa. \tag{1.22}$$

and

$$R_j = \frac{c_{j,j}}{\sum_j c_{i,j}}, \quad i, j = 1, 2, \ldots, \kappa, \tag{1.23}$$

respectively.

Once precision and recall for each class are determined, the F1-score per class can be computed as:

$$\text{F1} - \text{score}_i = 2\frac{P_i R_i}{P_i + R_i}, \quad i = 1, 2, \ldots, \kappa. \tag{1.24}$$

1.3.1.1 Classification of the System

In this case, we only evaluate if the system was classified correctly. A resume of the classifier performance, during the validation set, can be shown in the confusion matrix

$$\mathbf{C}_1 = \begin{bmatrix} 229 & 0 & 0 & 0 \\ 0 & 205 & 0 & 0 \\ 1 & 0 & 226 & 0 \\ 0 & 0 & 0 & 235 \end{bmatrix}, \tag{1.25}$$

where the sum of the elements in each row of $\mathbf{C}_1$ indicates the amount of true cases for each class in the following order, Lorenz, Rossler, Chua, and Chen. In the same manner, the sum per column indicates the amount of predicted cases for each class in the same order. A perfect classification would have a diagonal confusion matrix. In our case, as $\mathbf{C}_1$ is near diagonal, only one classification error was detected in the validation set, when a Chen system (row 3) was identified as a Lorenz system (column 1).

The same information is shown in a graphical representation in Fig. 1.4a, where true classes are defined by vertical labels and predicted values in horizontal labels are shown. We remark that the classification of the system was done using only time data from one state.

Table 1.2 Precision, recall, and F1-score metrics calculated for the system classification task on the validation dataset. The overall accuracy was $A = 0.9988$

Class	Precision	Recall	F1-score
Lorenz	0.9957	1.0000	0.9978
Rossler	1.0000	1.0000	1.0000
Chua	1.0000	0.9956	0.9978
Chen	1.0000	1.0000	1.0000

When only the system is classified, an overall accuracy of 0.9988 was reached. This percentage represents the ratio between the sum of the elements in the diagonal of $\mathbf{C}_1$ between the sum of all its elements. Table 1.2 shows the values for precision, recall, and F1-score for each class.

1.3.1.2 Classification of the System and State

A similar analysis is presented for the case when the classifier asserts the system and the state responsible for creating the data. In this case, the classifier has to choose between 12 classes possible as there are three states of four possible systems.

The confusion matrix for this case is

$$\mathbf{C}_2 = \left[\begin{array}{ccc|ccc|ccc|ccc}
70 & 0 & 0 & 0 & 0 & 0 & 0 & 0 & 0 & 0 & 0 & 0 \\
0 & 83 & 0 & 0 & 0 & 0 & 0 & 0 & 0 & 0 & 0 & 0 \\
0 & 0 & 76 & 0 & 0 & 0 & 0 & 0 & 0 & 0 & 0 & 0 \\
\hline
0 & 0 & 0 & 62 & 0 & 0 & 0 & 0 & 0 & 0 & 0 & 0 \\
0 & 0 & 0 & 14 & 64 & 0 & 0 & 0 & 0 & 0 & 0 & 0 \\
0 & 0 & 0 & 0 & 0 & 65 & 0 & 0 & 0 & 0 & 0 & 0 \\
\hline
0 & 0 & 0 & 0 & 0 & 0 & 69 & 9 & 0 & 0 & 0 & 0 \\
0 & 0 & 0 & 0 & 0 & 0 & 12 & 72 & 0 & 0 & 0 & 0 \\
0 & 0 & 1 & 0 & 0 & 0 & 0 & 0 & 64 & 0 & 0 & 0 \\
\hline
0 & 0 & 0 & 0 & 0 & 0 & 0 & 0 & 0 & 57 & 0 & 14 \\
0 & 0 & 0 & 0 & 0 & 0 & 0 & 0 & 0 & 0 & 75 & 1 \\
0 & 0 & 0 & 0 & 0 & 0 & 0 & 0 & 0 & 0 & 0 & 88
\end{array}\right], \tag{1.26}$$

where again true classes are defined by rows and predicted classes are identified as columns. A graphical representation with labels is presented in Fig. 1.4b. From $\mathbf{C}_2$ we note that, even there is only one system classification error, state classification errors were presented for the Rossler, Chua, and Chen systems. When the state of the system is classified, an overall accuracy of 0.9431 is reached.

Precision, recall, and F1-score were also computed for the system and state classification according to (1.22)–(1.24). Table 1.3 shows the values for precision and recall for each class.

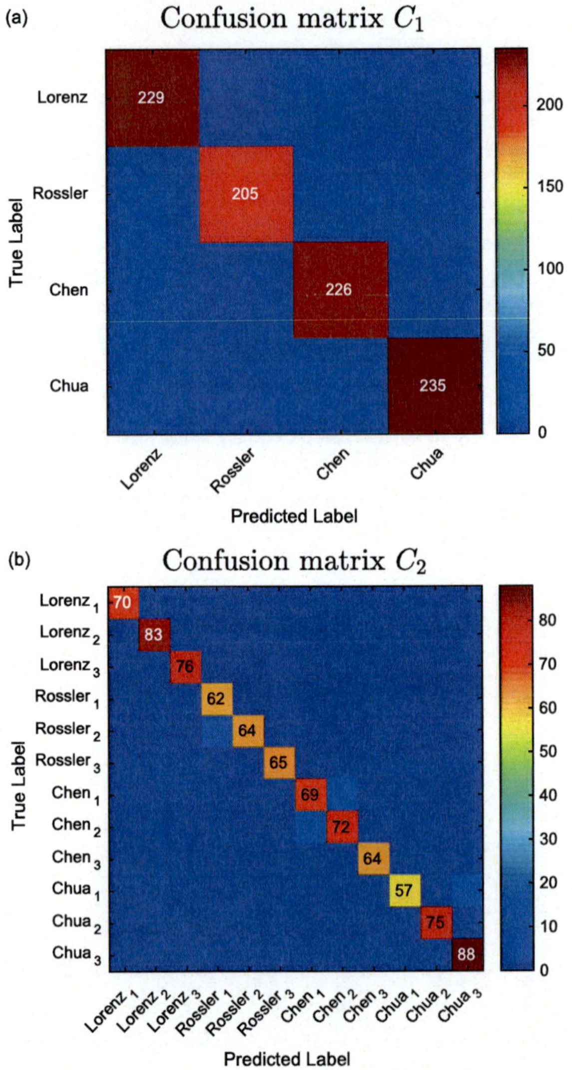

Fig. 1.4 Confusion matrix for the classification. (**a**) C_1 evaluates the system classification, (**b**) C_2 evaluates the case when the system and the state are classified

1.4 Conclusions and Future Work

In this study, we have introduced an approach for classifying chaotic dynamical systems and their states by analyzing time-domain data from a single state and employing signal processing techniques. The proposed methodology classified chaotic time series data from the Lorenz, Rossler, Chen, and Chua oscillators.

The results demonstrated a good performance in classifying both the chaotic system and its state. The confusion matrix and performance metrics, including accuracy, precision, recall, and F1-score, indicated the robustness and reliability

Table 1.3 Precision, recall, and F1-score metrics calculated for the system and state classification task on the validation dataset. The overall accuracy was $A = 0.9431$

Class	Precision	Recall	F1-score
Lorenz, state 1, $x_1(t)$	1.0000	1.0000	1.0000
Lorenz, state 2, $x_2(t)$	1.0000	1.0000	1.0000
Lorenz, state 3, $x_3(t)$	0.9870	1.0000	0.8985
Rossler, state 1, $x_4(t)$	0.8158	1.0000	0.8985
Rossler, state 2, $x_5(t)$	1.0000	0.8205	0.9014
Rossler, state 3, $x_6(t)$	1.0000	1.0000	1.0000
Chen, state 1, $x_7(t)$	0.8519	0.8846	0.8679
Chen, state 2, $x_8(t)$	0.8889	0.8571	0.8727
Chen, state 3, $x_9(t)$	1.0000	0.9846	0.9922
Chua, state 1, $x_{10}(t)$	1.0000	0.8028	0.8906
Chua, state 2, $x_{11}(t)$	1.0000	0.9868	0.9933
Chua, state 3, $x_{12}(t)$	0.8544	1.0000	0.9214

of our classification model. Additionally, results indicate that classifying the system itself is generally easier than classifying the specific state within it. When classification errors are present, they tend to occur between two states of the same system. This is possibly attributed to the symmetry present in the attractor.

As possible uses of this work, identifying chaotic data within these widely used systems can have significant implications, particularly in the context of breaking secure communication schemes. These schemes often employ a state-based approach, where one state is responsible for encoding the message, and another state, transmitted through a private channel, is used for synchronization between the transmitter and receiver. In the event of a successful attack that captures the state within the private channel responsible for synchronization, there is a potential vulnerability. To mitigate this risk, it is essential for secure communication systems to employ robust encryption and authentication measures.

In the future, researchers could explore the expansion of this approach to more complex dynamical systems and examine the impact of noise on classification accuracy, thus enhancing the practical applicability of the model for analysis of a more realistic scenario. In addition, since scalograms are essentially images, it would be worthwhile to explore the application of deep learning techniques, such as convolutional neural networks, commonly employed for image classification.

Finally, the proposed approach does not account for scaling, offset, and timescale variations in the analyzed data. In real-world applications, it is common to adjust states through scaling, introduce offsets, and generate data at different time intervals, which can potentially alter the system dynamics. At this point, the system lacks the capability to detect and adapt to these variations.

Acknowledgments All the authors acknowledge the research grant CF-2023-I-1110 from CONAHCYT (México) and the support from the "Departamento de Electrónica y Automatización" at the Facultad de Ingeniería Mecánica y Eléctrica, UANL.

Competing Interests This study was funded by CONAHCYT grant number CF-2023-I-1110.
The authors have no conflicts of interest to declare that are relevant to the content of this chapter.

References

1. Sergey P Kuznetsov. *Hyperbolic chaos*. Springer, 2012.
2. Eric Campos Cantón, Rodolfo de Jesús Escalante González, and Héctor Eduardo Gilardi Velázquez. *Generation of Self-Excited, Hidden and Non-Self-Excited Attractors in Piecewise Linear Systems: Some Recent Approaches*. World Scientific, 2023.
3. Andrew Fowler and Mark McGuinness. *Chaos: An Introduction for Applied Mathematicians*. Springer Nature, 2020.
4. Saureesh Das, Rashmi Bhardwaj, and Varsha Duhoon. Chaotic dynamics of recharge–discharge el niño–southern oscillation (enso) model. *The European Physical Journal Special Topics*, 232(1):217–230, 2023.
5. B Fernández-Carreón, JM Munoz-Pacheco, E Zambrano-Serrano, and OG Félix-Beltrán. Analysis of a fractional-order glucose-insulin biological system with time delay. *Chaos Theory and Applications*, 4(1):10–18, 2022.
6. Jing Luo and Xue Chen. Transmission synchronization of multiple memristor chaotic circuits via single input controller and its application in secure communication. *Integration*, 90:40–50, 2023.
7. Chunbo Xiu, Ruxia Zhou, Shaoda Zhao, and Guowei Xu. Memristive hyperchaos secure communication based on sliding mode control. *Nonlinear Dynamics*, 104(1):789–805, 2021.
8. Guodong Li, Yue Pu, Bing Yang, and Jing Zhao. Synchronization between different hyper chaotic systems and dimensions of cellular neural network and its design in audio encryption. *Cluster Computing*, 22:7423–7434, 2019.
9. Chih-Min Lin, Duc-Hung Pham, and Tuan-Tu Huynh. Synchronization of chaotic system using a brain-imitated neural network controller and its applications for secure communications. *IEEE Access*, 9:75923–75944, 2021.
10. Yuwei Yang, Jie Gao, and Hashem Imani. Design, analysis, circuit implementation, and synchronization of a new chaotic system with application to information encryption. *AIP Advances*, 13(7), 2023.
11. O. Garca-Seplveda, C. Posadas-Castillo, A. D. Cortés-Preciado, M. A. Platas-Garza, E. Garza-González, and A. G. Soriano-Sanchez. Synchronization of fractional-order Lü chaotic oscillators for voice encryption. *Revista Mexicana de Fsica*, 66(3):364–371, 2020.
12. Rodrigo Méndez-Ramírez, Adrian Arellano-Delgado, and Miguel Ángel Murillo-Escobar. Network synchronization of macm circuits and its application to secure communications. *Entropy*, 25(4):688, 2023.
13. Luis Gerardo de la Fraga, Brisbane Ovilla-Martínez, and Esteban Tlelo-Cuautle. Echo state network implementation for chaotic time series prediction. *Microprocessors and Microsystems*, 103(104950), 2023.
14. Shahrokh Shahi, Flavio H Fenton, and Elizabeth M Cherry. Prediction of chaotic time series using recurrent neural networks and reservoir computing techniques: A comparative study. *Machine learning with applications*, 8:100300, 2022.
15. Soroosh Solhjoo, Ali Motie Nasrabadi, and Mohammad Reza Hashemi Golpayegani. Classification of chaotic signals using hmm classifiers: Eeg-based mental task classification. In *2005 13th European Signal Processing Conference*, pages 1–4. IEEE, 2005.
16. Woo Seok Lee and Sergej Flach. Deep learning of chaos classification. *Machine Learning: Science and Technology*, 1(4):045019, 2020.
17. Shahrokh Shahi, Flavio H Fenton, and Elizabeth M Cherry. A machine-learning approach for long-term prediction of experimental cardiac action potential time series using an autoencoder and echo state networks. *Chaos: An Interdisciplinary Journal of Nonlinear Science*, 32(6), 2022.
18. Bowen Wang, Shuxian Lun, Ming Li, Xiaodong Lu, and Tianping Tao. Adaptive echo state network with a recursive inverse-free weight update algorithm. *Information Sciences*, 647:119436, 2023.

19. Rashmi Bhardwaj and Saureesh Das. Chaos control dynamics of cryptovirology in blockchain. *Cryptocurrencies and blockchain technology applications*, pages 129–148, 2020.
20. Jun Shi, Gong Chen, Yanan Zhao, and Ran Tao. Synchrosqueezed fractional wavelet transform: A new high-resolution time-frequency representation. *IEEE Transactions on Signal Processing*, 71:264–278, 2023.
21. Ahmed Cemiloglu, Licai Zhu, Sibel Arslan, Jinxia Xu, Xiaofeng Yuan, Mohammad Azarafza, and Reza Derakhshani. Support vector machine (svm) application for uniaxial compression strength (ucs) prediction: A case study for maragheh limestone. *Applied Sciences*, 13(4):2217, 2023.
22. Funda Cinyol, Uğur Baysal, Deniz Köksal, Elif Babaoğlu, and Sevinç Sarınç Ulaşlı. Incorporating support vector machine to the classification of respiratory sounds by convolutional neural network. *Biomedical Signal Processing and Control*, 79:104093, 2023.
23. TJ Tsai, Thomas Prätzlich, and Meinard Müller. Known-artist live song identification using audio hashprints. *IEEE Transactions on Multimedia*, 19(7):1569–1582, 2017.
24. Ervin Sejdić, Igor Djurović, and Jin Jiang. Time–frequency feature representation using energy concentration: An overview of recent advances. *Digital signal processing*, 19(1):153–183, 2009.

Chapter 2
Optimization of Echo State Neural Networks to Solve Classification Problems

Andres Cureño Ramírez and Luis Gerardo De la Fraga

2.1 Introduction

ESNNs belong to the class of recurrent neural networks. An ESNN have three main components: an input, a reservoir, and an output layer. The reservoir layer mapping the inputs into a high-dimensional space and a readout for pattern analysis from the high-dimensional states in the reservoir [1]. According to [1, 2], ESNNs can be used in machine learning applications of supervised learning such as classification and regression.

Nowadays, exists different state-of-the-art algorithms to perform classification tasks, such as the decision trees that classify instances by sorting them based on feature values, artificial neural networks (ANN) consist of large number of neurons joined together and their values need to be trained, and the support vector machines (SVM) that try to maximize the margin that separates two classes. In the work [3], the author mentioned that SVMs and neural networks tend to perform much better when dealing with multidimensions and continuous features and the decision trees with categorical and discrete features. Also the ANN and the SVM need more data to achieve a better accuracy but that also increase the time of the training.

In an ESNN, the size of the reservoir is fixed. For a classification task, the number of neurons on the input is equal to the number of features, and number of neurons in the output is equal to the number of classes. The weights of the neurons of the input and the reservoir are initialized randomly. In this work, we have used the minimal reservoir: the reservoir can be seen as a matrix of size $n \times n$, the diagonal entries of such matrix are initialized randomly and then permuted. Thus, this reservoir can be characterized by two integer numbers with the meaning of the seeds for the two

A. Cureño Ramírez · L. G. De la Fraga (✉)
Cinvestav, Computer Science Department, Mexico City, México
e-mail: andres.cureno@cinvestav.mx; fraga@cs.cinvestav.mx

E. Campos-Cantón et al. (eds.), *Complex Systems and Their Applications*,
https://doi.org/10.1007/978-3-031-51224-7_2

pseudorandom number generators, one for its random weights, and another for the random permutation of those weights. We will show that this idea allows us to obtain great classifiers based on ESNNs.

The weights of the output neurons in an ESNN are calculated with a simple method such as linear regression, which has the cost of a single matrix inversion. Thus, the major advantage of reservoir computing compared to other models of neural networks is its fast learning, resulting in a low training cost. A disadvantage of ESNNs is that it is necessary to tuning its five associated parameters.

One option is to manually tuning the reservoir parameters, but changes in several parameters at once often have different effects on the performance, and it is almost impossible to tell which parameter makes it worst or makes it better [4]. Rather than tuning by hand, other works focus on evolutionary optimization to optimize the parameters [5]. It has been tested a genetic algorithm to optimize the parameters, in [6] authors optimize four parameters and in [7] six parameters are optimized. Both works agree with the size of the reservoir as a parameter and how the weights of the reservoir are calculated, but it does not mention if they use a different seed to initialize the values of the reservoir and the input matrix, also these articles were proposed to make regression tasks, neither of these two references show the evaluation function.

In this work, we will also show that is possible to use a genetic algorithm (GA) and a pattern search algorithm (PSA) to train and to optimize an ESNN characterized by five variables, the two integer numbers for the seeds to build a reservoir, plus noise and leaking rate of the input, and the reservoir size. Noise and leaking rate input will be explained in detail in Sect. 2.2.

The rest of the chapter is organized as follows. In Sect. 2.2, the ESNN details are explained, including the description of the fitness function to optimize. In Sect. 2.3, the two heuristics GA and PSA are briefly explained. In Sect. 2.4, the results with some public database are shown and a discussion about this work is given. Finally, in Sect. 2.5, some conclusions are drawn.

2.2 Description of an ESNN

An ESNN has an input matrix W_{in}, a reservoir matrix W_{res}, and an output matrix W_{out}, as it is shown in the Fig. 2.1. The values of the matrices W_{in} and W_{res}, called also weights of the input and reservoir neurons, are randomly generated, while the values of the output matrix W_{out} are calculated by performing a pseudo matrix inversion. In an ESNN the gradient is not available to obtain its optimal values [2].

We have a set with five variables to optimize $\boldsymbol{x} = \{x_1, x_2, x_3, x_4, x_5\}$, variable $x_1 \in [0, 1]$ is the noise, variable $x_2 \in [0, 1]$ is the leaking rate, these variables affect the input before propagation through the network, variable $x_3 \in [5, 1000]$ is the size of the reservoir matrix W_{res}, variable $x_4 \in [0, 100]$ is the seed to initialize the weights of the reservoir matrix W_{res} and the input matrix W_{in}, and the variable

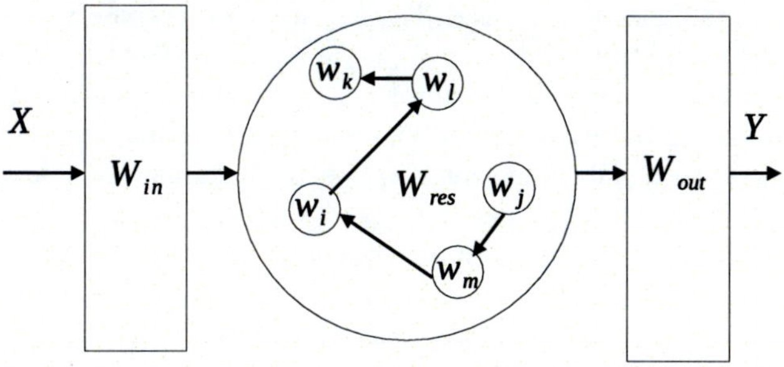

Fig. 2.1 Diagram of an ESNN for classification tasks

$x_5 \in [0, 100]$ is the seed to make a permutation of the reservoir matrix W_{res}. The leaking rate of the reservoir nodes can be regarded as the speed of the reservoir update dynamics dicretized in time. As we have solved a supervised classification task, the problem to optimize is to maximize the averaged accuracy of the ESNN using a fivefold cross-validation.

2.2.1 *Evaluation Function for Training an ESNN*

Let $X \in \mathbb{R}^{n \times c}$ be the matrix of the input features, and $Y \in \{0, 1\}^{n \times o}$ the target data, where n is the length of the data, c is the number of features, and o the number of classes.

We propose an evaluation function to train the ESNN that is shown in the Algorithm 1. Parameters x_3, x_4 and x_5 are coded as real numbers but only their integer parts are taken, as it is shown in lines 2–4 of Algorithm 1. The sign of the accuracy is negative because we use the software package in [8], which has a optimization procedure that only minimizes.

Algorithm 1 Evaluation function for training an ESNN

Input: X the samples, Y the target of the samples, c the number of features, o the number of classes, k the number of folds, x_1 noise variable, x_2 leaking rate variable, x_3 reservoir size variable, x_4 weights seed variable, and x_5 permutation seed variable.

Output: The negative of the accuracy average.

1: **procedure** EVALUATE_ESNN($X, Y, c, o, k, x_1, x_2, x_3, x_4, x_5$)
2: $x_3' = \lfloor x_3 \rfloor$
3: $x_4' = \lfloor x_4 \rfloor$
4: $x_5' = \lfloor x_5 \rfloor$
5: $W_{\text{in}}, W_{\text{res}} = \text{ESNN}(x_3', x_4', x_5')$ ▷ Initialization of the ESNN
6: accuracy = CrossValidation($X, Y, c, o, k, W_{\text{in}}, W_{\text{res}}, x_1, x_2, x_3', x_4'$)
7: **return** (−accuracy)
8: **end procedure**

We need to initialize the $W_{\rm in}$ and $W_{\rm res}$ matrices. We propose to use a diagonal matrix for the initilization of matrix $W_{\rm res}$, the reason of this idea is because it is easier to store it, its computations can be faster, and it is a regular matrix. We use for the initialization the Algorithm 2. Variable x_3' defines the size of the matrices, variable x_4' is the seed of random values for both matrices, and variable x_5' is the seed to create a permutation of the matrix $W_{\rm res}$, by shuffling first a vector of x_3' values with content $[0, 1, \ldots, x_3' - 1]$, as it is indicated in line 8 of Algorithm 2.

Algorithm 2 Initialization of the ESNN

Input: x_3' reservoir size, x_4' weights seed, and x_5' permutation seed, all these three variable are integer values.

Output: $W_{\rm in}$ input matrix, and $W_{\rm res}$ reservoir matrix.

1: **procedure** ESNN(x_3', x_4', x_5')
2:     seed(x_4')
3:     $W_{\rm in} \leftarrow$ rand()   ▷ $W_{\rm in} \in \mathbb{R}^{x_3' \times 2}$
4:     $W_{\rm res} \leftarrow 0$ ▷ $W_{\rm res} \in \mathbb{R}^{x_3' \times x_3'}$
5:     $d \leftarrow$ rand() $- 0.5$ ▷ $d \in \mathbb{R}^{x_3'}$
6:     $p \leftarrow [0, 1, \ldots, x_3' - 1]$
7:     seed(x_5')
8:     $p \leftarrow$ shuffle(p)
9:     **for** $i = 0 : x_3' - 1$ **do**
10:         $W_{\rm res}[\, i, p[i]\,] = d[\, p[i]\,]$
11:     **end for**
12:     $e \leftarrow$ maximum_eigen_value($W_{\rm res}$)
13:     $W_{\rm res} \leftarrow W_{\rm res}/e$
14:     **return** $W_{\rm in}$, $W_{\rm res}$
15: **end procedure**

Once the matrices $W_{\rm in}$ and $W_{\rm res}$ are initialized, we make a cross-validation with the data and the ESNN, which is described in the Algorithm 3. To obtain the accuracy, the input data X and its target values Y are split in k folds, and then one set of $k - 1$ folds is used as the training sets X_{train}, Y_{train}, and one set is used as the test sets X_{test} and Y_{test}. Then, we obtain the average of the accuracy of this results, and this is the accuracy obtained for that set of variables $\boldsymbol{x}$.

The training of the ESNN is described in the Algorithm 4, we can see that the variable x_1 just adds some noise to the features, but the variable x_2 controls how related are the features in each sample, and this is the feedback (or the echo) inside the ESNN. In line 10, we can see how $W_{\rm in}$ and $W_{\rm res}$ matrices are use to propagate each feature. The output of Algorithm 4 is the values of the elements of matrix $W_{\rm out} \in \mathbb{R}^{o \times (2 + x_3')}$, at a cost of a Moore–Penrose pseudoinverse. The size of $W_{\rm out}$ depends of the number of classes and the size of the reservoir. It is important to see that the values of this matrix change depending of the order of the data, that is why we use the cross-validation to measure the accuracy with different training and testing subsets of the same data. The inverse_activation() function on line 15 computes the inverse of the sigmoid function as it is described in the Eq. (2.1).

Algorithm 3 Cross validation algorithm

Input: X the samples, Y the target of the samples, c the number of features, o the number of classes, k the number of folds, W_{in} input matrix, W_{res} reservoir matrix, x_1 noise variable, x_2 leaking rate variable, x_3' reservoir size variable, and x_4' weights seed variable.

Output: Averaged accuracy of k folds.

procedure CROSSVALIDATION($X, Y, c, o, k, W_{\text{in}}, W_{\text{res}}, x_1, x_2, x_3', x_4'$)
    sum = 0
    for $i = 1 : k$ **do** ▷ k is the number of folds
        $X_{\text{test}}, Y_{\text{test}}, X_{\text{train}}, Y_{\text{train}}$ = folds(X, Y, k, i)
        $W_{\text{out}} = \text{FIT}(X_{\text{train}}, Y_{\text{train}}, c, o, W_{\text{in}}, W_{\text{res}}, x_1, x_2, x_3', x_4')$
        $Y_p = \text{PREDICT}(X_{test}, c, W_{\text{res}}, W_{\text{in}}, W_{\text{out}}, x_1, x_2, x_3', x_4')$
        sum = sum + accuracy(Y_p, Y_{test})
    end for
    return sum/k ▷ Returns the averaged accuracy of k folds
end procedure

Algorithm 4 Fit the data (calculates W_{out} matrix)

Input: X the samples, Y the target of the samples, c the number of features, o the number of classes, W_{in} input matrix, W_{res} reservoir matrix, x_1 noise variable, x_2 leaking rate variable, x_3' reservoir size variable, and x_4' weights seed variable.

Output: W_{out} output matrix.

procedure FIT($X, Y, c, o, W_{\text{in}}, W_{\text{res}}, x_1, x_2, x_3', x_4'$)
    $n = \text{length}(X)$
    $v \in \mathbb{R}^{x_3'}$
    $V \leftarrow 0$ ▷ $V \in \mathbb{R}^{(2+x_3')\times(c\times n)}$
    $Y_T \in \mathbb{R}^{o\times(c\times n)}$
    for $i = 0 : n - 1$ **do** ▷ Analize each sample in the dataset
        $v \leftarrow [0]$
        for $j = 0 : c - 1$ **do** ▷ Analize each feature in the sample
            $u = X[i, j]$
            $T = W_{\text{in}} \cdot [1, u]^T + W_{\text{res}} \cdot v$ ▷ Progate each feature in the network
            $v = v \cdot (1 - x_2)$
            $r = \text{rand}(x_4') - 0.5$
            $v = v + x_2 \cdot \tanh(T + (r \cdot x_1))$ ▷ Connection between the features
            $V[:, i \cdot c + j] = [1, u, v^T]^T$
            $Y^{\text{T}}[:, i \cdot c + j] = \text{inverse\_activation}(Y[i, :])^T$ ▷ Set the target values
        end for
    end for
    $W_{\text{out}} = Y^{\text{T}}\, V^{+}$ ▷ $V^{+} \in \mathbb{R}^{(c\times n)\times(2+x_3')}$
    return W_{out} ▷ $W_{out} \in \mathbb{R}^{o\times(2+x_3')}$
end procedure

$$f(x) = \ln \frac{0.01 + 0.98x}{0.99 - 0.98x} \tag{2.1}$$

The Algorithm 5 does the prediction for a given input, this is the algorithm that we use to measure the accuracy of the training. We can see that we need the three matrices W_{in}, W_{res}, and W_{out}, and the values of variables x_1, x_2, x_3' and x_4'. So, it is important to store those matrices and variable values. If we analyze the Algorithm 2, the matrices are created with the set of variables $\boldsymbol{x}$, and to obtain W_{out} we also need the values for all the elements in the set of parameters. Hence, we can describe the whole ESNN with the set of variables $\boldsymbol{x}$ and the W_{out} matrix. In Fig. 2.2, it is the flowchart of this algorithm for one sample.

Algorithm 5 Predict

Input: X the samples, c the number of features, W_{in} input matrix, W_{res} reservoir matrix, W_{out} output matrix, x_1 noise variable, x_2 leaking rate variable, x_3' reservoir size variable, and x_4' weights seed variable.

Output: Y_p the predicted class for each sample.

procedure PREDICT($X, c, W_{\text{res}}, W_{\text{in}}, W_{\text{out}}, x_1, x_2, x_3', x_4'$)
  $n = \text{length}(X)$
  for $i = 0 : n-1$ **do** ▷ Analize each sample in the dataset
    $v \leftarrow [0]$ ▷ $v \in \mathbb{R}^{x_3'}$
    for $j = 0 : c-1$ **do** ▷ Analize each feature in the sample
      $u = X[i, j]$
      $T = W_{in} \cdot [1, u]^T + W_{res} \cdot v$ ▷ Progate each feature in the network
      $v = v \cdot (1 - x_2)$
      $r = \text{rand}(x_4') - 0.5$
      $v = v + x_2 \cdot \tanh(T + (r \cdot x_1))$ ▷ Connection between the features
      $V[:, j] = [1, u, v^T]^T$ ▷ $V \in \mathbb{R}^{(2+x_3') \times (c)}$
    end for
    $y = W_{out} V$ ▷ $y \in \mathbb{R}^{o \times c}$
    $\bar{y} = \text{mean}(y)$ ▷ $y \in \mathbb{R}^{o \times 1}$, is the mean of the rows
    $Y_p[n, :] = \text{argmax}(\bar{y}^T)$ ▷ $Y_p \in \mathbb{R}^{n \times o}$
  end for
  return Y_p
end procedure

2.3 Heuristics GA and PSA

2.3.1 Genetic Algorithm

The genetic algorithm was initially conceived by Holland as a means of studying adaptive behaviour, and they have been considered as function optimization methods [9], their parameters consist of the number and range of the variables, the size of the population, maximum number of generations, operators of crossover, mutation, selection, the evaluating functions and the restrictions, the pseudocode it is presented in the Algorithm 6.

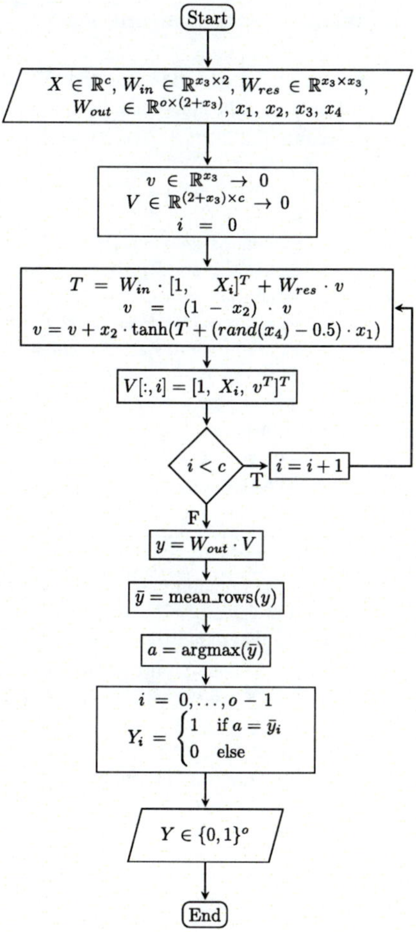

Fig. 2.2 Flowchart of the ESNN's prediction for classification tasks, given a sample, the W_{in}, W_{res}, and W_{out} matrices and the noise and leaking rate parameters

2.3.2 *Pattern Search Algorithm*

Pattern search is a direct search routine for minimizing a function $f(x)$ of several variables $\boldsymbol{x}$, the direct methods do not use derivatives. The argument $\boldsymbol{x}$ is varied until the minimum of $f(x)$ is obtained [10]. In each iteration, it takes the point that minimizes $f(x)$ or shrinks the step if there is no better point, until the algorithm cannot make more steps. The pseudocode for this algorithm is presented in the Algorithm 7.

Algorithm 6 Genetic algorithm

Input: P_c crossover probability, P_m mutation probability, G_{max} maximum number of generations, P_{size} population size, and $f(x)$ function to evaluate.

Output: I_{best} the individual that minimizes the function.

procedure GENETICALGORITHM(P_c, P_m, G_{max}, P_{size}, $f(x)$)
  $gen = 0$
  P=InitializePopulation(P_{size}, x) ▷ x is the value of each individual
  while $gen < G_{max}$ **do**
    Evaluate P in $f(x)$
    I_{best} = Elitism(P) ▷ Retain the best individual
    Selection ▷ Select the best individuals
    P_n = Crossover(P_c) ▷ Make the crossover of the best individuals with a probability P_c
    P_n = Mutation(P_m) ▷ Make the mutation of each individual with a probability P_m
    $P = P_n$
    $gen = gen + 1$
  end while
  return I_{best}
end procedure

Algorithm 7 Pattern search

Input: $f(x)$ function to evaluate, x starting point, δ exploration move, λ minimum step size, and ρ reduction value

Output: x that minimizes the function.

procedure PATTENSEARCH($f(x)$, x, δ, λ, ρ)
  while $\delta > \lambda$ **do** ▷ λ is the minimun step size and δ is the initial step size
    Evaluate $f(x)$
    Make a exploratory move $x_e = x \mp \delta$
    if $f(x_e) < f(x)$ **then**
      $x = x_e$ ▷ Set a new base point
    else
      $\delta = \delta \times \rho$ ▷ Reduce the exploratory step
    end if
  end while
  return x
end procedure

2.4 Results

We used ten different datasets to test our evaluation function, the datasets are described in the Table 2.3. The samples in the datasets are normalized within the range $[-1, 1]$. Also some datasets have categorical values, and the target values of the samples need to be changed to the dimensions of Y as described in Sect. 2.2.1.

We use three algorithms to train the ESNN, the first one is a Manual Algorithm that is described in the Algorithm 8, in this case we set the variables x_1, x_2, and x_4 as constants and the variable x_3 has a vector of values in every value we make e experiments and in every evaluation we generate p random permutations of the matrix W_{res} and store the best matrix. For the data presented in the Table 2.4, we use

$e = 5$, $p = 20$, and 13 different values for x_3, this is why the number of evaluations are 1300. After the e evaluations, we obtain the mean value of the accuracy and the best value of the accuracy is the value presented in Table 2.4. This algorithm could be seen as a semi-exhaustive way of training the ESNN, because we need to analyze how is the accuracy for different size of the reservoir to get the best accuracy.

Algorithm 8 Algorithm for manual adjustment of reservoir size

Input: X the samples, Y the target of the samples, c the number of features, o the number of classes, k the number of folds, e the number of experiments, and p the number of random permutations.

Output: The average accuracy and a set of W_{in} input matrix and W_{res} reservoir matrix.

procedure MANUAL_ALGORITHM(X, Y, c, o, k, n, p)
    $x_1 = 0.0$
    $x_2 = 1.0$
    $x_4 = 2$
    $r = [5, 10, 25, 50, 75, 100, 150, 200, 250, 300, 500, 750, 1000]$
    mean $\in \mathbb{R}^{13} \leftarrow 0$
    for $i = 1 : 13$ **do**
        $x_3 = r[i]$
        sum = 0
        for $j = 1 : e$ **do** ▷ Do e experiments
            accuracy = 0
            for $l = 1 : p$ **do** ▷ Generate p random permutations
                $x_5 = \leftarrow$ rand()
                $W_{\text{in}}, W_{\text{res}} = \text{ESNN}(x_3, x_4, x_5)$
                $a = \text{CrossValidation}(X, Y, c, o, k, W_{\text{in}}, W_{\text{res}}, x_1, x_2, x_3, x_4)$
                if $a >$accuracy **then**
                    accuracy = a
                    $W'_{\text{res}} = W_{\text{res}}$
                    $W'_{\text{in}} = W_{\text{in}}$
                end if
            end for
            sum = sum + accuracy
        end for
        mean[i] = sum/5
        save($W'_{\text{in}}, W'_{\text{res}}$) ▷ Save the matrices to a file
    end for
    return mean
end procedure

The other two algorithms are implemented in the library pymoo [8], the parameters of the genetic algorithm [11] are presented in the Table 2.1, let P_{size} be the population size, C the type of crossover, P_c the probability of crossover, M the type of mutation, and P_m the probability of mutation. The parameters of the pattern search algorithm [12] are presented in the Table 2.2, let δ be the value for the exploration move and ρ the reduction value, in these algorithms we use the whole set $\boldsymbol{x}$ described in Sect. 2.2.1 and we introduce the variable x_5 to look for the best permutation seed like a variable and make it more deterministic. And we use

Table 2.1 Parameters for the genetic algorithm

P_{size}	Representation	C	P_c	M	P_m	Selection
70	Real	SBX	0.5	PM	0.2	Binary tournament

Table 2.2 Parameters for the pattern search algorithm

δ	ρ	Step size
0.25	0.5	1

Table 2.3 Datasets

Dataset	Length	Features	Classes	Source
Balance scale	625	4	3	[13]
Banana	5300	2	2	[14]
Banknote authentication	1372	4	2	[13]
Breast tissue	106	9	6	[13]
Cardiotocography	2126	22	3	[13]
Diabetic retinopathy	1151	19	2	[13]
Image segmentation	2315	19	7	[13]
Pima Indian Diabetes	762	8	2	[15]
Seeds	210	7	3	[13]
Wine	178	13	3	[13]

the evaluation function in the Algorithm 1. In Table 2.4 is the comparison between the accuracy obtain using the three different algorithms to train the ESNN. All the tests were performed on two different computers, computer A has the following specifications: Intel i7-13700, 16 GB RAM, and Ubuntu 22.04, and computer B has the following specifications: Intel Xeon E5-2658 v2, 256 GB RAM, and Fedora 30 (Workstation Edition).

Observe that the accuracy obtained by the three algorithms is similar, also in some cases the values for the size of the reservoir are almost the same. But the values for the noise, leaking rate, and the seeds are different. It is important to remark the number evaluations and the time (minutes), here it is the main difference between these algorithms. In all the cases thee PSA makes less evaluations and time than the manual and genetic algorithm. In Figs. 2.3 and 2.4, we analyze the results obtained, we compare the values for the size of the reservoir, leaking rate, and noise, and if the point is more red, it has higher accuracy, and if it is more blue, it has lower accuracy. The first thing to notice is that the search space is bigger in the genetic algorithm, we can observe this in Fig. 2.4, also there are three well-defined sets of points with a higher accuracy, and if we see Fig. 2.3, it is possible to see that space search is lower than the genetic algorithm, but the set of points with the highest accuracy matches one of the three sets of points of the genetic algorithm. We cannot affirm that this is the global maxima, because the other two sets in the genetic algorithm look like they have a similar accuracy.

In Fig. 2.5, we evaluate the size of the reservoir against the accuracy, the orange curve is the result of manual algorithm, the red dot is the best result found by the

Table 2.4 Results comparison (see the text for details of the computers)

Dataset	Alg.	x_1	x_2	x_3	x_4	x_5	Accuracy	Evals.	Time (min)	Comp.
Balance scale	GA	0.58	0.78	207	13	81	0.84	700	12	
	PSA	0.20	0.73	173	0	56	0.84	120	2	A
	Manual	0.00	1.00	200	2	–	0.84	1300	17	
Banana	GA	0.49	0.98	478	31	51	0.90	700	76	
	PSA	0.58	0.49	395	96	26	0.90	111	8	A
	Manual	0.00	1.00	150	2	–	0.89	1300	68	
Banknote authentication	GA	0.51	0.55	503	7	47	1.00	700	36	
	PSA	0.76	0.52	224	21	62	0.99	111	4	A
	Manual	0.00	1.00	500	2	–	0.99	1300	33	
Breast tissue	GA	0.30	0.03	37	14	69	0.74	770	4	
	PSA	0.55	0.12	76	29	71	0.70	151	1	A
	Manual	0.00	1.00	100	2	–	0.71	1300	8	
Cardiotocography	GA	0.86	0.29	987	32	16	0.86	1050	1600	
	PSA	0.52	0.29	999	81	92	0.91	119	198	B
	Manual	0.00	1.00	1000	2	–	0.89	1300	935	
Diabetic retinopathy	GA	0.74	0.85	961	77	60	0.76	840	227	
	PSA	0.35	0.60	924	37	4	0.76	110	48	A
	Manual	0.00	1.00	1000	2	–	0.75	1300	134	
Image segmentation	GA	0.17	0.04	986	85	82	0.94	910	1341	
	PSA	0.77	0.04	1000	81	67	0.94	170	270	B
	Manual	0.00	1.00	1000	2	–	0.90	1300	890	
Pima Indian diabetes	GA	0.17	0.17	508	82	14	0.78	1820	89	
	PSA	0.89	0.74	589	11	47	0.76	129	9	A
	Manual	0.00	1.00	75	2	–	0.76	1300	36	
Seeds	GA	0.59	0.51	61	21	16	0.92	1400	8	
	PSA	0.76	0.59	99	25	62	0.91	118	1	A
	Manual	0.00	1.00	300	2	–	0.91	1300	11	
Wine	GA	0.98	0.91	926	14	61	0.96	980	47	
	PSA	0.40	0.83	866	25	10	0.97	182	10	A
	Manual	0.00	1.00	750	2	–	0.94	1300	16	

genetic algorithm, and the blue dot is the best result found by the pattern search algorithm. The curve obtained by the manual algorithm lets us see the behavior of the accuracy of this dataset, and it is important to mention that every dataset generates a different behavior, in Fig. 2.6, it is the behavior of the accuracy for the Image Segmentation dataset, and we can observe that is completely different. But we can remark that the higher point in the curve is highly related with the best solutions found by the genetic and pattern search algorithms, even when the best values for the set of variables $\boldsymbol{x}$ are different.

The library pymoo needs an end criterion to stop the algorithms, we propose that if the value of the standard deviation of the minimum accuracy of the last ten

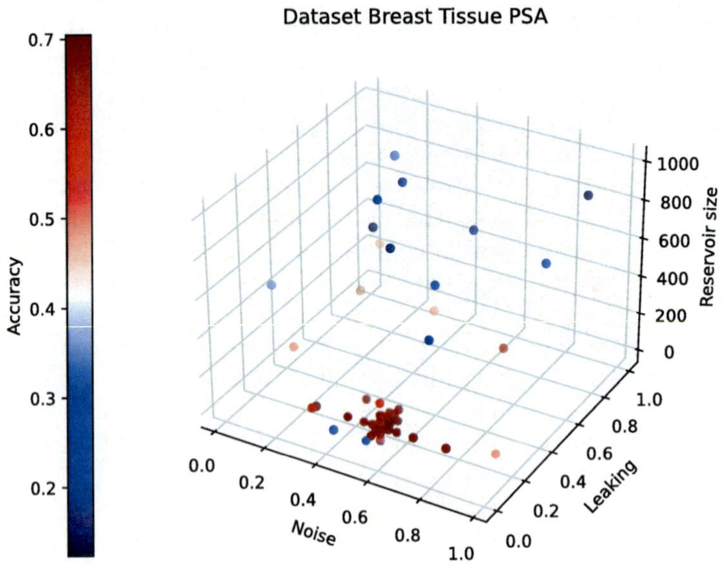

Fig. 2.3 Comparison of the evaluations made by the pattern search algorithm

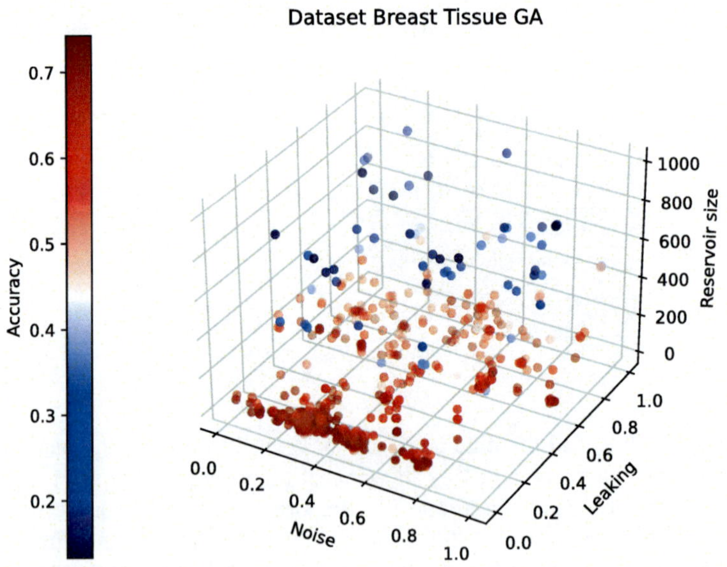

Fig. 2.4 Comparison of the evaluations made by the genetic algorithm

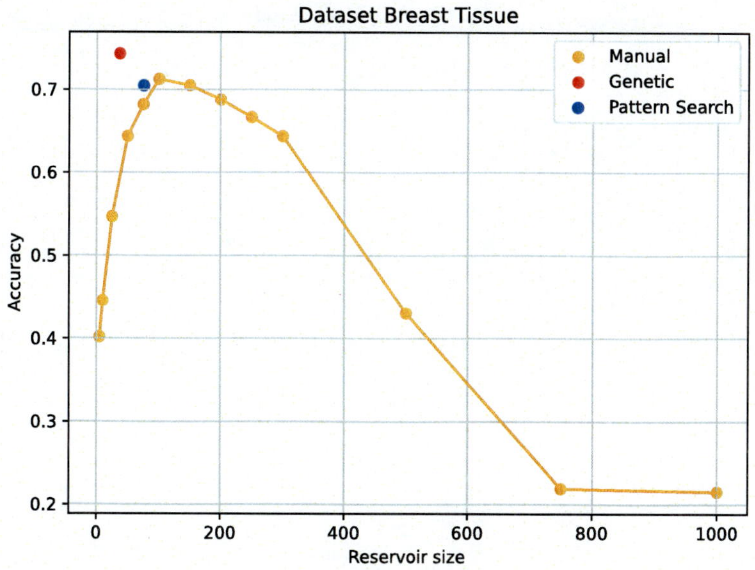

Fig. 2.5 Comparison between the results obtained by the manual algorithm, the genetic algorithm, and the pattern search algorithm for the Breast Tissue dataset

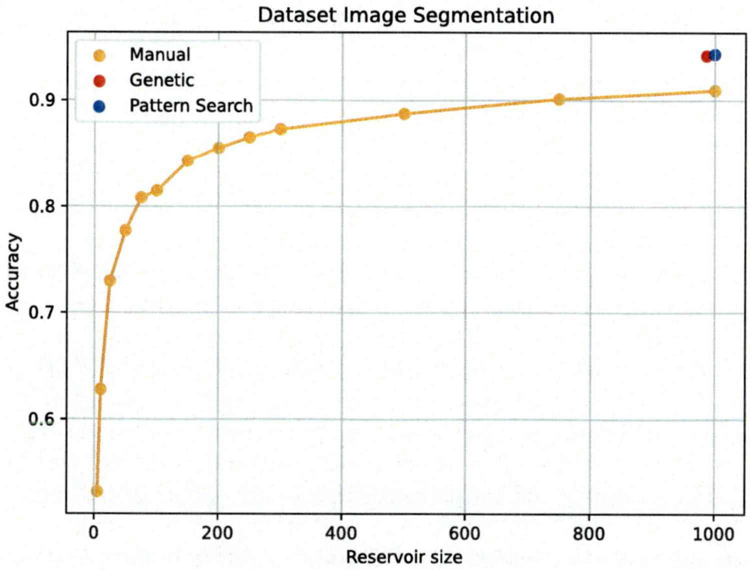

Fig. 2.6 Comparison between the results obtained by the manual algorithm, the genetic algorithm, and the pattern search algorithm for the Image Segmentation dataset

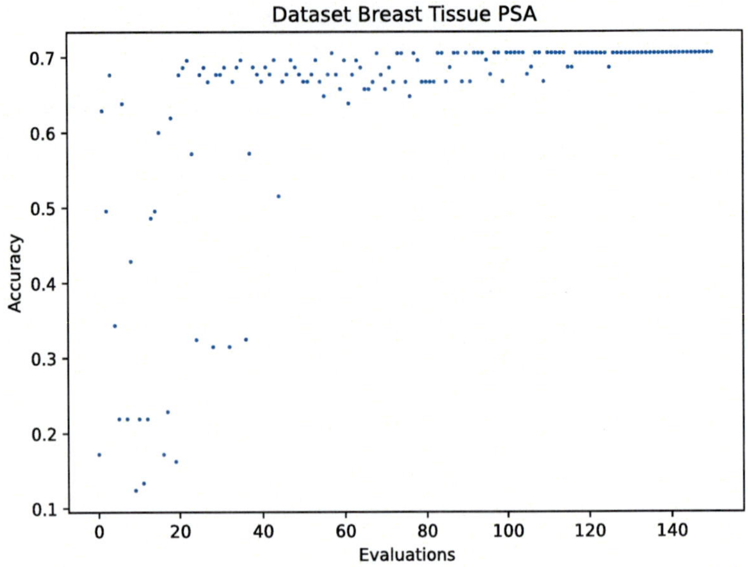

Fig. 2.7 Convergence graph of the pattern search algorithm for the Breast Tissue dataset

generations is lower than 0.001 stops the algorithm. In Figs. 2.7 and 2.8, we can see the differences between the number of evaluations between the genetic algorithm and the pattern search algorithm, but even so, their accuracy is very similar.

In Sect. 2.2.1, we mentioned that the cost of the training is to obtain the pseudoinverse of matrix, but we can analyze in Algorithm 4 the dimension of the matrix V^+. Lets take the example for the features and results of the Image Segmentation dataset, it has 2315 samples, 19 features, and 7 classes, for the manual algorithm we obtained a size of the reservoir of 1000 and the number of samples is 1852, this is because of the fivefold, then the size of V^+ is (35188×1002), after obtaining the V^+ we multiply with the Y_T matrix that has a dimension for this example of (7×35188) and we need to compute this five times because of the cross-validation. So, we can see that the cost to compute the output matrix W_{out} is directly related with the number of samples, the number of features, the number of classes, and the size of the reservoir. Therefore, although it is only a pseudoinverse, the number of operations grows depending on size of the dataset. This can be seen in Table 2.4 in the column of time, the datasets with more features like Cardiotocography and Image Segmentation took longer to finish the ESNN training, on the other hand, we can also see that Diabetic Retinopathy and Wine datasets also have a lot features, but the length of the dataset is much smaller, so also the time of the training is shorter and something similar happens with the Banana dataset, but in this case the number of features is small and the length is large.

We can remark that the search space of the permutation depends of the size of the reservoir, but when we make the experiments using Algorithm 8, we did not

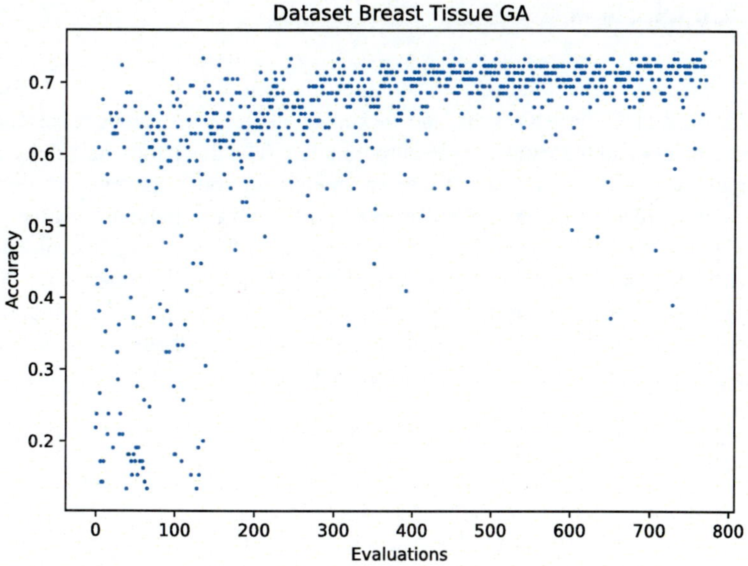

Fig. 2.8 Convergence graph of the genetic algorithm for the Breast Tissue dataset

know which was the best value for e and p, even that the value for p was really low for higher sizes of the reservoir, we obtain a good accuracy with only 20 random permutations. It it also not easy to specify the values for the noise and leaking rate, because as we can see in Table 2.4 their values change a lot depending on the input dataset. Therefore, we use the genetic algorithm and pattern search algorithm to optimize the parameters depending on the input dataset.

It is important to mention that in some cases the accuracy just starts to change after two decimal points, and that increases the number of evaluations. It would be better to only just look for the first two decimals as a stopping criterion, maybe it will reduce the number of evaluations.

Even that this ESNN used for classifications tasks does not have a feedback between the samples, as we mentioned before, the output matrix W_{out} will depend on the order of the samples in the input dataset.

It is not easy to choose randomly the best value for any of the ESNN's parameters in $\boldsymbol{x}$, even if we set some values as constants as it is presented in Algorithm 8, because every dataset needs different parameters, and its behavior is different.

In the results from Table 2.4, we can observe that in some datasets the accuracy is almost the same for GA, PSA, and manual results, but the reservoir size is smaller, this leaves us with the question if we can minimize the reservoir size and maximize the accuracy and treat it now as a biobjective optimization problem.

Another important thing is that maybe the time of the training can be optimized by changing the implementation to another language and also using GPUs.

2.5 Conclusions

We proposed an evaluation function and three algorithms to train an ESNN for classifications tasks with great results, and we show that an ESNN can be represented by five parameters and the values of the output matrix. The five parameters determinate the input and reservoir matrices. These parameters are the input noise and leakage, the size of the reservoir, and two integer seeds, one for the initialization of the weight values in the input and reservoir matrices, and another for shuffling the values in the reservoir. We used a minimum reservoir of a single neuron per each row. We can conclude that is possible to train an ESNN in systematic way, but it is important to remark that the ESNN for classification problems is just another tool of machine learning. We should compare different methods and choose the one that fits more to our needs.

We would like to extend the use of ESNNs to solve computer vision problems in real time.

References

1. Gouhei Tanaka, Toshiyuki Yamane, Jean Benoit Héroux, Ryosho Nakane, Naoki Kanazawa, Seiji Takeda, Hidetoshi Numata, Daiju Nakano, and Akira Hirose. Recent advances in physical reservoir computing: A review. *Neural Networks*, 115:100–123, 2019.
2. Ömer Faruk Ertuğrul. A novel randomized machine learning approach: Reservoir computing extreme learning machine. *Applied Soft Computing*, 94:106433, 2020.
3. Sotiris B Kotsiantis, Ioannis Zaharakis, P Pintelas, et al. Supervised machine learning: A review of classification techniques. *Emerging artificial intelligence applications in computer engineering*, 160(1):3–24, 2007.
4. Mantas Lukoševičius. A practical guide to applying echo state networks. In *Neural Networks: Tricks of the Trade: Second Edition*, pages 659–686. Springer, 2012.
5. Chenxi Sun, Moxian Song, Shenda Hong, and Hongyan Li. A review of designs and applications of echo state networks. *arXiv preprint arXiv:2012.02974*, 2020.
6. Mao Cai, Xingming Fan, Chao Wang, Linlin Gao, and Xin Zhang. Research for parameters optimization of echo state network. In *2018 11th International Symposium on Computational Intelligence and Design (ISCID)*, volume 2, pages 177–180. IEEE, 2018.
7. Aida A Ferreira and Teresa B Ludermir. Evolutionary strategy for simultaneous optimization of parameters, topology and reservoir weights in echo state networks. In *The 2010 International Joint Conference on Neural Networks (IJCNN)*, pages 1–7. IEEE, 2010.
8. J. Blank and K. Deb. pymoo: Multi-objective optimization in python. *IEEE Access*, 8:89497–89509, 2020.
9. Agoston E Eiben and James E Smith. *Introduction to evolutionary computing*. Springer, 2015.
10. Robert Hooke and Terry A Jeeves. "direct search" solution of numerical and statistical problems. *Journal of the ACM (JACM)*, 8(2):212–229, 1961.
11. J. Blank and K. Deb. Ga: Genetic algorithm. https://pymoo.org/algorithms/soo/ga.html, 2023. Accessed: 2023-07-09.
12. J. Blank and K. Deb. Pattern search. https://pymoo.org/algorithms/soo/pattern.html, 2023. Accessed: 2023-07-09.
13. UCI. Machine Learning Repository. https://archive.ics.uci.edu/, 2023. Accessed: 2023-07-09.
14. Banana. https://www.openml.org/, 2013. Accessed: 2023-07-09.
15. Pima Indians Diabetes Database. https://www.kaggle.com/datasets/, 2016. Accessed: 2023-07-09.

Chapter 3
Deep Learning in the Expansion of the Urban Spot

Eduardo Jiménez López

3.1 Introduction

Deep learning is one of the lines of artificial intelligence that performs data abstraction and characterization. Unlike machine learning, which requires human intervention in the process, deep learning transforms data and learns abstract features on its own [1]. In spectral satellite images with numerous bands, deep learning techniques are used to determine which bands are the most outstanding or contribute the most to the process of generating high-resolution images. They also use deep learning as a classifier that can perform high-precision comparison of images or image layers [2].

Within Artificial Neural Networks (ANNs) is deep learning that is used as a classifier of spatial information (i.e., places in geographic space). Efficiently configure the layers, size, and shape of the pixel [3]. In this work, ANNs linked to Cellular Automata (CA) can model the growth of the urban area, trying to determine which layers or factors that make up the image contribute the most or are most important in the growth of the city.

The ANN model offers important conceptual and operational advantages to urban planning. It allows simulating multiple city expansion dynamics, controlling the following factors: i. The functional relationship of the land with the transportation network (accessibility); ii. Physical variables of the environment that facilitate or hinder real estate development (suitability); and iii. Planning policies (zoning) [4].

In this work, we propose the search for the factor that contributes the most to the growth of the urban area, where each factor is managed as a layer with pixels of different values. For example, for the accessibility layer, the pixels located closest to

E. Jiménez López (✉)
El Colegio Mexiquense A.C., Ex hacienda Santa Cruz de los Patos S/N, Cerro del Murcielago, Mexico

E. Campos-Cantón et al. (eds.), *Complex Systems and Their Applications*,
https://doi.org/10.1007/978-3-031-51224-7_3

the main communication routes will have a value equal to 1 and the pixels far from the communication routes will have a value very close to 0 (i.e., pixels are assigned a threshold value) [5].

The model considers the possibility that an importance weight (weight) is added to each factor or layer. This allows, for example, to make an area more attractive where one wants to observe its growth behavior in the coming years (i.e., to propose growth scenarios for the urban area). The idea is to give the user control in the simulation of the model: include the voice of experts regarding where the city will grow, where the value of the land changes, or where real estate developments are most likely to be built (including privileged information from businessmen and officials) [6].

One type of ANN is the multilayer perceive, which is made up of input layers, hidden layers, and output layer [7]. It is composed of a series of neurons that are responsible for receiving, processing, classifying, and sending data to other neurons based on a unidirectional flow of information that is activated by receiving information from the previous layer, weighting it, adding it, and based on a function of activation, the neuron outputs occur, that is, it learns and classifies the most important factor in the growth of the urban area [8].

The way ANN is used in this work, a stage is added to the model where the data obtained are converted into spatial data. To determine where and when these data affect or benefit urban expansion. CAs are systems that change in time and space, they are dynamic systems. The space is divided into a finite number of cells (pixels or cells), and each cell can take a finite number of states (in our example, there are two: built and unconstructed). Cell values are updated individually separately. That is, the cells are updated every certain period, for example, it is the ticking of a clock, defined by the analyst (specified in the transition instructions). This update depends on the present value of the cell, as well as the values of its two neighboring pixels [9, 10].

In our work, the transition rules reflect the spatiotemporal behavior of the process to be simulated (e.g., urban expansion) and apply to any system. Each pixel transitions from one state to another based on transition rules, its state, and the state of its neighbors [11]. In urban expansion modeling, transition rules are linked to the behavior of land use change and factors that drive or slow down growth. All this allows CAs to generate alternative scenarios of urban expansion [12].

We add some interesting components to the transition rules, such as the transition probability of each cell. This probability results from synthesizing each pixel: the neighborhood effect (whether its neighbors are built or not), accessibility to primary communication routes, suitability for real estate development (flat areas are usually more attractive than areas with steep slopes), and municipal zoning that determines which areas are susceptible to construction and which areas are restricted. We also add a stochastic factor to avoid over-determinism [13].

The ANN model offers important conceptual and operational advantages to urban planning. It allows simulating multiple city expansion dynamics, controlling the following factors: i. The functional relationship of the land with the transportation network (accessibility); ii. Physical variables of the environment that facilitate or hinder real estate development (suitability); and iii. Planning policies (zoning). We look for the factors that contribute the most to the growth of the urban area, where each factor is managed as a layer with pixels of different values. For example, for the accessibility layer, the pixels located closest to the main communication routes will have a value equal to 1 and the pixels far from the communication routes will have a value very close to 0 (i.e., pixels are assigned a threshold value).

The writing is organized into four sections. The first presents an introduction to the topic of study and its importance in urban planning. The second section details the design methodology of the proposed model and exposes the city being analyzed. In the third section, the results obtained are presented, and finally in the fourth section, the conclusions that highlight significant points of the research are presented.

3.2 Methodology

Urban growth does not occur randomly, but rather results from general actions that allow the formulation of models that explain its behavior [14]. Those actions that drive urban growth, whether geographical, social, or economic, are defined in the literature as driving forces or factors, which are considered in urban planning studies and are part of urban growth models [15].

The urban area of Toluca in recent decades has become a more dynamic area of urban growth in the center of the Mexican Republic. This dynamism has also caused changes in urban growth patterns, with important consequences for the urban structure. The urban area contains a large amount of protected land such as national parks. Examples of these areas are the Nevado de Toluca volcano or the Sierra Morelos park. This represents a significant constraint on growth, although the natural attractions of such areas often lead to increased interest in developing the surrounding sites of the Toluca urban area. Consequently, the City of Toluca is considered an interesting place to study the dynamics and growth of the City [16].

The model proposed in this work requires three stages: Data Preparation, Artificial Neural Networks, and Cellular Automata Model, as shown in Fig. 3.1.

3.2.1 Data Preparation

This stage consists of adapting the satellite images of the City of Toluca, coming from the Landsat 8 satellite, whose repository is located in [17]. Once the image is downloaded, an attempt is made to reduce the errors produced by the capture. The types of corrections are atmospheric, radiometric, and geometric [18]. The data that are analyzed have characteristics of the georeferencing type. Combinations between the spectral bands are possible, which allow working with areas of greatest interest, urban area, vegetation, or restrictive areas [19].

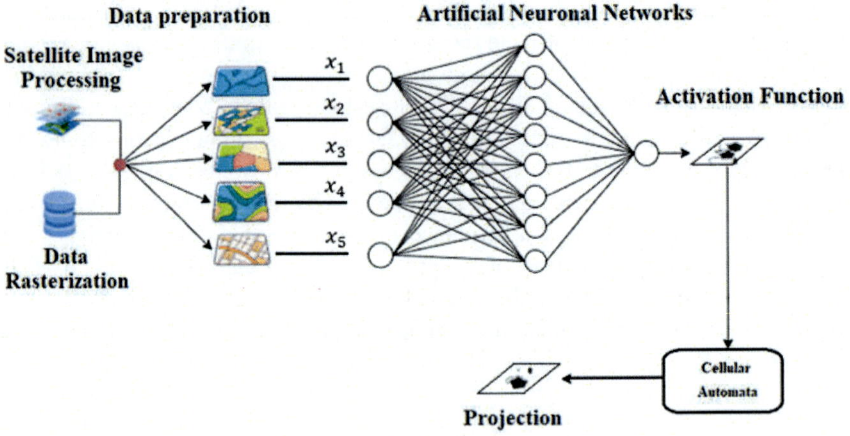

Fig. 3.1 Proposed model of growth of the Urban area

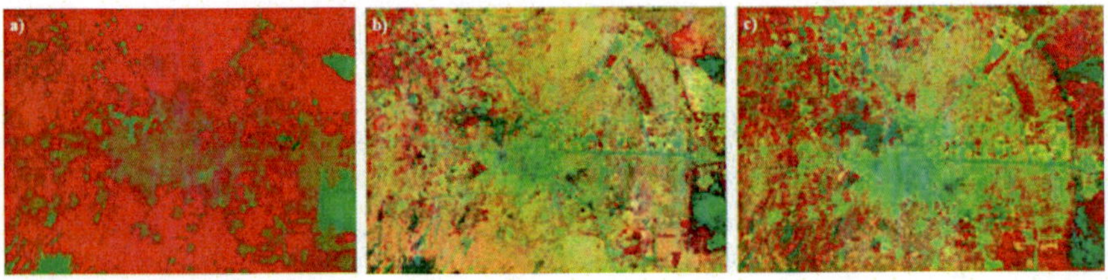

Fig. 3.2 Work layers

Geographic Information Systems (GISs) are a useful tool for preparing data that are analyzed such as images from satellites. They are defined as a database management system capable of capturing, storing, retrieving, analyzing, and displaying spatially defined data [20].

The representation of geographical areas in digital formats and their characteristics are in vector and raster formats [21]. The vector format represents geographic space with geometric figures [22]. Spatial information and model attribute information are linked using an identification number provided to each feature on a map. The raster format is made up of a matrix of cells or pixels organized in rows and columns (i.e., pixel matrix), in which each cell contains a value that represents information, such as temperature or type of terrain. The source of raster data is digital aerial photographs, satellite images, digital images, or even scanned maps [23].

We work with four layers in raster format: accessibility, suitability, restrictions, and neighborhood. Accessibility determines the space where growth is most likely due to the factor of proximity to the main communication routes, see Fig. 3.2a. Suitability is the areas where it cannot be built due to restrictions by the authorities, such as water restrictions, wetlands, bodies of water, rivers, banks, water currents, canals, see Fig. 3.2b.

The base restrictions are the areas where you cannot build, because they are protected natural areas, shopping centers, markets, aerodrome, cemetery, median, school, medical care center, bodies of water, waste deposit, building, sports facility or recreational, treatment plant, storage warehouses, bus station. Industrial facility, government palace, plaza, dam, archaeological features, electrical substation, fuel tank, water tank, temples, among others, see Fig. 3.2c. The neighborhood is a very important component for the proposed model, because it seeks to have a deep knowledge of the neighborhood dynamics in the city and thus find the best spatialization rule for the data.

3.2.2 *Artificial Neural Network (ANN)*

Models and simulations are used to understand, test, and experiment with theoretical methodologies related to complex processes. Using models and simulations, different plans are studied and possible scenarios are examined [24]. In this work, a multilayer perceptron ANN is used that can be described with input layers, hidden layers, and output layer, composed of a series of neurons that are responsible for receiving, processing, and sending data to other neurons from a flow of information that is activated by receiving information from the previous layer, weighting it, adding it, and then from an activation function, the outputs of the neuron are produced, the prediction [25].

The input layers are called x_n, where n is the number of input variables to the ANN, which correspond to the explanatory variables. The connection weight from the input layer to the hidden layer is denoted as w_{ni}, the input n and the hidden neuron i. The connection weight between the hidden layer and the output layer is called the connection weight w_{ik} where the hidden neuron i and the output neuron k. With an output layer, a neuron with an output value equal to 1. The activation functions are represented by the summation symbol and an activation function, which is the sigmoid function. See Fig. 3.3.

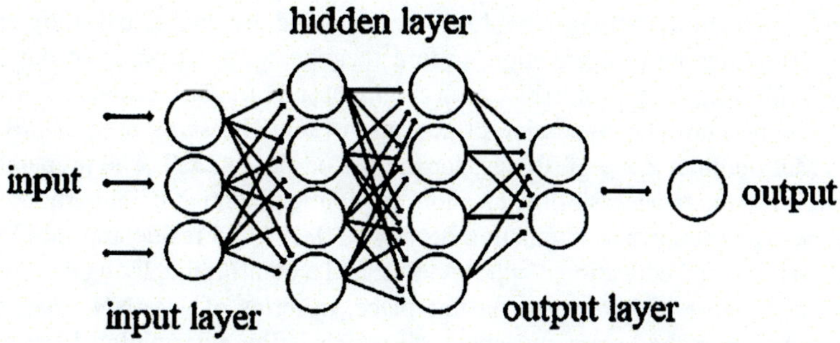

Fig. 3.3 Multilayer perceptron neural network with supervised learning

An example of supervised learning is shown in Fig. 3.3 and is carried out through backpropagation, a linear perceptron generalization with a least squared error algorithm. Learning occurs by changing the weights of the connections after each campaign on each node that processes it, based on the minimum expected error of the result. For each node i in each input n, the error is determined as $e_i(n) = d_i(n) - y_i(n)$, where d is the value of the map in our case, and the value produced by the perceptron is y. Corrections are made to the weights w_{ik} in each node, minimizing the error at the output of each node and in the output layer, given by Eq. 3.1.

$$\epsilon(n) = \frac{1}{2}\sum_i e_i^2(n). \tag{3.1}$$

The gradient decrease is determined by Eq. 3.2.

$$\Delta w_{ik}(n) = -\mu \frac{\partial \epsilon(n)}{\partial \varphi_i(n)} y_i(n). \tag{3.2}$$

Where μ is the learning rate, which is selected so that the weights reach a sufficiently fast response without producing variations. φ_i is the induced local field, calculated by the derivative, and which varies by itself. For each node, the derivative can be simplified shown in Eq. 3.3.

$$-\frac{\partial \epsilon(n)}{\partial \varphi_i(n)} = -\epsilon_i(n)\phi'(\varphi_i(n)). \tag{3.3}$$

Where ϕ' is the derivative of the function, which does not vary. For the change of weights in the hidden layer, therefore in the hidden nodes, it is shown that the derivative is relevant in Eq. 3.4.

$$-\frac{\partial \epsilon(n)}{\partial \varphi_i(n)} = \phi'(\varphi_i(n)) \sum_k -\frac{\partial \epsilon(n)}{\partial \varphi_i(n)} w_{ki}(n). \tag{3.4}$$

The connections between neurons are modeled by numerical weights that establish the degree of correlation, in turn these weights are adjusted during the network training stage, and then each neuron sends to the neurons of the next layer a value known as the value of activation [26]. The study area of this work is the Metropolitan Zone of Toluca, for the period 2013–2020. The probability of growth is related to accessibility, location, economic potential, neighborhood of the territory, topography, and population density [27]. A good residential location has better medical and educational opportunities, and a comfortable living environment that attracts more people to live in this place, in terms of an ANN what are the weights between the layers and above all which is the determining layer for the expansion. This population growth triggers an increase in the probability of growth in the city.

An advance proposed in this work is the management of ANN with MATLAB, which has tools and functions to manage large datasets specific to city maps. The software offers specialized toolboxes for working with neural networks. Data collection is the first step in ANN modeling, which is performed outside of the Deep Learning Toolbox software. In this case, the data to be worked on are prepared with GIS, mainly the union of spectral bands from satellite images. The software used are QGIS and Python, since they are free software and reduce operating costs.

A network object is to store all the information that defines a neural network that is made with the Deep Learning Toolbox software. This utility describes the basic components of a neural network, showing how they are created and stored in the network object. The configuration consists of organizing the network in a way that is compatible with the problem to be solved, defined by the map data in our case. Once the network is configured, you must adjust the network parameters, called weights, which optimize network performance. This adjustment process is called network training. Configuration and training require that the network receive data in the best possible way.

3.2.3 Data Spacialization

The next stage in the model proposed in this work is the spatialization of the data where the CA is used, see Fig. 3.1. The CA uses regular grids (i.e., pixels on maps) to represent the geographical space. The same neighborhood size or shape is generally used to identify those cells that make up the neighborhood. In CAs, a top-down analysis process is usually adopted to simulate patterns and processes of urban expansion; this proposed model includes a stochastic component, which simulates spontaneous growth as it happens in real life [28, 29].

The neighborhood is one of the most important factors in the CA, it is defined by the internal dynamics of the model, and the effects are observed in the local result [16]. The transition rules that determine the dynamics of the neighborhood can be deterministic or probabilistic, simple or highly elaborate. Deterministic rules start from the way of defining the neighborhood between cells. In a one-dimensional (i.e., linear) space, each cell for this work has two close neighbor cells with which it shares a border on the sides. They are altered from one moment to the next in a discrete time, through neighborhood rules that convert the cells into another configuration. There are different types of neighborhood analysis. Wolfran's rule 219 is proposed, which in many works is detected to be the best rule that simulates the growth of the urban area and especially the city of Toluca, it consists of the cells that surround the central cell as a configuration of tower [16, 28, 30].

With a parameter that we obtain from ANN, areas or sets of pixels that have a high probability of growth are determined. If the value is low, it introduces a disturbance into the model (areas where it cannot be built), areas where there are construction restrictions, inadequate accessibility, a slope greater than 30° on the land, or inappropriate neighborhoods. While if the value is high, areas that stand out above the others can be built without any problem. It has no problems with restrictions, very close accessibility to the main communication routes, a slope of less than 30°, and an adequate neighborhood for expansion. All the factors mentioned can be added to the ANN model and checked with CA. See Fig. 3.4.

3.3 Results

The classification of occupied and empty space gives a sample of the growth of the urban area in a given period of time. The ANN model allows us to know the transition probabilities that an area may have so that construction is carried out or not in it, that the pixels remain in one or another classification of occupied or empty space. A simple and efficient way to discretize the categories is to encode/filter as 1 to occupied space and 0 to empty space, Fig. 3.4.

The transition probability for the city of Toluca in 2013 is shown in Fig. 3.4a, five circles are observed in the following areas. In the north of the city of Toluca, the main growth area is in the following municipalities of Almoloya de Juárez and Temoaya. To the Northeast, the growth is in the municipalities of Lerma, Temoaya, Villa Cuauhtémoc, Xonacatlén, and Santa María Zolotepec. To the Southeast, the growth is in Metepec and San Mateo Atenco. To the west of the city of Toluca in the Municipality of Zinacantepec and finally to the Northwest in the Municipality of Almoloya de Juárez. The highlight of this analysis is that there are two areas that stand out from the others, the growth of Almoloya de Juárez and the Lerma Area with Temoaya, since the probability of growth in these areas is above the other areas, which may be due to the change in use of agricultural land to urbanized land.

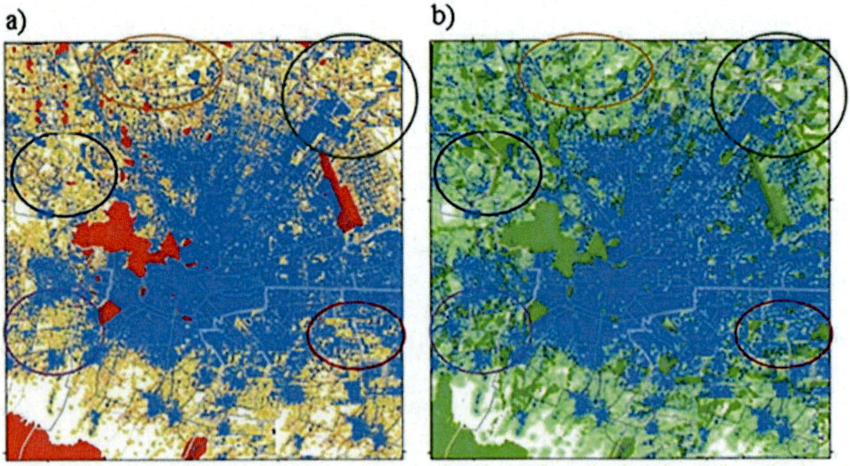

Fig. 3.4 Results obtained with ANN of the transition probability of the Metropolitan area of Toluca in 2013 and 2020

For 2020, the same phenomenon is observed in the same areas of the city of Toluca with an increase in the surfaces of the same areas marked with circles, this can be seen in Fig. 3.4b, which are the first results of the ANN in his training stage.

The proposed neural network is the combination of several layers and neurons, the connection between them, and the software used, there is no optimal structure for all applications. Determining the type of ANN and then its architecture is the best path for design. With the MATLAB Toolboxes, several ANN architectures were tested, varying the parameters of hidden layers, neurons in hidden layers, layer activation function, activation function in the output layer, and optimization until obtaining a loss of less than 25% in the architecture. With this result, it can be said that the network design is supervised.

The design architecture is determined by the number of layers, neurons, and type of network to implement. The ANN has been run on a computer with an Intel Core *i*7 3.1 GHz processor, 16 GB RAM, Windows 10 64-*bit* operating system, for a period of 23 days, testing a total of 601 architectures with 550 epochs for the metropolitan area of Toluca. Once the best architecture is defined, it is trained again for 1500 epochs, these results are shown in Table 3.1.

The explanation of the notation seen in Table 3.1, for the values (6, 9, 9, 13), should be interpreted as an architecture that consists of four hidden layers, with six neurons in the first layer, nine in the second layer, nine in the third layer, and thirteen in the last layer. The highest pressure is reached in the first 1000 epochs. The activation of hidden layers is carried out with the hyperbolic tangent function that transforms the entered values to a scale $(-1, 1)$, where high values asymptotically have 1 and very low values tend asymptotically to -1. For the activation output layer, the Sigmoid function is used because it exists between (0 and 1). It is especially used for models where we have to predict the probability as an outcome. Since the probability of anything exists only between the range of 0 and 1. Finally, the accuracy and pressure of the multilayer perceptron neural network its highest value are 81.12%.

Table 3.1 shows the results of each of the simulations and effectiveness of the proposed model. First, the validation technique used is shown. Followed by the percentage of efficiency of agreement for the base simulation results with the model projection (how closely they resemble each other at the pixel level), which is simply the percentage of cells that coincide spatially in both maps. In the following columns, the weights that were added to our simulation. Main effects and total effects of each parameter are provided for the comparison methods.

Table 3.1 ANN architecture results for metropolitan areas of the Toluca

Variable	Toluca
Hidden layers	4
Nodes in hidden layers	(6, 9, 9, 13)
Hidden layers activation	Hyperbolic tangent
Output layers activation	Sigmoid
Accuracy and precision	81.12%

In the final validation, after having entered all the necessary data into the proposed model, it is validated again with an image comparison/position technique. The technique is the Jaccard index, also considered a similarity coefficient, it compares the equality between two sets of samples. In this work, it is adapted for the comparison of images of the urban area that by its nature contains a large amount of data or pixels [31]. The index can be defined as the size of the intersection of two sets and the size of the union 3.5.

$$I_j = \frac{T_{11}}{T_{21} + T_{12} + T_{11}}. \quad (3.5)$$

In set A, the objects found in its domain are named T_{21}. In set B, the objects that are in its domain are labeled T_{12}. Objects found at the junction of the two sets are classified as T_{11}. Everything outside of these we label as T_{22}. When the inequality between the maps is total, the value of the Jaccard index is equal to 0.0, if the two maps have the same concordance position the value of the index is equal to 1.0. The results obtained in this work in what was done in the ANN have a value of equality in the index equal to 0.79, according to the accuracy and precision (see Table 3.1), obtained by the network model in the software is good results, very close in their values.

The classification of occupied and empty space gives us a sample of the growth of the urban area in a given period of time, the urban area is coded 1, for occupied space and 0 for empty space. In the city of Toluca for 2013, the pixels that have a value of 1 are 1,210,423. In 2020 the pixels in value of 1 are 2,002,030, an increase of 791,607 in 7 years, almost an increase of 40%.

With the use of Simulink tools in MATLAB, the projection is carried out with the help of the proposed model and the comparison indicators are estimated, which validates the multilayer perceptron neural network with a neighborhood rule that spatializes the data in the growth in the urban stain. The network is trained to make a projection of the urban area using the map of Toluca in 2013 as a reference to sketch the map in 2020 and thus make the comparison with the real map of 2020. Figure 3.5a shows the real map of Toluca in 2020. The simulation of the growth of the urban area for 2020 after going through the different processes shown in this work is shown in Fig. 3.5b.

To be certain of what is reported in this work and verify the results of the network, an evaluation of the training is carried out with the Jaccard comparison index, which is carried out in a separate mode. Toluca recorded a Jaccard of 0.81, which indicates a high capacity of the model to distribute the pixels on the map with high efficiency. The value of the comparison indicators is good, the entire model as a whole allows us to distribute the pixels with less uncertainty in the expansion of the urban area for the Toluca valley in a period of seven years. The multilayer deperceptron neural network could be an important instrument for urban planners.

The files and images used are massive and highly complex, but QGIS was able to handle all this information. Satellite images of the city of Toluca contain around 4 million pixels. Considering the magnitude of the information, the complexity

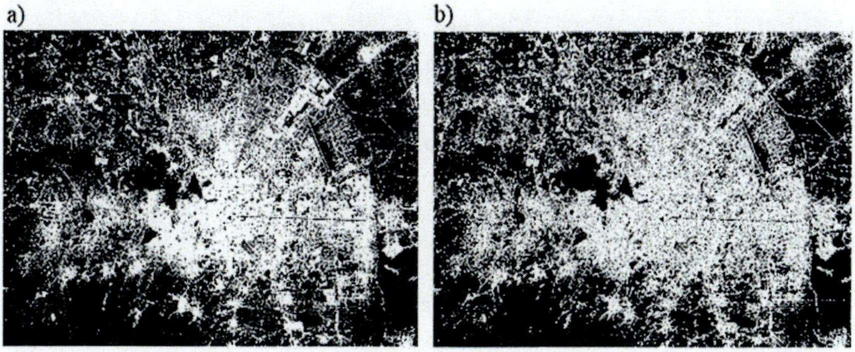

Fig. 3.5 Real maps and projections of the city of Toluca in 2020

of the calculations, the use of the neural network, the estimation of the urban comparison/expansion indexes, and the automatic generation of the simulation map, we can consider that the performance in the Perceptron Neural Network model Multilayer is fast and efficient in projection. On average, this model requires 18 days to generate the results and projections of the city, which translates into good results in the application.

With the certainty that the proposed model is efficient and has good results, we can generate a projection of the city of Toluca for 2027, Fig. 3.6. This map shows a balance in the values that changed from 0 to 1; therefore, it can be said that in a period of 7 years, the speed of growth of the city of Toluca was high, but this result needs to be verified with tests that are not discussed in this work. It is worth mentioning that mathematically this assertion is possible, but geographically it is more difficult for it to occur.

Figure 3.6 highlights the projection image of Toluca with areas that restrict the model used. With a visual inspection, we can realize that there are three areas that maintain their physiognomy almost without modifications up to 2027. In the center of the city of Toluca, you can see the Cerro de la Teresona, which is a national park and nature reserve. In the south, we see the second area where accelerated growth is not seen due to the Nevado de Toluca volcano. In the airport area to the north-east, in the municipality of Lerma, there is land destined for the Toluca airport, but in its surroundings there is a high number of buildings due to the logistical advantages that the airport provides.

3.4 Conclusions

In this work, an innovative technique was adopted to explore the growth of the urban area in the city of Toluca, artificial neural networks programmed in Matlab. The objective was to show how this growth-projection technique works, taking as

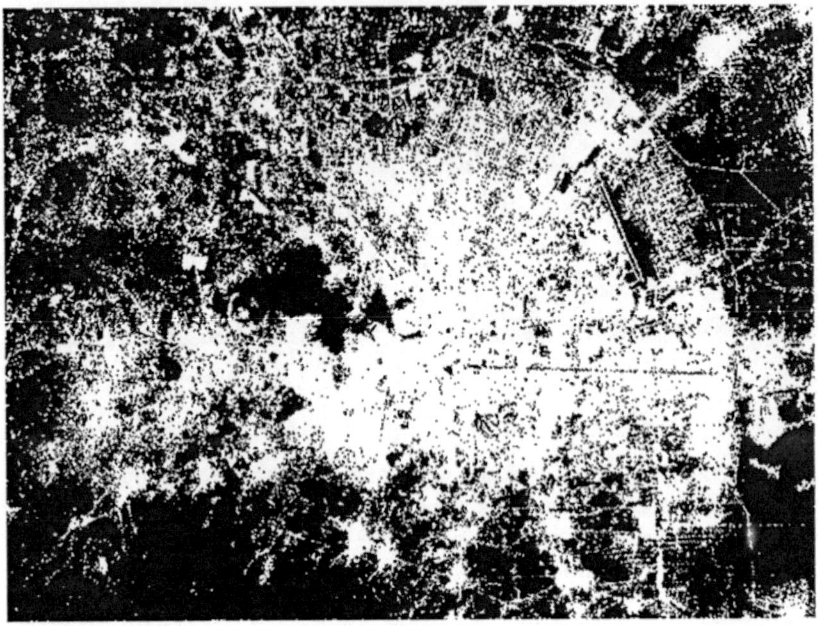

Fig. 3.6 Map of Toluca in the 2027 projection, use of the spatial multilayer perceptron neural network model

an example the Metropolitan Area of Toluca, which can be said to have a very high growth speed for being a city with more than 1 million inhabitants. The results show that the analysis approach with Neural Networks offers valuable information and an alternative vision to the traditional approach to city growth, it focuses mainly on the growth of the urban area with pixels in binary maps based on rasters.

The multilayer perceptron neural network efficiently performs spatiotemporal analysis, which is very difficult to find in techniques that project maps or perform growth on them. The power of the neural network is used to generate various resources that are used in urban planning. How to generate transition potentials in a given space, determine what factors promote growth determined by the weights in the network and an adequate projection after supervised training.

The union of a geographic information system with neural networks leads us to generate a tool programmed in MATLAB, with several programming variants in Code and others with graphic tools such as those provided by simulink. This software automates analysis processes and generates an interesting tool that can be used in different areas of knowledge.

The map comparison techniques with the Jaccard Index show us that the results are accurate. We continue to try to improve the variable over which we have no control, which is obtaining the maps for the analysis, everything depends on having the same capture conditions in the shot, but this cannot be achieved easily. To overcome all this, the factors that form the maps try to correct the differences in the neural network and where there is still a lot of work to be done and where we will focus our attention to overcome this obstacle.

With the results shown by the neural networks programmed in MALAB and GIS, we can say that it is possible to build a predictive model. It can be concluded that the combination of mathematical tools, programming and GIS in the analysis, supervision, and control of urban phenomena constitutes a very useful tool. If maps or images with good resolution, both spatial and temporal, are available, very precise information can be extracted for processing. The greater their precision, they greatly facilitate obtaining indices and show a correct projection of the cities under study.

Data Availability

The data used to support the findings of this study are included within the article.

Conflicts of Interest

The authors declare that there are no conflicts of interest regarding the publication of this work.

References

1. Yann LeCun, Yoshua Bengio, and Geoffrey Hinton. Deep learning. *Nature*, 521(7553):436–444, 2015.
2. Konstantinos Makantasis, Konstantinos Karantzalos, Anastasios Doulamis, and Nikolaos Doulamis. Deep supervised learning for hyperspectral data classification through convolutional neural networks. *In 2015 IEEE international geoscience and remote sensing symposium (IGARSS)*, pages 4959–4962, 2015.
3. Yu Yu, Tianran Jia, and Xuemei Chen. The how and where of plant micro rna's. *New Phytologist*, 216(4):1002–1017, 2017.
4. Xinxin Wu, Xiaoping Liu, Dachuan Zhang, Jinbao Zhang, Jialyu He, and Xiaocong Xu. Simulating mixed land-use change under multi-label concept by integrating a convolutional neural network and cellular automata: A case study of huizhou, china. *GIScience & Remote Sensing*, 59:609–632, 2022.
5. Shi Li, Jianping Chen, and Jie Xiang. Applications of deep convolutional neural networks in prospecting prediction based on two-dimensional geological big data. *Neural computing and applications*, 32:2037–2053, 2020.
6. Nithila Devi, B Sridharan, and Soumendra Kuiry. Impact of urban sprawl on future flooding in chennai city, india. *Journal of Hydrology*, 574:486–496, 2019.
7. Norbert Agana and Abdollah Homaifar. A deep learning based approach for long-term drought prediction. *In SoutheastCon 2017. IEEE*, pages 1–8, 2017.

8. Aneta Poniszewska-Maranda, Daniel Kaczmarek, Natalia Kryvinska, and Fatos Xhafa. Studying usability of ai in the iot systems/paradigm through embedding nn techniques into mobile smart service system. *Computing*, 101:1661–1685, 2019.
9. Al Adamatzky. *Cellular Automata: A Volume in the Encyclopedia of Complexity and Systems Science*. Springer, 2018.
10. Weixing Zhang, Weidong Li, Chuanrong Zhang, Dean Hanink, Yueyan Liu, and Ruiting Zhai. Analyzing horizontal and vertical urban expansions in three east asian megacities with the ss-co mcrf model. *Landscape and urban planning*, 177:114–127, 2018.
11. Charles Newland, Holger Maier, Aaron Zecchin, Jeffrey Newman, and Hedwig van Delden. Multi-objective optimisation framework for calibration of cellular automata land-use models. *Environmental modelling & software*, 100:175–200, 2018.
12. Yongjiu Feng and Xiaohua Tong. Incorporation of spatial heterogeneity-weighted neighborhood into cellular automata for dynamic urban growth simulation. *GIScience & Remote Sensing*, pages 1–22, 2019.
13. M. Munthali, S. Mustak, A. Adeola, J. Botai, S. Singh, and N. Davis. Modelling land use and land cover dynamics of dedza district of malawi using hybrid cellular automata and markov model. *Remote Sensing Applications: Society and Environment*, 17:100276, 2020.
14. Guztavo Buzai. Crecimiento urbano y potenciales conflictos entre usos del suelo en el municipio de luján (provincia de buenos aires, argentina): Modelado espacial 2016-2030. *Cuadernos geográficos de la Universidad de Granada*, 57(1):155–176, 2018.
15. M. De la Luz Hernández-Flores, E. Otazo-Sánchez, M. Galeana-Pizana, E. Roldán-Cruz, R. Razo-Zárate, C. ... González-Ramírez, and A. Gordillo-Martínez. Urban driving forces and megacity expansion threats. study case in the mexico city periphery. *Habitat International*, 64:109–122, 2017.
16. Eduardo Jiménez-López, Carlos Garrocho, and Tania Chávez. Autómatas celulares en cascada para modelar la expansión urbana con áreas restringidas. *Estudios demográficos y urbanos*, 36(3):778–823, 2021.
17. Earthexplorer. Repositorio online: https://earthexplorer.usgs.gov. 2020.
18. Heileen Arias, Rodolfo Zamora, and Christian Bolaños. Metodología para la corrección atmosférica de imágenes aster, rapideye, spot 2 y landsat 8 con el módulo flaash del software envi. *Revista Geográfica de América Central*, 53(2):39–59, 2014.
19. Edgar Jardón, Eduardo Jiménez, and Marcelo Romero. Spatial markov chains implemented in gis. *2018 International Conference on Computational Science and Computational Intelligence (CSCI)*, pages 361–367, 2018.
20. NCGIA. Fundamental research in geographic information analysis.
21. Paul Longley, Michael Goodchild, David Maguire, and David Rhind. *Geographic information systems and science*. John Wiley & Sons, 2005.
22. Cheryl Jones. *Geographical information systems and computer cartography*. Routledge, 2014.
23. Jonathan Campbell and Michael Shin. *Essentials of Geographic Information Systems*. Computer Sciences Commons, 2011.
24. M. Kaviari, F.and Mesgari and H. Seidi, E.and Motieyan. Simulation of urban growth using agent-based modeling and game theory with different temporal resolutions. *Cities*, 95:102387, 2019.
25. M. Gardner and S. Dorling. Artificial neural networks (the multilayer perceptron)âa review of applications in the atmospheric sciences. *Atmospheric environment*, 32(14-15):2627–2636, 1998.
26. Zulifqar Ali, Ijaz Hussain, Muhammad Faisal, Hafiza Mamona, Tajammal Hussain, Muhammad Yousaf, 1Alaa Mohamd, and Showkat Hussain. Forecasting drought using multilayer perceptron artificial neural network model. *Advances in Meteorology*, 2017.
27. Cong Ou, Jianyu Yang, Zhenrong Du, Pengshan Li, and Dehai Zhu. Simulating multiple land use changes by incorporating deep belief network into cellular automata: A case study in beijing-tianjinhebei region, china. *AGILE 2018 Lund*, pages June 12–15, 2018.

28. Eduardo Jiménez, Carlos Garrocho, and Tania Chávez. Modelando la expansión urbana con autómatas celulares: Aplicación de la estación de inteligencia territorial (christaller)®. *Geografía y Sistemas de Información Geográfica (GeoSIG)*, 11:1–26, 2018.
29. Eduardo Jiménez-López and Luis López-Rivera. Artificial neural networks in the application of the growth of the urban sprawl. *Pädi Boletín Científico de Ciencias Básicas e Ingenierías del ICBI*, 11(21):109–119, 2023.
30. Eduardo Jiménez López. Inverse filter in the growth of urban sprawl with cellular automata model. *Complex Systems and Their Applications: Second International Conference (EDIESCA 2021*, pages 231–247, 2022.
31. Eduardo Jiménez López. Cadenas de markov espaciales para simular el creccimiento del área metropolitana de toluca, 2017-2031. *Economa, sociedad y territorio*, 19(60):109–140, 2019.

Part II
Biomedical Advancements in Rehabilitation Through Complex Systems

Chapter 4
Solving Inverse Kinematics Problem for Manipulator Robots Using Artificial Neural Network with Varied Dataset Formats

Rania Bouzid, Jyotindra Narayan, and Hassène Gritli

4.1 Introduction

Robotics kinematics plays a pivotal role in various domains of robotics research. Understanding how robots navigate and orient in space is essential for developing effective control systems and algorithms [10, 25]. Such knowledge empowers researchers to enhance a robot's mobility, enabling them to plan trajectories for seamless movement and navigation [2, 12, 25]. Understanding kinematics has far-reaching effects, spanning various domains, including the deployment of robots in industrial environments and outer space exploration. Researchers and engineers have proposed several algorithms and techniques to tackle inverse kinematics problems for various robot types, including robotic arms, humanoid robots, and mobile robots [2]. Beyond its significance in general robotics, kinematics finds significant applications in specialized domains, like medical robots, where precision and stability are paramount for conducting intricate surgical procedures [24, 29]. Inverse kinematics is essential in many applications, most notably in robot motion planning, trajectory generation, and interactive control [1, 20, 22]. This concept

R. Bouzid (✉)
Higher Institute of Information and Communication Technologies, University of Carthage, Borj Cedria, Tunis, Tunisia
e-mail: rania.bouzid@istic.ucar.tn

J. Narayan
Mechatronics and Robotics Laboratory, Department of Mechanical Engineering, Indian Institute of Technology Guwahati, Guwahati, Assam, India

H. Gritli
Higher Institute of Information and Communication Technologies, University of Carthage, Borj Cedria, Tunis, Tunisia

Laboratory of Robotics, Informatics and Complex Systems (RISC Lab – LR16ES07), National Engineering School of Tunis, University of Tunis El Manar, Tunis, Tunisia

E. Campos-Cantón et al. (eds.), *Complex Systems and Their Applications*,
https://doi.org/10.1007/978-3-031-51224-7_4

finds particular significance in tasks that involve a robot's interaction with objects within its environment, encompassing object grasping, regardless of their diverse shapes and sizes, and executing intricate movements necessitating utmost precision in positioning and orientation [1, 25, 26].

Inverse kinematics equations pose greater challenges compared to forward kinematics equations [8, 11, 17, 24]. Solving inverse kinematics (IK) problems is generally more complex than solving forward kinematics (FK) problems. In forward kinematics, the task is determining the end-effector pose given the joint angles or positions. This can often be accomplished using straightforward geometric calculations. However, inverse kinematics involves finding the joint angles or positions corresponding to a desired end-effector pose. This requires solving a system of nonlinear equations, which can be computationally intensive and involve complex relationships and constraints. As the degrees of freedom (DoF) of a robotic arm increase, the complexity of the IK problem grows, and finding a solution becomes more challenging. Numerous scholars and research institutions have extensively studied inverse kinematics methods in robotics [3, 11, 16, 17, 22, 23, 25, 26, 31]. These methods can be categorized into four types: geometric, analytical, numerical, and intelligent, as shown in Fig. 4.1. Xu et al. [31] proposed an analytical algorithm for the kinematics inverse problem in palletizing robotics. Although this algorithm exhibits practicality to a certain extent, it often encounters the challenge of multiple analytical results corresponding to a single attitude, making it difficult to determine a unique solution. Empirical observation reveals that implementing the algorithm in robotics offers versatility but complicates the design procedure. Frequently, achieving the solution involves multiple iterations, leading to inefficiencies. In the study by Anschober et al. [3], the authors introduce an effective geometric algorithm for addressing the inverse kinematics challenge of a 6R robot manipulator intended for

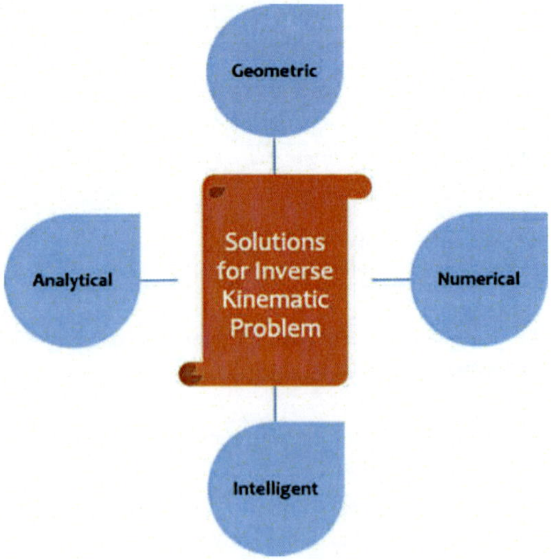

Fig. 4.1 Different methods for solving the inverse kinematics (IK) problem of manipulator robots

use with embedded control hardware. Additionally, the research delves into aspects like singularities, limits, workspace, and general solvability of the manipulator. In a recent work by Petelin et al. [26], an analytical relationship between platform and controlled coordinates is presented for a new 4-DoF kinematically redundant planar parallel grasping manipulator. However, such solutions become computationally expensive in the case of non-closed-form problems.

A closed-form solution might not be attainable in certain instances, necessitating numerical or iterative techniques, eventually leading to increased computational intricacy. Furthermore, in their work, Mahajan et al. [18] introduced a neural network that autonomously controls the manipulator's movements, eliminating the requirement for supervision. The network is trained using unsupervised learning techniques for a two-degrees-of-freedom (DoF) manipulator. The research paper primarily centers on two critical tasks for the robotic arm: drawing a circular trajectory and catching a ball. The research by Kshitish et al. [6] utilized the LM algorithm to train the proposed network for a specific number of epochs. The study addresses the challenge of solving the inverse kinematics problem of a six-degree-of-freedom system using an Artificial Neural Network (ANN). Additionally, the research highlights the importance of a substantial amount of training datasets for the ANN to achieve satisfactory accuracy levels. Hang and Savkin [15] proposed the utilization of competitive neural networks for addressing the inverse kinematics challenge in the field of robotics, specifically focusing on mechanical arm grabbing. Other studies, as [14], explored the use of a neural network (NN) to solve the inverse kinematics problem (IKP) for a planar 3-link manipulator. The trained NN is tested by having the manipulator perform square and triangle motions within its workspace. The Levenberg–Marquardt (LM) algorithm is employed to train the NN for the IKP solution. Multilayered neural networks (NNs) are employed to independently learn the inverse kinematics of serial redundant manipulators [30], regardless of the structure of the evaluation function. The training data used in this method consist of the manipulators' postures, endpoints, and evaluation values. Another study by Gao [9] demonstrates that the proposed inverse kinematics algorithm, which utilizes an enhanced BP neural network, performs better than traditional inverse solution algorithms when addressing the kinematics inverse problem in six-degree-of-freedom Robotics. Moreover, Aravinddhakshan et al. proposed in [4] a neural network–based solution for pick-and-place operations of a 5-DoF manipulator with RRRPR configuration. Using supervised learning, they train a neural network that accurately predicts the inverse kinematic solution with low MSE. The researchers successfully conducted a pick-and-place simulation within the manipulator's structure, optimizing the path planner for the shortest path. Lately, Singla et al. [29] proposed a neuro-fuzzy approach to solve the inverse kinematics of patient-side medical manipulators; however, they have only considered the fixed step size dataset.

Based on the previous research, it becomes evident that a shortage of studies comparing various approaches for configuring datasets and hyperparameters within neural networks to address the issue of inverse kinematics in robot manipulators. Our work presents a significant contribution by introducing an ANN-based intelligent method tailored to tackle the inverse kinematics of a two-joint robotic manipulator. The primary goal of the ANN is to deduce the joint variables based on the input spatial coordinates and orientations of the manipulator's end effector. To accomplish this, the proposed network utilizes deep learning methodologies enabling rapid convergence while undergoing training. During training, a key focus is to attain a minimal Mean Squared Error (MSE) and training error, ideally driving them toward zero. The MATLAB toolbox is utilized for training the ANN, employing three distinct datasets: a random step-size dataset, a fixed step-size dataset, and a sinusoidal signal-based dataset with varying frequencies. The testing and verification stages are conducted to evaluate the reliability of the trained ANN in minimizing approximation errors and accurately estimating the inverse kinematics. The research showcases the effectiveness of the trained ANN in successfully solving the inverse kinematics problem for the 2-DoF manipulator. The network's ability to minimize approximation errors and provide accurate estimations while experimenting with various hyperparameters is validated, underscoring its potential as a dependable tool in robotic kinematics.

4.2 Inverse Kinematics of Manipulator Robots

Inverse Kinematics (IK) is a significant concept in robotics that determines joint configurations (angles or positions) required for a robotic manipulator to achieve a specific end-effector position and orientation in space. In simple terms, inverse kinematics allows us to find the joint angles or coordinates needed to move a robot's end effector to a desired location and orientation. To better understand inverse kinematics, let's contrast it with forward kinematics (Fig. 4.2):

1. *Forward Kinematics*: Given the joint angles or coordinates of a robot's joints, forward kinematics calculates the position and orientation of the robot's end effector in space.
2. *Inverse Kinematics*: In contrast, inverse kinematics works the other way around. It inputs a desired end-effector position and orientation and calculates the corresponding joint angles or coordinates to achieve that desired pose.

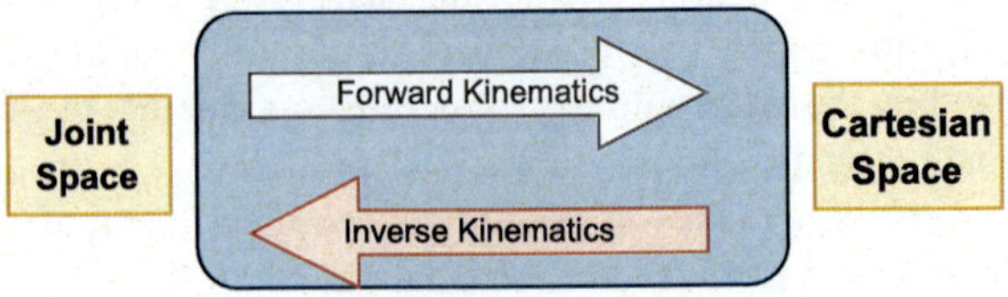

Fig. 4.2 Relationship between the forward and inverse kinematics

4.2.1 Manipulator Model

In this work, we have considered a 2-DoF model that comprises three links, namely l_0, l_1, and l_2, which correspond to the lengths of the different segments in the robot arm (Fig. 4.3). For our specific scenario, the lengths are set as follows: $l_0 = 1$ $[m]$, $l_1 = 2$ $[m]$, and $l_2 = 3$ $[m]$. The forward kinematics equations for the robotic arm are derived through the geometrical method. The well-known Denavit Hartenberg (DH) convention can be used for more complex joint configurations [7]. Taking into account the lengths of the links and the joint angles, the following Eqs. (4.1) and (4.2) can be used to determine the position of the end effector in the X-Y plane for our 2-DoF manipulator robot:

$$X = l_0 + l_1 \cos(\theta_1) + l_2 \cos(\theta_1 + \theta_2), \tag{4.1}$$

$$Y = l_1 \sin(\theta_1) + l_2 \sin(\theta_1 + \theta_2). \tag{4.2}$$

4.2.2 Problem Formulation

In the present work on inverse kinematics for manipulator robots, we are utilizing a 2-DoF model. This model enables us to calculate the joint angles required to position the end effector at specific Cartesian coordinates. The model is designed to take two inputs, representing the Cartesian position coordinates X and Y, and it generates two outputs, θ_1 and θ_2, representing the angles of the joint rotations, as illustrated in Fig. 4.4. This approach is essential for precisely planning the robot's movements, allowing it to perform tasks accurately. This problem becomes particularly complex due to the nonlinearity and intricacy of the mathematical equations that govern the relationship between joint angles and the resulting end-effector pose.

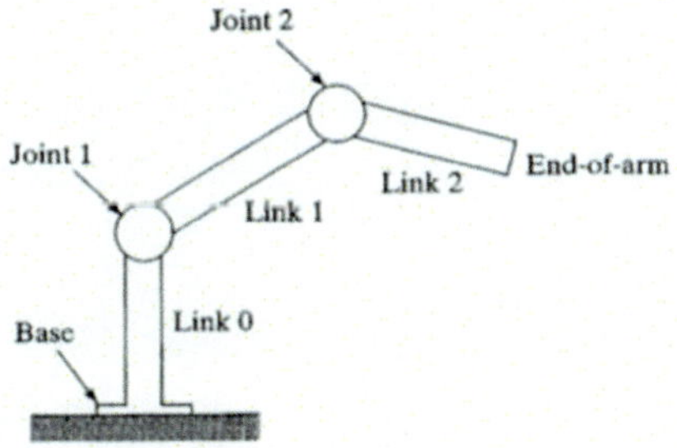

Fig. 4.3 Joints and links of the proposed 2-DoF manipulator [28]

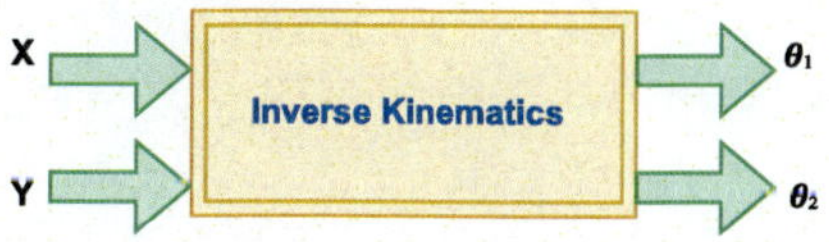

Fig. 4.4 Schema of the 2-DoF inverse kinematics

4.3 Artificial Neural Networks for Solving IK Problems

4.3.1 An Overview of Artificial Neural Networks (ANNs)

Artificial neural networks play a crucial role in the realm of artificial intelligence (AI), machine learning (ML), and deep learning systems (DL), as shown in Fig. 4.5. Inspired by the structure and functioning of biological neural networks within the human brain, ANNs serve as computational models. ANNs are designed to process and interpret intricate patterns or input data, facilitating predictions and decisions based on this input. These networks comprise interconnected nodes, akin to artificial neurons, structured into layers: the input layer, hidden layer(s), and output layer being the fundamental ones [32].

Figure 4.6 illustrates an ANN architecture featuring two hidden layers. In each layer, neurons receive input signals, perform calculations, and transmit outcomes to the subsequent layer. The connections between neurons are defined by weights that undergo adjustments throughout the learning process to optimize the network's overall performance. In the training phase of an ANN, a substantial dataset comprising input–output pairs is introduced to the network. Employing the provided inputs, the network generates predictions, contrasting these projected outputs against the genuine outputs derived from the dataset. The variance between predictions and actual outputs is gauged using a loss function, and the training's aim revolves around minimizing this disparity. This objective is achieved through diverse algorithms, which systematically fine-tune the connection weights to reduce the overall loss.

Applying ANN in inverse kinematics involves training the network with a dataset comprising known joint angles and their corresponding end-effector poses. Through training, the ANN learns to approximate the inverse kinematics function, effectively creating a mapping between the input (desired end-effector pose) and the output (joint angles). Once trained, the ANN can take a new desired end-effector position as

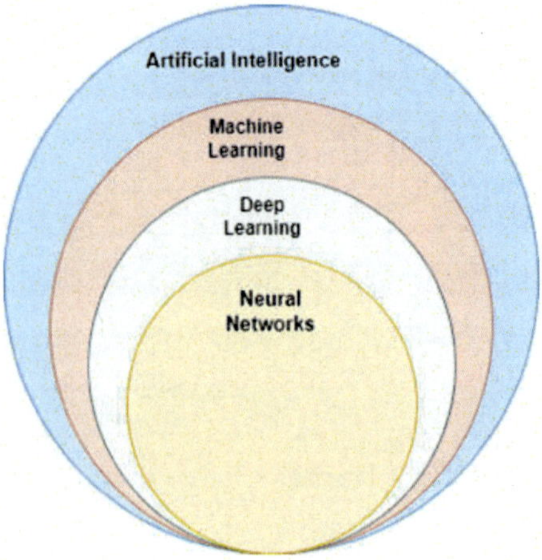

Fig. 4.5 Relationship between AI, ML, DL, and neural networks

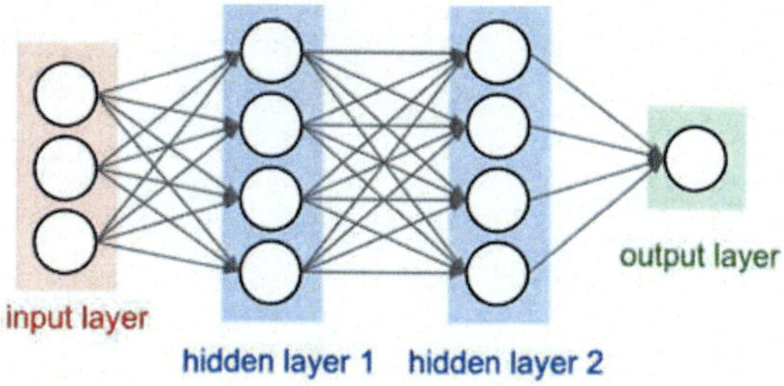

Fig. 4.6 ANN architecture representation with two hidden layers [5]

input and accurately predict the corresponding joint angles. This prediction process is usually achieved using optimization algorithms like gradient descent to minimize the loss function that quantifies prediction errors. Integrating ANNs into inverse kinematics empowers robots to efficiently determine joint angles for desired end-effector positions, enhancing motion planning and control. ANNs' adaptive and streamlined approach improves efficiency and accuracy compared to traditional methods.

4.3.2 Proposed ANN Architectures

During the training process of the ANN, three distinct datasets were employed to assess its performance across various input data types. The first dataset utilized a fixed step size, while the second one incorporated a random step size. The third dataset was generated from a sinusoidal signal-based dataset with varying frequencies. This diverse selection of datasets allowed for a comprehensive evaluation of the ANN's capabilities. It is crucial to acknowledge that the parameters employed in the training process were used arbitrarily. This implies that numerous possible combinations of these parameters exist based on our choices. The results obtained through our experimentation have been promising, shedding light on valuable insights. Nonetheless, it is essential to recognize the potential for getting different values and potentially more effective models through additional exploration and experimentation. The present work considers an ANN architecture with a different number of hidden layers, as shown in Fig. 4.7.

4.3.2.1 ANN with a Fixed Step-Size Dataset

The first dataset was utilized for training ANNs with a fixed step-size dataset, which consists of diverse input–output pairs designed to evaluate the ANN's performance. To construct this dataset, we vary θ_1 and θ_2 with a step size of 0.02 between $-\pi$ and π, generate many Cartesian coordinate combinations using forward kinematic formulations from Eqs. (4.1) and (4.2). The input dataset is Cartesian coordinates (X and Y), and the output dataset is respective joint angles denoted as (θ_1, θ_2). By

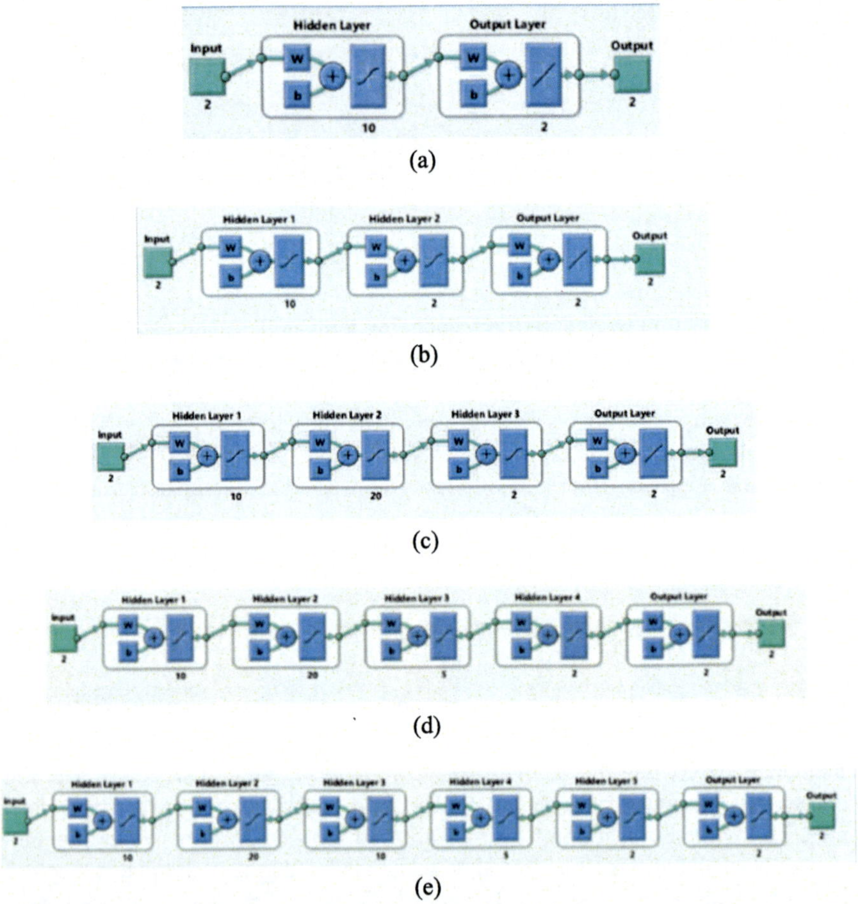

Fig. 4.7 Different ANN architectures with a different number of hidden layers. (**a**) ANN architecture with one hidden layer. (**b**) ANN architecture with two hidden layers. (**c**) ANN architecture with three hidden layers. (**d**) ANN architecture with four hidden layers. (**e**) ANN architecture with five hidden layers

constructing diverse combinations of inputs and outputs in the dataset, our primary objective was to effectively enhance the ANN's capacity to deduce accurate joint angles from various end-effector positions.

4.3.2.2 ANN with a Random Step-Size Dataset

The second dataset utilized in our study is the "random step-size dataset," designed to introduce unpredictability and variability into the input data. The primary purpose

Table 4.1 Parameters used for the sinusoidal signal based dataset

Parameter	Signal 1	Signal 2
Frequency [Hz]	1.5	10
Phase [rad]	0	$\pi/4$
Amplitude [rad]	π	π
Angle [rad]	$[-\pi\pi]$	$[-\pi\pi]$
Number of samples	1000	1000

of this dataset was to assess the ANNs ability to handle unforeseen patterns and evaluate its robustness under such circumstances. To create the random dataset for the 2-DoF robot's inverse kinematics, we employed the rand function to generate random values for the joint angles as shown in expression (4.3). Each data point in the dataset consists of two joint angles, θ_1 and θ_2, randomly selected from a uniform distribution. We applied a scaling and shifting technique to the random values to ensure that these joint angles fall within the desired range. Specifically, we multiplied the random values by 2π and subtracted π, resulting in joint angles ranging from $-\pi$ to π. The following expression gives the computation of the random dataset:

$$\theta = \text{rand(size, 2)} \times 2\pi - \pi. \tag{4.3}$$

Subsequently, we iterated through the forward kinematics equations for each data point, utilizing the randomly generated joint angles. This process allowed us to obtain a diverse and random dataset representing various combinations of Cartesian coordinates and their corresponding joint angles. Each data entry within the dataset depicted a distinct amalgamation of joint angles and their corresponding Cartesian coordinates. This arrangement allowed for a randomized exploration and examination of the robot's end effector positions.

4.3.2.3 ANN with a Sinusoidal Signal-Based Dataset

In our research, we utilized a third dataset that focused on sinusoidal signals with varying frequencies. This dataset played a crucial role in evaluating the capabilities of the ANN when dealing with different frequency patterns. We wanted to understand how well the ANN could grasp and predict the diverse cyclic patterns present in the input data. This was essential to help the ANN learn and generalize from the information. The details of the sinusoidal signals are provided in Table 4.1.

4.4 Results and Discussions

We conducted a series of experiments to explore how neural networks perform when they have varying numbers of hidden layers. Specifically, we compared networks with a single hidden layer to those with five hidden layers, as depicted in Sect. 4.3.2. To evaluate their performance, we employed several metrics, including the Mean Squared Error (MSE). Our primary focus was to investigate the impact of increasing the number of hidden layers on the network's ability to learn and represent intricate patterns in the data. By carefully analyzing the outcomes, we aimed to gain valuable insights into the benefits and potential drawbacks of using a deeper architecture with multiple hidden layers. Additionally, in this section, we will present the effectiveness of the proposed Artificial Neural Networks (ANNs) architecture by employing various optimizers like Levenberg–Marquardt (LM) [27], Bayesian regularization (BR) [13], and scaled conjugate gradient (SCG) [19].

4.4.1 ANN with the Levenberg–Marquardt (LM) Algorithm

When evaluating the training of Artificial Neural Networks (ANN) through the utilization of the Levenberg–Marquardt (LM) algorithm across various datasets—fixed, random, and sinusoidal—it becomes apparent that their performance displays notable fluctuations. The results are shown in Fig. 4.8. Regarding the fixed step-size data, the ANN model achieved its best validation MSE of 2.2574 at epoch 224. This particular architecture comprised five hidden layers with neuron configurations of (10, 20, 10, 5, 2). On the other hand, when training the ANN with random step-size data, it achieved its highest validation MSE of 2.1573 at epoch 13. For this dataset, the ANN architecture involved a single hidden layer with 10 neurons. Lastly, training the ANN with a sinusoidal signal-based dataset resulted in the best validation MSE of 3.5612 at epoch 998. Similarly, this model employed a single hidden layer with 10 neurons, as shown in Table 4.2.

In this scenario, we divided the dataset into three parts: model training, validation, and testing. Specifically, 70% of the dataset was utilized to train the model, allowing it to learn from a substantial portion of the available data. During the training process, 15% of the dataset served as the validation set, enabling us to fine-tune hyperparameters and monitor the model's performance. Finally, the remaining 15% of the dataset was reserved for the final testing phase. This subset of data served as a crucial evaluation set, as it contained unseen data the model had not encountered before. Testing the model on this unseen data provided valuable insights into its generalization ability and performance on new, previously unseen instances.

After analyzing the performance of these models, it becomes apparent that the random data produced the lowest MSE, indicating a superior fit to produce the target output as shown in Fig. 4.9. This outcome suggests that the random data

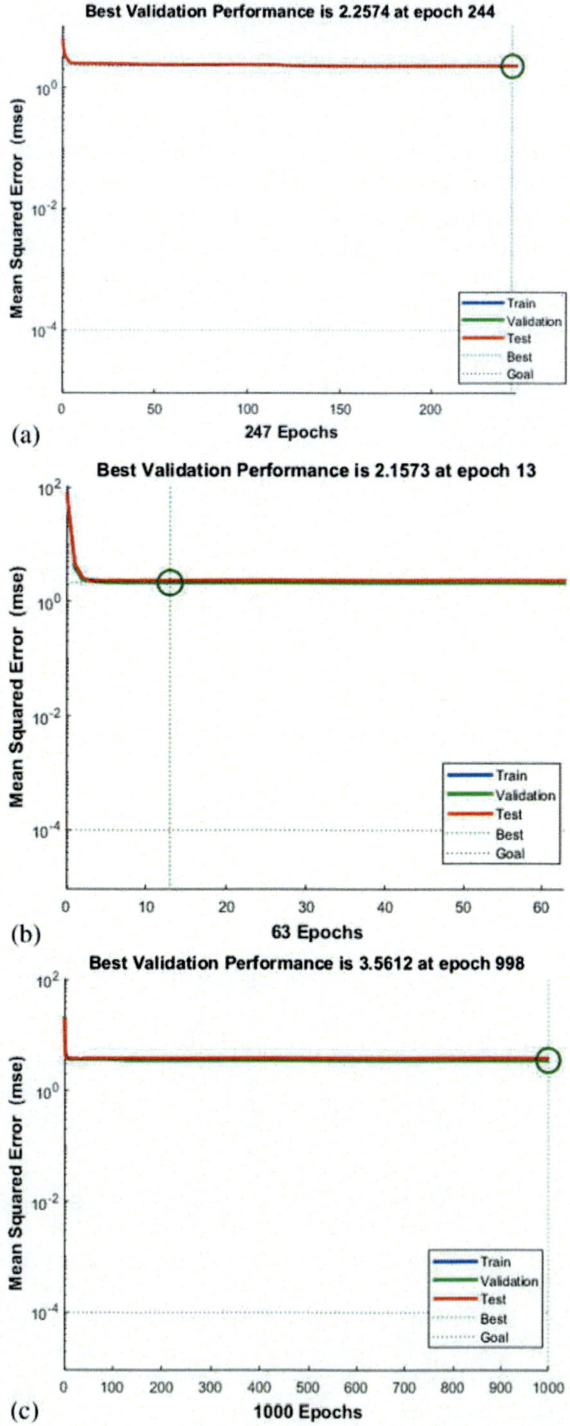

Fig. 4.8 Best ANN's performances with the LM algorithm. (**a**) Lowest MSE for fix step-size data. (**b**) Lowest MSE for random step-size data. (**c**) Lowest MSE for sinusoidal signal-based data

Table 4.2 ANN's performance results with the LM algorithm for three different datasets

Number of hidden layers with associated number of neurons	Fixed dataset		Random dataset		Sinusoidal dataset	
	MSE	Epoch	MSE	Epoch	MSE	Epoch
1 hidden layer (10)	2.4153	127	2.1573	13	3.5612	998
2 hidden layer (10,2)	2.4059	523	2.3055	9	4.0517	20
3 hidden layer (10,20,2)	2.3110	412	2.2923	7	4.4504	27
4 hidden layer (10,20,5,2)	2.2746	152	2.2993	13	4.5774	7
5 hidden layer (10,20,10,5,2)	2.2574	244	2.1970	9	4.3239	4

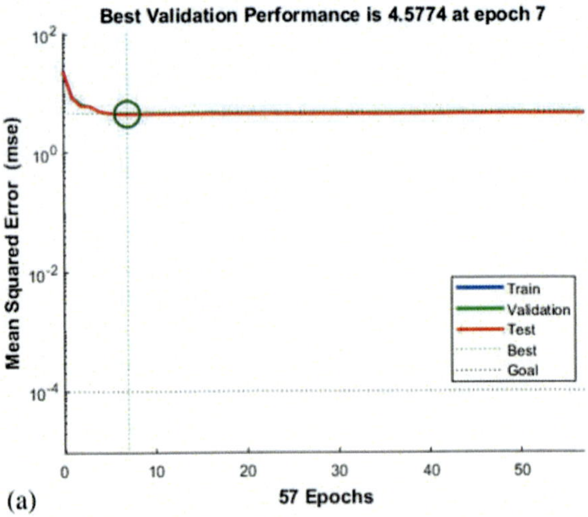

(a)

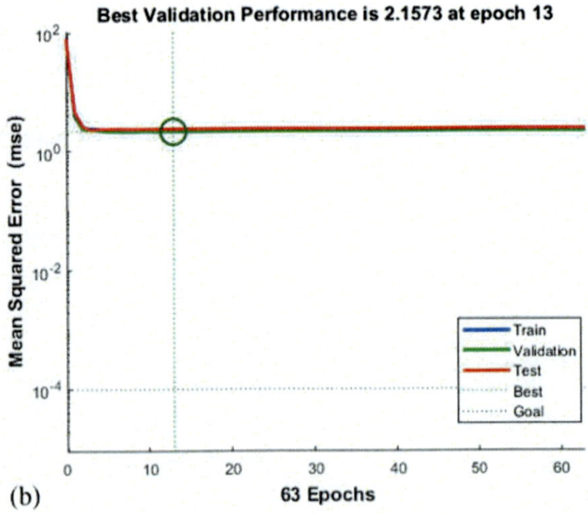

(b)

Fig. 4.9 Highest and lowest MSE with the LM algorithm. (**a**) Highest MSE (Sinusoidal, 3 hidden layers). (**b**) Lowest MSE (Random, 1 hidden layer) offered a more predictable and regular pattern for the LM algorithm. Conversely, the sinusoidal data exhibited a higher MSE, indicating a more difficult learning task for the network. The complex nature of sinusoidal data posed challenges for the LM algorithm in accurately capturing the underlying patterns, resulting in a comparatively higher error. In the case of the fixed data, the LM algorithm demonstrated competitive performance, yielding a relatively low MSE. This result underscores the algorithm's capability to handle complex data patterns and its adaptability to varied input characteristics. Overall, the results highlight the strengths and weaknesses of the LM algorithm with different datasets, showcasing its ability to excel in situations where the data exhibit regularity and predictability while also managing reasonably well with more intricate patterns.

The error histogram, represented with 20 bins, visually represents error distribution within a dataset. The x-axis denotes the error range, while the y-axis illustrates the frequency or count of errors falling within each bin. One notable observation emerges when comparing the distribution of errors in the histogram with the highest MSE to the histogram with the lowest MSE. The histogram with the highest MSE exhibits a broader range of errors than the one with the lowest MSE, as shown in Fig. 4.10. This indicates that the model with the highest MSE displays a wider spread of prediction errors, including some significantly larger errors. The increased range of errors in the highest MSE histogram suggests that the model encountered challenges in accurately predicting certain data points, leading to more significant discrepancies between the predicted and actual values. Consequently, the model's performance was relatively poorer in these instances, possibly indicating difficulties in capturing complex patterns or nuances in the data. In contrast, the histogram with the lowest MSE presents a narrower range of errors, implying that the model achieved higher accuracy and precision in its predictions. The smaller magnitude of errors indicates that the model more successfully approximated the target values, resulting in a more concentrated distribution of errors.

4.4.2 ANN with the Bayesian Regularization (BR) Algorithm

Next, we examined the performance of an ANN trained using the BR algorithm on various datasets: fixed, random, and sinusoidal. The outcomes showed notable network performance variations, as illustrated in Fig. 4.11. When the ANN was trained on the fixed dataset, we achieved its best MSE of 2.2680 at epoch 232. The model architecture comprised three hidden layers with neuron configurations of (10, 20, 2). The BR algorithm demonstrated its efficacy in preventing overfitting and enhancing generalization, resulting in effective training on the fixed data. On the other hand, training the ANN with random data led to the best MSE of 2.1830 at epoch 169. The model architecture comprised five hidden layers. Furthermore, we trained the ANN on sinusoidal data, which produced the best MSE of 3.0553 at epoch 46. The model architecture for this scenario involved three hidden layers.

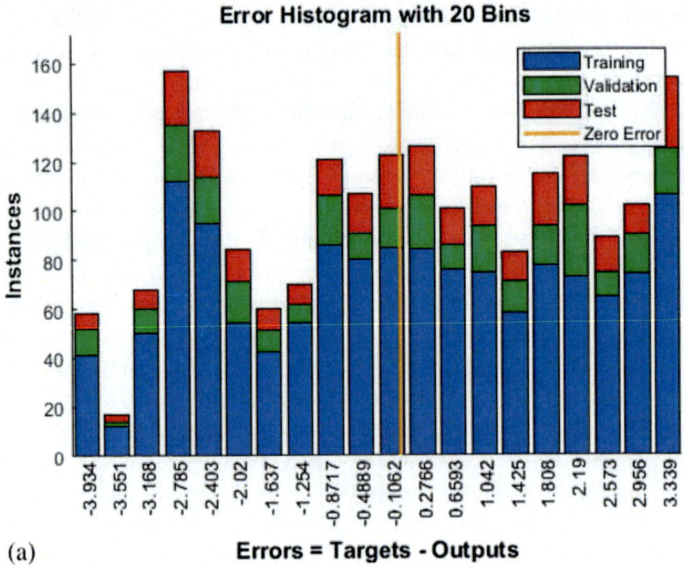

(a)

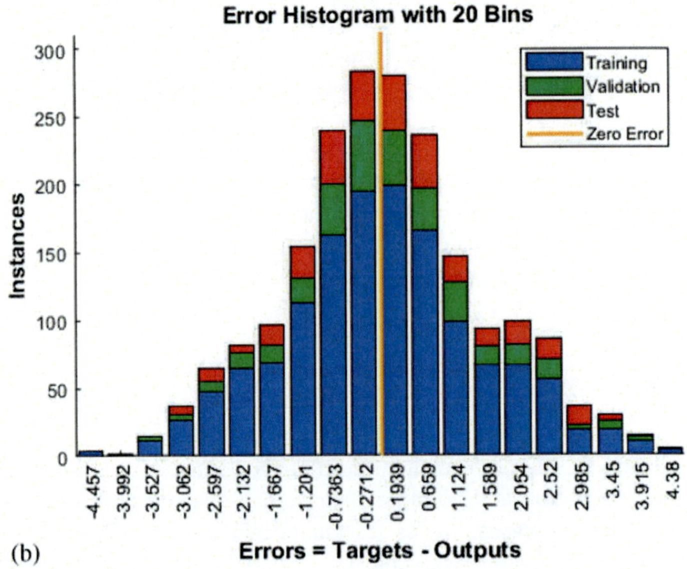

(b)

Fig. 4.10 Error histogram of the highest and lowest MSE. **(a)** Error histogram of the highest MSE. **(b)** Error histogram of the lowest MSE

When using the BR algorithm, we split the dataset into 80% for training and 20% for testing purposes. The unique aspect of the BR algorithm is that it doesn't demand a separate validation dataset. Instead, the validation process is seamlessly integrated during the training phase to tune the hyperparameters effectively. This

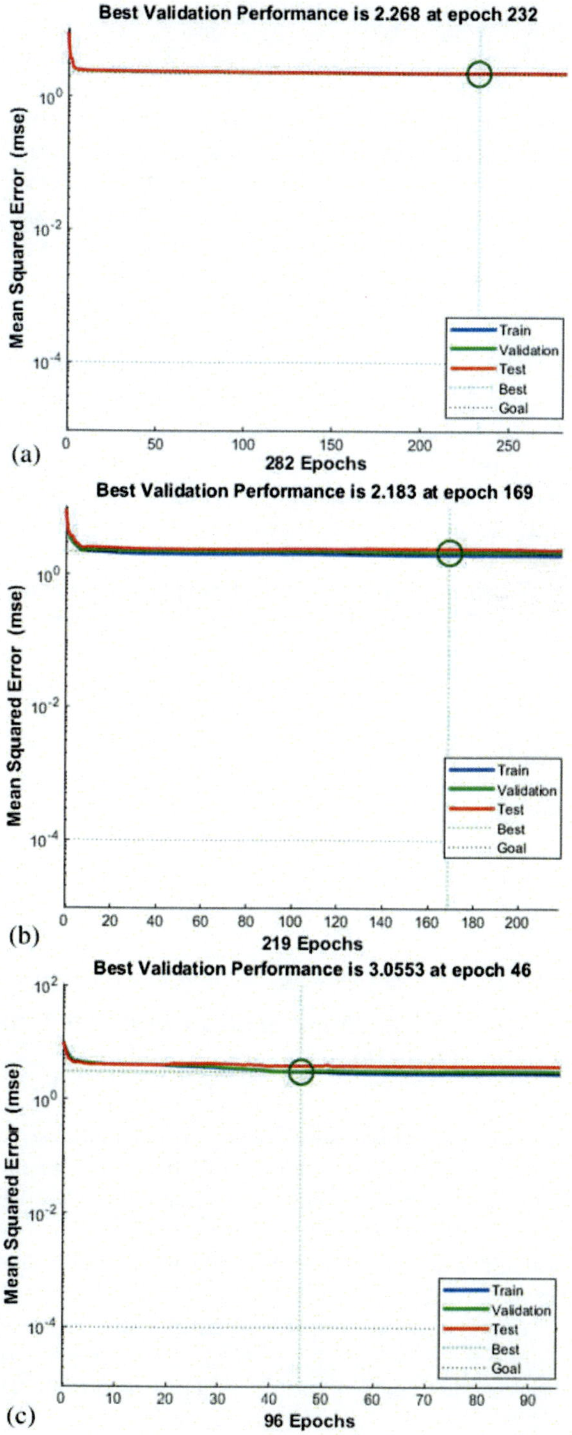

Fig. 4.11 Best ANN's performances with the BR algorithm. (**a**) Lowest MSE for fix step-size data. (**b**) Lowest MSE for random step-size data. (**c**) Lowest MSE for sinusoidal signal-based data.

Table 4.3 ANN's performance results with the BR algorithm for different datasets

Number of hidden layers with associated number of neurons	Fixed dataset		Random dataset		Sinusoidal dataset	
	MSE	Epoch	MSE	Epoch	MSE	Epoch
1 hidden layer (10)	2.3998	344	2.2384	27	3.8994	16
2 hidden layer (10,2)	2.3746	113	2.4714	108	3.3756	65
3 hidden layer (10,20,2)	2.2680	232	2.444	40	3.0553	46
4 hidden layer (10,20,5,2)	2.3475	361	2.3908	7	3.0902	70
5 hidden layer (10,20,10,5,2)	2.3108	85	2.1830	169	3.0693	100

feature simplifies the training process and ensures optimal parameter configuration without the need for an explicit validation set. Once again, the BR algorithm played a vital role in regularization, leading to optimal performance on the sinusoidal data, as summarized in Table 4.3. After examination of the model performances, it becomes apparent that the random data resulted in the lowest MSE. This outcome suggests that the random data provided a more predictable pattern, as shown in Fig. 4.12.

When comparing the error histograms of training with the Bayesian Regularization (BR) algorithm on random data and sinusoidal data, a notable distinction emerges shown in Fig. 4.13. The error histogram for random data demonstrates superior performance compared to that of sinusoidal data. The random data error histogram exhibits a narrower spread of errors, indicating a more precise and accurate model fit. In contrast, the sinusoidal data error histogram displays a wider range of errors, suggesting that the model struggled to capture the complexities of the sinusoidal pattern effectively. This comparison underscores the advantage of using random data over sinusoidal data when training with the BR algorithm, as it results in a model with improved predictive capabilities and reduced prediction errors.

4.4.3 ANN with the Scaled Conjugate Gradient (SCG) Algorithm

The SCG algorithm is well regarded for its impressive convergence rate and ability to handle complex optimization problems. In our investigation, we compared the training of an Artificial Neural Network (ANN) using the SCG algorithm on various datasets: fixed data, random data, and sinusoidal data. The results displayed variations in their performance, which are visualized in Fig. 4.14. The ANN attained its best Mean Squared Error (MSE) of 2.3364 for the fixed data at epoch 986. The model architecture consisted of four hidden layers with respective neuron configurations of (10, 20, 5, 2). On the other hand, training the ANN with random data yielded the best MSE of 2.2419 at epoch 38. The model architecture, in this

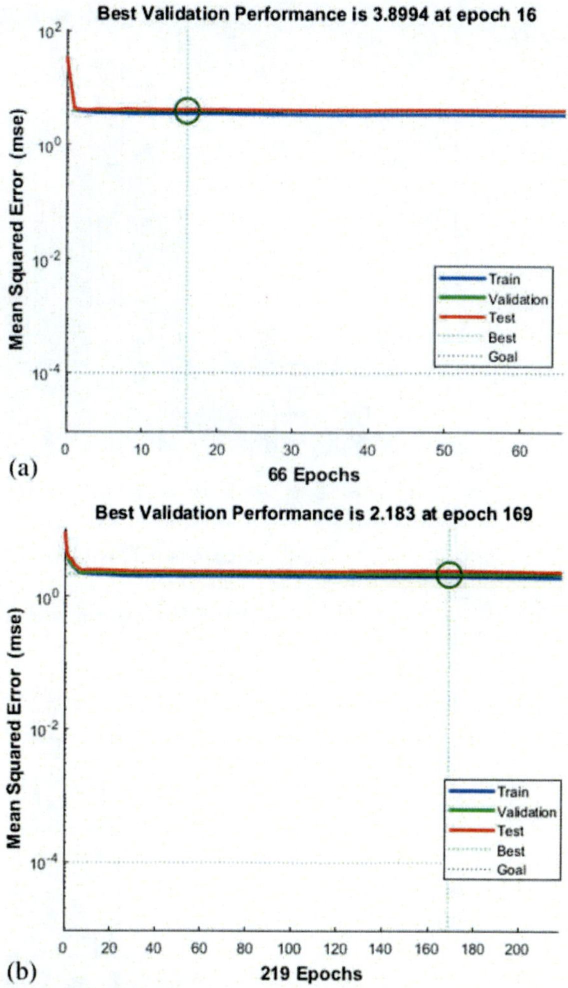

Fig. 4.12 Highest and lowest MSE with the BR algorithm. (**a**) Highest MSE (Sinusoidal, 1 hidden layer). (**b**) Lowest MSE (Random, 5 hidden layers)

case, was simpler, comprising two hidden layers with configurations of (10, 2). Additionally, we trained the ANN on sinusoidal data, resulting in the best MSE of 3.8386 at epoch 314. The model architecture for this scenario included four hidden layers, as documented in Table 4.4.

In Fig. 4.15, the lower MSE of the random data implies that the predictions closely align with the true values, indicating a better fit than the sinusoidal signal. However, it is important to mention that the sinusoidal signal may have been chosen based on its periodicity and inherent patterns. When utilizing the SCG algorithm, we divided the dataset into 70% for training, 15% for validation, and 15% for testing sets, allowing for robust evaluation of the network's performance.

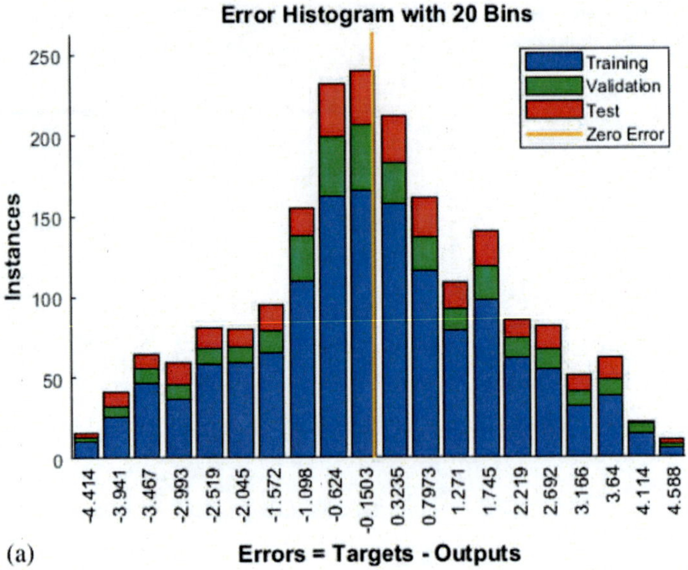

(a)

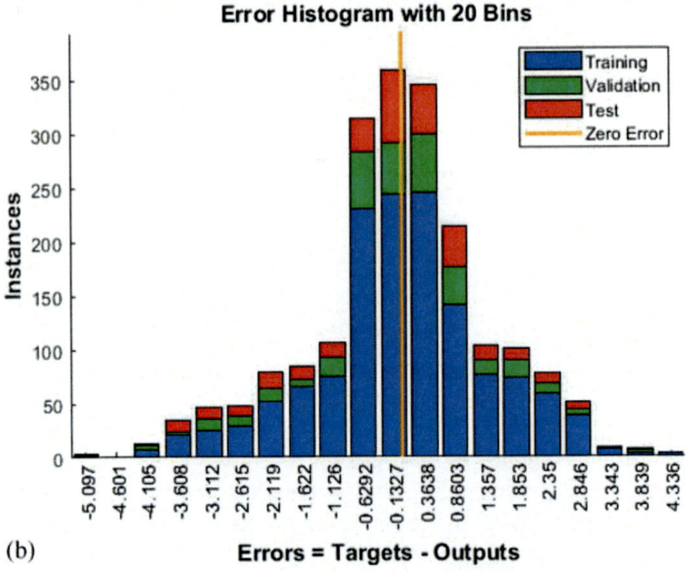

(b)

Fig. 4.13 Error histogram of the highest and lowest mean squared error (MSE). (**a**) Error histogram of the highest MSE. (**b**) Error histogram of the lowest MSE

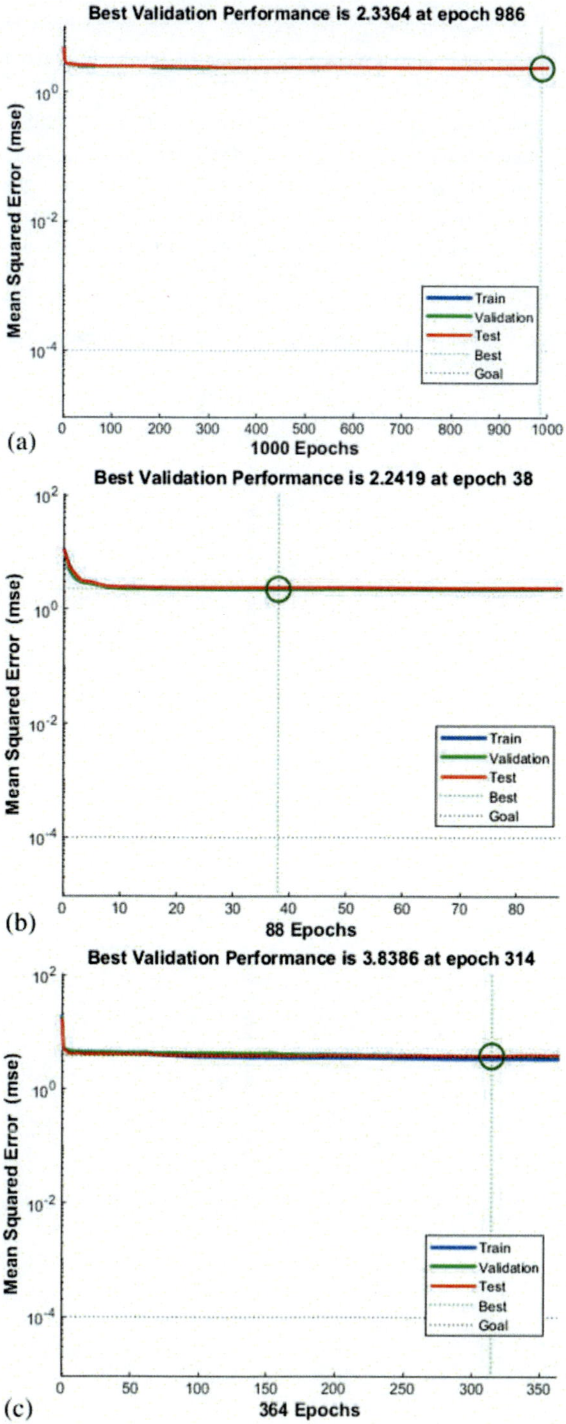

Fig. 4.14 Best ANN's performances with the SCG algorithm. (**a**) Lowest MSE for fix step-size data. (**b**) Lowest MSE for random step-size data. (**c**) Lowest MSE for sinusoidal signal-based data

Table 4.4 ANN's performance results with the SCG algorithm for different datasets

Number of hidden layers with associated number of neurons	Fixed dataset		Random dataset		Sinusoidal dataset	
	MSE	Epoch	MSE	Epoch	MSE	Epoch
1 hidden Layer (10)	2.4132	604	2.2681	12	3.9165	35
2 hidden layer (10,2)	2.4892	436	2.2419	38	4.1159	15
3 hidden layer (10,20,2)	2.3364	783	2.6728	22	4.1439	138
4 hidden layer (10,20,5,2)	2.3364	986	2.7028	33	3.8386	314
5 hidden layer (10,20,10,5,2)	2.3396	1000	2.4799	82	4.3179	18

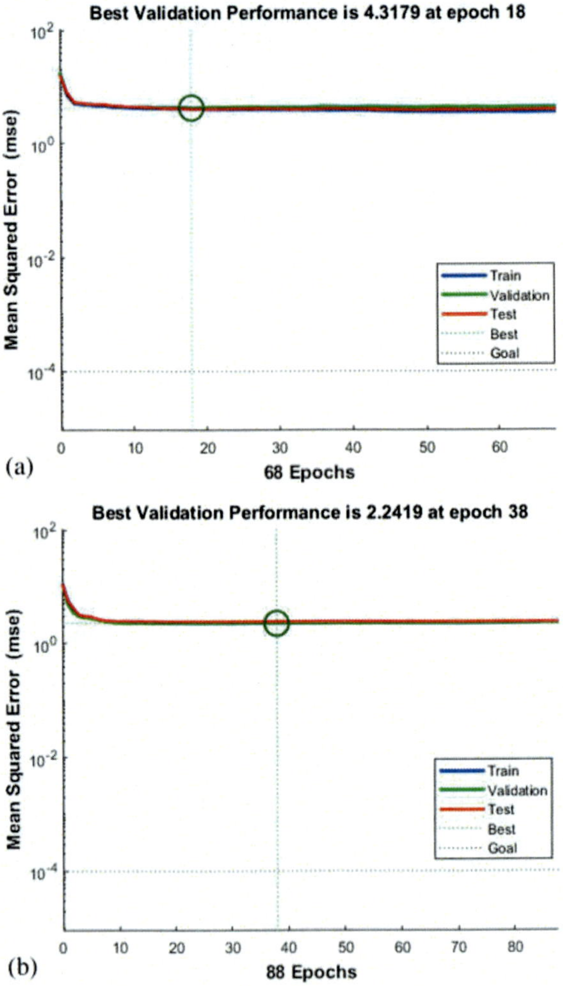

Fig. 4.15 Highest and lowest mean squared error (MSE). (**a**) Highest MSE (Sinusoidal, 5 hidden layers). (**b**) Lowest MSE (Random, 2 hidden layers)

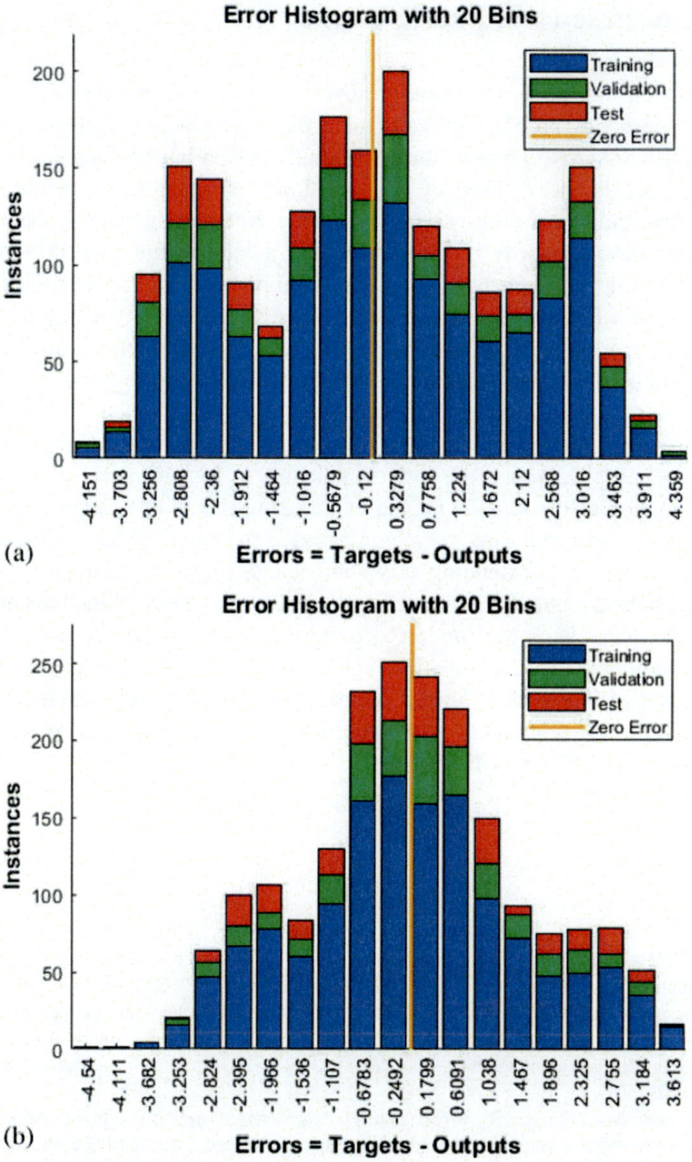

Fig. 4.16 Error histogram of the highest and lowest mean squared error (MSE). (**a**) Error histogram of the highest MSE. (**b**) Error histogram of the lowest MSE

In this context, the higher error histogram corresponding to the highest Mean Square Error (MSE) implies a prevalence of samples with larger prediction errors compared to the error histogram associated with the lowest MSE. Figure 4.16 indicates that the neural network faced challenges in accurately predicting the output for those specific samples, leading to elevated errors in those cases.

4.5 Conclusion and Future Works

In conclusion, our paper presented a novel and effective approach to solving the inverse kinematics problem of a 2-DoF robotic manipulator using an Artificial Neural Network (ANN) architecture. The study delved into the potential of ANNs, utilizing three distinct datasets of joint positions: fixed step size, random step size, and a sinusoidal signal with varying frequency. Notably, the results unequivocally underscore the superiority of random data in achieving the best training performance, offering a significant insight into the importance of dataset selection when training ANNs for complex tasks. The ANN's remarkable efficiency in detecting joint angle errors is a pivotal contribution, as it facilitates precise control of robotic arms, particularly in intricate joint configurations, thus promising a leap in the quality and precision of robotics applications.

This research opens new horizons for solving intricate inverse kinematics problems using neural computation, with the potential to drive advancements in robotics and automation across various industries. Looking ahead, this chapter thoughtfully outlines future research directions, emphasizing the need for dataset enhancement to ensure robustness, extending this approach to higher DoF manipulators, and exploring hybrid methods to further improve accuracy. These directions are pivotal in guiding future work and underscore the chapter's contribution to the ongoing progress in robotics and automation technology. In addition, the present work can be extended and introduced in the control part of manipulator robots using some nonlinear control techniques [12] and also to be applied for medical robotics like exoskeleton systems for pediatric gaits [21, 25].

References

1. Abbas, M., Narayan, J., Dwivedy, S.K.: Simulation analysis for trajectory tracking control of 5-DOFs robotic arm using ANFIS approach. In: 2019 5th International Conference On Computing, Communication, Control And Automation (ICCUBEA), pp. 1–6 (2019)
2. Abbas, M., Narayan, J., Dwivedy, S.K.: A systematic review on cooperative dual-arm manipulators: modeling, planning, control, and vision strategies. International Journal of Intelligent Robotics and Applications **7**(4), 683–707 (2023). https://doi.org/10.1007/s41315-023-00292-0
3. Anschober, M., Edlinger, R., Froschauer, R., Nüchter, A.: Inverse kinematics of an anthropomorphic 6r robot manipulator based on a simple geometric approach for embedded systems. Robotics **12**(4), 101 (2023)
4. Aravinddhakshan, S., Apte, S., Akash, S.M.: Neural network based inverse kinematic solution of a 5 DOF manipulator for industrial application. Journal of Physics: Conference Series **1969**(1), 012010 (2021)
5. Dabbura, I.: Coding neural network - forward propagation and backpropagtion. https://towardsdatascience.com/coding-neural-network-forward-propagation-and-backpropagtion-ccf8cf369f76. Published in Towards Data Science. Accessed: 2023-09-30
6. Dash, K.K., Choudury, B.B., Senapati, S.K.: A inverse kinematic solution of a 6-dof industrial robot using ANN. Indian Journal of Scientific Research **15**(2), 97–101 (2017)
7. Denavit, J., Hartenberg, R.S.: A kinematic notation for lower-pair mechanisms based on matrices. Journal of Applied Mechanics **22**(2), 215–221 (1955)

8. Duka, A.V.: Neural network based inverse kinematics solution for trajectory tracking of a robotic arm. Procedia Technology **12**, 20–27 (2014)
9. Gao, R.: Inverse kinematics solution of robotics based on neural network algorithms. Journal of Ambient Intelligence and Humanized Computing **11**(12), 6199–6209 (2020)
10. Ho, M.K., Griffiths, T.L.: Cognitive science as a source of forward and inverse models of human decisions for robotics and control. Annual Review of Control, Robotics, and Autonomous Systems **5**(1), 33–53 (2022)
11. Ibarra-Pérez, T., Ortiz-Rodríguez, J.M., Olivera-Domingo, F., Guerrero-Osuna, H.A., Gamboa-Rosales, H., Martínez-Blanco, M.d.R.: A novel inverse kinematic solution of a six-DOF robot using neural networks based on the taguchi optimization technique. Applied Sciences **12**(19) (2022)
12. Jenhani, S., Gritli, H., Carbone, G.: Comparison between some nonlinear controllers for the position control of Lagrangian-type robotic systems. Chaos Theory and Applications **4**(4), 179–196 (2022)
13. Kayri, M.: Predictive abilities of Bayesian regularization and Levenberg–Marquardt algorithms in artificial neural networks: a comparative empirical study on social data. Mathematical and Computational Applications **21**(2), 20 (2016)
14. Lathifah, N., Handayani, A.N., Herwanto, H.W., Sendari, S.: Solving inverse kinematics trajectory tracking of planar manipulator using neural network. In: 2018 International Conference on Information and Communications Technology (ICOIACT), pp. 483–488. IEEE (2018)
15. Li, H., Savkin, A.V.: An algorithm for safe navigation of mobile robots by a sensor network in dynamic cluttered industrial environments. Robotics and Computer-Integrated Manufacturing **54**, 65–82 (2018)
16. Liu, H., Zhou, W., Lai, X., Zhu, S.: An efficient inverse kinematic algorithm for a puma560-structured robot manipulator. International Journal of Advanced Robotic Systems **10**(5), 236 (2013)
17. Luo, S., Chu, D., Li, Q., He, Y.: Inverse kinematics solution of 6-DOF manipulator based on multi-objective full-parameter optimization PSO algorithm. Frontiers in Neurorobotics **16** (2022)
18. Mahajan, A., Singh, H., Sukavanam, N.: An unsupervised learning based neural network approach for a robotic manipulator. International Journal of Information Technology **9**, 1–6 (2017)
19. Møller, M.F.: A scaled conjugate gradient algorithm for fast supervised learning. Neural networks **6**(4), 525–533 (1993)
20. Mueller, A.: Modern robotics: Mechanics, planning, and control [bookshelf]. IEEE Control Systems Magazine **39**(6), 100–102 (2019)
21. Narayan, J., Abbas, M., Patel, B., Dwivedy, S.K.: Adaptive RBF neural network-computed torque control for a pediatric gait exoskeleton system: an experimental study. Intelligent Service Robotics **232**(7), 726–732 (2023). https://doi.org/10.1007/s11370-023-00477-3
22. Narayan, J., Banerjee, S., Kamireddy, D., Dwivedy, S.K.: Fuzzy membership functions in anfis for kinematic modeling of 3r manipulator. In: C.M. Hussain, P. Di Sia (eds.) Handbook of Smart Materials, Technologies, and Devices: Applications of Industry 4.0, pp. 1101–1119. Springer International Publishing, Cham (2022)
23. Narayan, J., Jhunjhunwala, S., Gupta, M., Dwivedy, S.K.: Backpropagation neural network based design of a novel sit-to-stand exoskeleton at seat-off position for paraplegic children. In: 2020 6th International Conference on Control, Automation and Robotics (ICCAR), pp. 546–552 (2020)
24. Narayan, J., Singla, A.: ANFIS based kinematic analysis of a 4-DOFs SCARA robot. In: 2017 4th International Conference on Signal Processing, Computing and Control (ISPCC), pp. 205–211 (2017)
25. Narayan, J., Singla, E., Soni, S., Singla, A.: Adaptive neuro-fuzzy inference system–based path planning of 5-degrees-of-freedom spatial manipulator for medical applications. Proceedings of the Institution of Mechanical Engineers, Part H: Journal of Engineering in Medicine **232**(7), 726–732 (2018)

26. Petelin, D., Fomin, A., Laryushkin, P., Fomina, O., Carbone, G., Ceccarelli, M.: Design, kinematics and workspace analysis of a novel 4-DOF kinematically redundant planar parallel grasping manipulator. Machines **11**(3), 319 (2023)
27. Ranganathan, A.: The levenberg-marquardt algorithm. Tutoral on LM algorithm **11**(1), 101–110 (2004)
28. Savatekar, R.D., Dum, A.A.: Design of control system for articulated robot using leap motion sensor. International Research Journal of Engineering and Technology **3**(3), 1407–1417 (2016)
29. Singla, A., Narayan, J., Arora, H.: Investigating the potential of redundant manipulators in narrow channels. Proceedings of the Institution of Mechanical Engineers, Part C: Journal of Mechanical Engineering Science **235**(18), 3723–3736 (2021)
30. Takatani, H., Araki, N., Sato, T., Konishi, Y.: Neural network-based construction of inverse kinematics model for serial redundant manipulators. Artificial Life and Robotics **24**, 487–493 (2019)
31. Xu, W., Mu, Z., Liu, T., Liang, B.: A modified modal method for solving the mission-oriented inverse kinematics of hyper-redundant space manipulators for on-orbit servicing. Acta Astronautica **139**, 54–66 (2017)
32. Zou, J., Han, Y., So, S.S.: Overview of artificial neural networks. In: D.J. Livingstone (ed.) Artificial Neural Networks: Methods and Applications, pp. 14–22. Humana Press, Totowa, NJ (2009)

Chapter 5
CRNN-Based Classification of EMG Signals for the Rehabilitation of the Human Arm

Sami Briouza, Hassène Gritli, Nahla Khraief, and Safya Belghith

5.1 Introduction

Electromyography (EMG) signal processing occupies a pivotal role in the dynamic landscape of biomedical engineering. These signals, which serve as electrical manifestations of muscle activity [1], hold the key to unlocking new frontiers in arm rehabilitation. By harnessing the insights embedded within EMG signals, the field stands poised for groundbreaking advancements in understanding motor control mechanisms and devising tailored interventions for the rehabilitation of the human arm.

The human arm, a cornerstone of functional independence, can be subject to impairment due to a diverse array of factors, encompassing injuries, traumatic incidents, and neuromuscular disorders [2]. To restore arm functionality, individuals traverse the path of rehabilitation, a journey frequently navigated through subjective assessments and generic exercise regimens. The potential for elevating the efficacy of this rehabilitation process lies in the precise utilization of EMG signals.

At the heart of this endeavor lies the accurate classification of EMG signals [2–4]. Such classification empowers a personalized and targeted approach to arm rehabilitation, grounded in an individual's unique neuromuscular dynamics. The transformational potential of this precision cannot be overstated: it holds the

S. Briouza · N. Khraief · S. Belghith
Laboratory of Robotics, Informatics and Complex Systems (RISC Lab – LR16ES07), National Engineering School of Tunis, University of Tunis El Manar, Tunis, Tunisia
e-mail: samibriouza@gmail.com

H. Gritli (✉)
Higher Institute of Information and Communication Technologies, University of Carthage, Borj Cedria, Tunis, Tunisia

Laboratory of Robotics, Informatics and Complex Systems (RISC Lab – LR16ES07), National Engineering School of Tunis, University of Tunis El Manar, Tunis, Tunisia

E. Campos-Cantón et al. (eds.), *Complex Systems and Their Applications*,
https://doi.org/10.1007/978-3-031-51224-7_5

capacity to revolutionize the outcomes of rehabilitation interventions, fostering a harmonious synergy between patient needs and therapeutic strategies [5, 6].

In the absence of robust signal classification, the landscape of arm rehabilitation is plagued by the specter of suboptimal outcomes. The lack of discernible progress becomes a recurring motif, eroding the efficacy of interventions and inducing frustration among patients and therapists alike. The urgency to transcend these limitations calls for the integration of advanced classification methodologies, particularly those borne from the realms of machine learning (ML) and deep learning (DL) [7–13].

Over the years, various methodologies have been deployed for the classification of electromyography signals in the context of arm rehabilitation. Traditional methods typically encompass a preliminary stage of feature extraction from EMG signals, delineating these features into Time Domain (TD) [14], Frequency Domain (FD) [15], and time–frequency domain [16, 17] categories. Subsequent stages entail the usage of machine learning algorithms, including support vector machines [6, 18], k-nearest neighbors [10], random forest [11, 19, 20], and artificial neural network (auto-AN) [21]. While these techniques have demonstrated commendable outcomes, a conspicuous drawback lies in their dependence on handcrafted features.

Handcrafted features, albeit meticulously crafted, fall short in capturing the intricate temporal tapestry woven within EMG signals. This deficiency becomes especially pronounced in the context of the human arm's nuanced and diverse motor actions. The reliance on predefined features poses a formidable barrier to adapting to the ever-evolving intricacies of muscle activity patterns, constraining the adaptability and robustness of traditional classification techniques.

In the current phase of our research, our primary focus revolves around the development of diverse machine and deep learning models tailored for the classification of Electromyography (EMG) signals related to arm movements. Our goal is to create robust and high-performing models that can accurately discern and interpret these signals. The ultimate aim of our work is to seamlessly integrate these well-established models into the control system of an exoskeleton robot. This innovative combination serves a vital role in the rehabilitation of the human arm. By employing the capabilities of the exoskeleton robot, we can provide personalized assistance or resistance to individuals based on their specific needs and conditions. This adaptability holds immense potential for optimizing the rehabilitation process.

5.2 Literature Review

Contemporary interactions between humans and machines, as well as clinical and biomedical uses, along with the analysis of movement, are areas where gesture classification from EMG signals is commonly applied [4, 5]. Surface electromyography (sEMG) signals are generated upon muscle contractions and can be captured by positioning electrodes on the skin's surface of an individual [22, 23].

This section will be used to provide an overview of the existing literature for EMG signal classification. The different approaches used to perform the classification task will be highlighted, their strengths, weaknesses, and contributions will be presented.

5.2.1 Traditional EMG Signal Classification Approaches

For the purpose of categorizing sEMG signals, diverse classification methods have been employed, including the k-Nearest Neighbor (kNN) approach [24], the Support Vector Machine (SVM) classifier [6], and the Random Forest (RF) technique [10, 11, 19, 25].

In [9], numerous models were assessed for the prediction of knee angles and subsequently evaluated. A set of 15 features was extracted, and the authors employed Principal Component Analysis (PCA) to reduce the feature count. The study involved testing several classifiers, namely Naive Bayes (NB), Linear Discriminant Analysis (LDA), k-Nearest Neighbor (k-NN), and Support Vector Machine (SVM). The study conducted by Dhindsa et al. [9] established that, upon assessing various models through different evaluation techniques, the Support Vector Machine (SVM) exhibited superior performance, as demonstrated in their findings.

Moreover, in [24], the authors employed both Support Vector Machine (SVM) and k-Nearest Neighbor (kNN) algorithms for classifying distinct movements utilizing EMG signals. They individually utilized multiple features to evaluate the impact of each feature on the classification process. Hence, the overall conclusion drawn from the study was that the Support Vector Machine (SVM) exhibited notably greater accuracy when compared to k-Nearest Neighbor (kNN) for the selected features in use.

In [26], the study involved a comparison of decision tree algorithms for denoising purposes. The authors employed MSPCA and DWT for feature extraction. The algorithms utilized included CART, C4.5, and random forest. Among these, the random forest algorithm exhibited the most favorable performance across various evaluation techniques.

Authors in [27] used the DWT for feature extraction. Three features were selected, namely: Maximum Absolute Value (MAV1), Mean Absolute Value (MAV2), and Root Means Square (RMS). For the classification, the SVM and the ANN were used. The accuracy of SVM was higher in case of AAA feature vector [27].

In [28], the authors used the time-domain technique to extract features, namely: Waveform Length (WL), Mean Absolute Value (MAV), Root Mean Square (RMS), among others. Then, to reduce the number of feature extractions, they used an Extremely Randomized Tree (ERT). They used multiple classifiers and found that the decision tree performed the best with a classification accuracy of 100%, 99%, and 99%, respectively, on training, testing, and validation dataset.

In [9], authors evaluated the performance of multiple classifiers for knee angle prediction. 15 features were extracted, and the PCA technique was used for feature reduction. In addition, four classifiers were examined including LDA, NB, k-NN, and SVM. The SVM method demonstrated the best results compared with multiple performance evaluation techniques.

In [27], the DWT method was employed to extract features. Specifically, three features were chosen: Maximum Absolute Value (MAV1), Mean Absolute Value (MAV2), and Root Mean Square (RMS). For the classification task, Support Vector Machine (SVM) and Artificial Neural Network (ANN) models were utilized. Notably, in the case of the AAA feature vector, the SVM method exhibited a superior level of accuracy.

Authors in [8] centered their study around wrist and hand movements. The researchers opted for combined time-domain descriptors to select features and employed a deep neural network for classification. They contrasted these outcomes with those from other classification methods. The findings showcased better performance in the majority of cases and demonstrated similarity with the efficacy of the Support Vector Machine (SVM) technique.

These traditional techniques have given important insights and reached good results for the classification of EMG signals. Nevertheless, they inherent some limitations especially with capturing the temporal dynamics of the data. This leads to more focus going to more advanced techniques.

5.2.2 *Emergence of Deep Learning in EMG Classification*

In recent years, Convolutional Neural Networks (CNNs) have gained substantial popularity and widespread adoption in the field of classifying surface electromyography (sEMG) signals. CNNs, originally designed for image analysis, have proven to be highly effective in handling sequential data like sEMG signals due to their ability to automatically extract meaningful features and capture intricate patterns from raw data.

Authors in [29] employed a short-time Fourier transform to create frequency feature maps. The signals were converted into images to serve as inputs for the CNN model. The CNN utilized encompassed nine layers. Although maximum pooling is the conventional choice for the pooling layer, mean pooling was opted in this investigation due to the inherent noise in EMG signals. A leaky rectified linear unit was employed as the activation function in the convolution layer, while the fully connected layer utilized a sigmoid activation function.

In [30–32], the authors harnessed the potential of CNNs in their study, leading to findings that significantly underscored the reliability and robustness of employing these architectural frameworks. The results not only validated the effectiveness of utilizing CNNs but also underscored their potential as a promising avenue for achieving accurate and dependable classifications in the context of their research.

Authors in [33] developed a featureless method that was employed for EMG pattern recognition, utilizing two different CNN architectures. The first one consisted of 5 layers, including a convolution with 32 feature maps using 3×3 filters, max pooling for the pooling layer, and a fully connected layer with a softmax function. On the other hand, the second architecture included an additional convolution and pooling layer, with the convolution layer having 16 feature maps. In both architectures, the Rectified Linear Unit (ReLU) activation function was applied after the convolution layer. For the first model, the mean accuracy among healthy subjects reached 86.18%, while the second architecture achieved a higher accuracy of 88.04%. When focusing on amputee subjects, the first model achieved an accuracy of 57.40%, whereas the second architecture showed an improved accuracy of 60.34%. As a result, the second architecture exhibited better results in both scenarios. Nevertheless, it is worth noting that the increased complexity of the second architecture led to longer training times, despite its superior performance.

In [34], the authors commenced their approach by initiating data preprocessing using the sliding window method. The dataset was subsequently partitioned into training, validation, and test sets. The validation set served the purpose of curbing potential overfitting of the network. The network architecture comprised an input layer, followed by three convolution layers, two max pooling layers, and a fully connected layer. Four distinct inputs were experimented with, namely: signals with 150 ms windowing, signal Fast Fourier Transform (FFT), signal Root Mean Square (RMS), and signal Intrinsic Mode Functions (IMFs). Stochastic Gradient Descent with momentum was employed, initiating with a learning rate of 0.01, while the maximum number of epochs was set at 6. Among the different input variations, the CNN utilizing signal IMFs displayed superior outcomes post standard training, achieving a validation accuracy of approximately 93.55% and a corresponding test accuracy of 93.70%. Notably, when applying cross-validation, closely aligned results were observed, with an average test accuracy of around 95.90%.

Authors in [22] did not use an initial preprocessing step involving windowing was utilized, as the intention was to transform EMG signals into images directly, thereby avoiding manual feature extraction. The architecture was designed with two distinct components: a feature extractor and a classifier. The feature extractor comprises two blocks. The first block consists of five convolutional layers and two max pooling layers. The second block mirrors the structure of the first block but with variations in the filter kernel size of the first three convolution layers. Importantly, these two blocks operate in parallel and do not influence each other. Their outputs are combined and serve as input for the subsequent classifier. The classifier itself consists of three fully connected layers, with the output layer using the softmax activation function. When subjected to testing, this architecture yielded promising results. Specifically, it achieved a classification accuracy of 83.79% when tested on a single exercise and 78.86% when evaluated across three distinct exercises. Remarkably, this architecture exhibited superior performance when compared to other testing scenarios.

Others chose to employ time series models like Long Short-Term Memory (LSTM) and Recurrent Neural Networks (RNNs). In [35], the authors used an

LSTM-RNN architecture for gesture recognition. The proposed model achieved high accuracy with a testing accuracy of 87.29%. The model was also small enough, which allows its use in a microcontroller. Reducing the model's complexity while maintaining good accuracy was the main goal for this work. The authors in [36] worked on online classification of EMG signals for different gestures on the DualMyo and NinaPro DB5.

In [37], the authors conducted a comprehensive presentation and comparative analysis of a range of studies focused on EMG signal analysis, employing diverse deep learning techniques. The outcomes garnered from these studies exhibited great promise in enhancing EMG signal processing. However, it is noteworthy that despite their promising potential, these techniques have not yet found widespread adoption in commercial applications. Additionally, while the results were encouraging, certain limitations were observed, indicating the need for further refinement and performance enhancements before these methodologies can be effectively integrated into practical commercial use.

5.3 Adopted Dataset: Ninapro DB2

The Ninapro Dataset is a publicly available database containing 10 different EMG datasets [25]. For our use case, we chose the Ninapro DB2. In this section, the Ninapro DB2 will be presented.

The Ninapro DB2 contains EMG signals of 40 intact subjects, where we have 29 males and 11 females, the age of the subjects is between 23 and 45 years. Most of the subjects are right-handed (35), the last five are left-handed subjects. In this dataset, we have the EMG data of 50 different movements. These movement are divided into three types of exercises.

Firstly, we have exercise-B movements. This exercise contains the data of 17 different movements. These movements present the basic movements of the wrist and the fingers, these movements are performed in free space. To see the 17 movements of exercise B, check Fig. 5.1. Secondly, we have exercise C. This exercise contains 23 movements. These movements are grasping and functional movements. In these movements, external objects were used such as a water bottle, a pencil, a CD, and others. The movements can be seen in Fig. 5.2. Lastly, we have exercise D. This exercise contains nine motions. The movements in this exercise represent different force patterns. The nine movements can be seen in Fig. 5.3.

Each movement in the dataset was performed by all subjects. Every one of the movements was repeated 10 times. Each of these repetitions was done for 5 seconds, with a 3 second rest in between each repetition. Twleve electrodes were used for the data collection. A sample of the retrieved signals can be seen in Fig. 5.4.

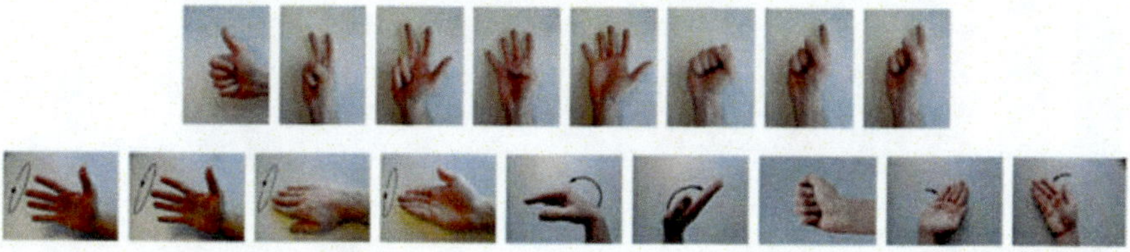

Fig. 5.1 Exercise B gestures [38]

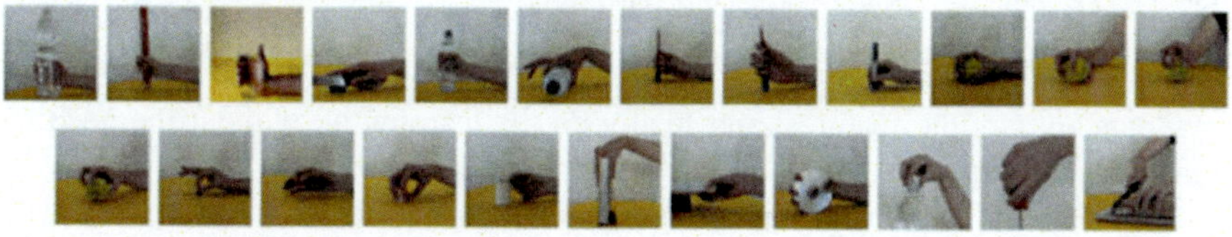

Fig. 5.2 Exercise C gestures [38]

Fig. 5.3 Exercise D gestures [38]

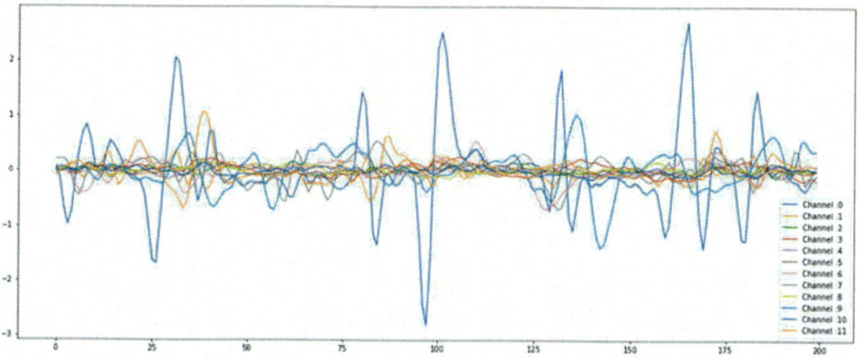

Fig. 5.4 A sample of the signal showing the 12 channels

5.4 The Proposed Classification Method: Convolutional Recurrent Neural Network (CRNN)

For this work and then for the classification of the sEMG signals, we used a Convolutional Recurrent Neural Network (CRNN) architecture, presented by Fig. 5.5. Its main advantage is that it seamlessly uses the power of convolution to capture spatial features from the data and uses the recurrent ones for capturing temporal dynamics. Another important and powerful advantage is that the learning using this architecture is end-to-end, meaning that the model would start from the

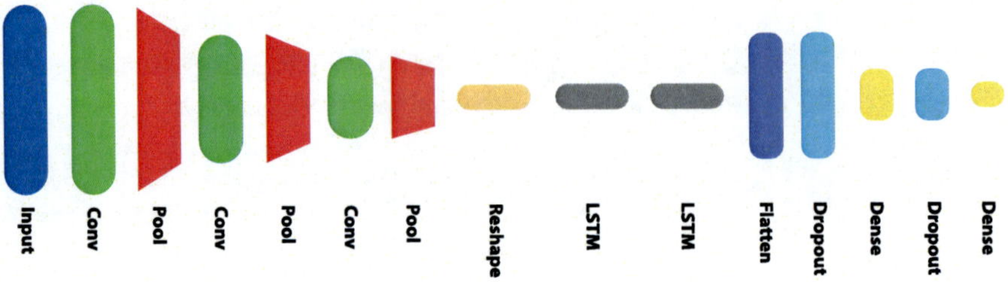

Fig. 5.5 The proposed CRNN model's different layers

raw data, extract the features, and finally do the learning all without the need for manual feature extraction, selection, or engineering.

The different parameters and hyperparameters of the different layers in the adopted CRNN model, illustrated in Fig. 5.5, are provided in Table 5.1. The adopted model contains three convolution layers; each of them is followed by a max pooling layer to reduce the spatial dimensions. The output of the third max pooling layer is reshaped using the Reshape layer to convert it to a 3D tensor. Then, two bidirectional LSTM layers are applied. The first LSTM layer returns sequences, which is then passed to the second LSTM layer, also returning sequences. This architecture enables the model to capture temporal dependencies in the data.

The output of the second LSTM layer is flattened using the Flatten layer to transform it into a 1D tensor. Two Dropout layers are introduced for regularization purposes, with a dropout rate of 0.5. The first Dropout layer is applied after the Flatten layer, and the second Dropout layer is applied after the first fully connected (dense) layer. The final dense layer has the number of units equal to the specified number of classes, and it uses a softmax activation function to produce class probabilities.

For the training process of the CRNN, the optimization algorithm "Adam" was used, which is an optimizer that helps in adapting the learning rate during the training. As a loss function, we opted for the "categorical_crossentropy". This loss function is used of multiclass classification tasks. We run the model for 50 epochs with a batch size of 128.

Now, let us take a closer look at how the dataset is divided among the different sets: the training set, the validation set, and the testing set. The data are divided into a 70% for the training, 15% for the validation, and 15% for the testing.

This breakdown provides insights into the distribution of samples and allows us to assess the representation of each subject within each set. In Table 5.2, we have the number of samples per subject in different set for exercise B. As we can see, each subject, identified by the labels "1" to "10", has a specific count of samples allocated for training, validation, and testing data. As we can see, the training set comprises the majority of the data as it serves as the foundation for training our classification model. The distribution for exercises C and D can be seen in Tables 5.3 and 5.4, respectively.

Table 5.1 Parameters of the different layers in the adopted CRNN

Type of layer	Units	Kernel	Activation function
Input		None	None
Convolutional	32	(20, 3)	ReLu
Pooling		(10, 1)	None
Convolutional	64	(3,3)	ReLu
Pooling		(3, 2)	None
Convolutional	32	(3, 3)	ReLu
Pooling		(2, 2)	None
Reshape			
LSTM	256		
LSTM	256		
Flatten			
Dropout	0.5		
Dense	128		Relu
Dropout	0.5		
Dense	Number of classes		Softmax

Table 5.2 Data distribution for exercise B

Subject	Training data	Validation data	Testing data
1	2584	554	554
2	3125	669	670
3	2981	638	639
4	3985	854	854
5	2524	540	541
6	3369	722	722
7	4217	903	904
8	2638	565	565
9	2573	552	551
10	2839	609	608

Table 5.3 Data distribution for exercise C

Subject	Training data	Validation data	Testing data
1	3858	826	827
2	4804	1030	1029
3	5598	1199	1200
4	5750	1233	1232
5	3631	778	778
6	3907	838	837
7	5443	1167	1166
8	4737	1015	1015
9	5090	1091	1091
10	5676	1217	1216

Table 5.4 Data distribution for exercise D

Subject	Training data	Validation data	Testing data
1	1898	406	407
2	1345	289	288
3	1551	333	332
4	1938	415	415
5	1672	359	358
6	2092	448	448
7	2247	481	482
8	1712	366	367
9	2206	473	473
10	2248	481	482

5.5 Experimental Results

In this section, we will present the results achieved using the proposed CRNN approach. For exercise B, which has 17 movements representing isometric, isotonic, and basic wrist movements, the highest testing accuracy achieved was 100% for subjects 1, 3, 6, and 9, and the lowest was found to be 97.45% for the seventh subject. The average testing accuracy for all subjects is 99.46%. More detailed results are found in Table 5.5, where the Training, Testing, and validation accuracy can be seen for all subjects.

For exercise C where we have the different grasping and functional movements, we achieved a highest testing accuracy of 99.7% for the first and second subjects with the lowest being and 82.27% for the eighth subject. The average testing accuracy achieved using the CRNN is 95.89%. The full results can be found in Table 5.6.

And finally, we have the results for exercise D, where we are classifying nine finger flexion and abduction movements. For this one, we achieved a 100% accuracy for eight out of the 10 subjects while having a 98% for the sixth subject and a 99.79% for the 10th one. The rest of the results are presented in Table 5.7.

The results from exercise C exhibit some noticeable inconsistencies, particularly in the cases of the seventh and eighth subjects. These inconsistencies might be attributed to a couple of factors. Firstly, the higher number of movements involved in this exercise could contribute to increased variability in the data. Secondly, it is possible that the data quality itself is a contributing factor to the observed inconsistencies. To gain a more comprehensive understanding of these inconsistencies and their underlying causes, conducting further assessments could prove insightful. Expanding the testing to a broader pool of subjects or incorporating additional datasets could provide a clearer perspective on the performance and reliability of the models.

We have worked on the classification of EMG signals using multiple techniques such as the random forest method where we achieved an average testing accuracy of 75.19% using the RMS, MAV, WL, and ZC features after testing different features

Table 5.5 Classification results for each subject using the proposed CRNN model for exercise B

Subject/Accuracy	Training (%)	Testing (%)	Validation (%)
1	100	100	100
2	99.104	99.55157	99.85
3	99.76518	100	99.84
4	99.32246	99.76581	99.65
5	98.65293	99.63031	100
6	99.22826	100	100
7	94.9253	97.45293	96.79
8	96.20925	98.40708	97.52
9	99.84454	100	100
10	98.94329	99.8358	99.51
Average	98.59952	99.46435	99.316

Table 5.6 Classification results for each subject using the proposed CRNN model for exercise C

Subject/Accuracy	Training (%)	Testing (%)	Validation (%)
1	97.82271	99.76	99.63724
2	98.66778	99.71	99.90282
3	97.37406	98.92	99.24938
4	96.17391	99.11	98.62125
5	93.25255	96.66	97.68637
6	97.00537	99.52	99.16368
7	79.55172	86.45	85.00428
8	71.09985	82.27	84.33498
9	93.37918	98.26	98.25848
10	91.75476	98.27	98.19228
Average	91.60819	95.893	96.00507

Table 5.7 Classification results for each subject using the proposed CRNN model for exercise D

Subject/Accuracy	Training (%)	Testing (%)	Validation (%)
1	100	100	100
2	100	100	100
3	99.93553	100	100
4	99.484	100	100
5	100	100	100
6	96.65392	98.44	98.21429
7	99.86649	100	100
8	99.94159	100	100
9	98.6854	100	99.78858
10	99.19929	99.79	100
Average	99.37662	99.823	99.80029

combinations [11]. We also used models such as the CNN [12], and the CNN-SVM architecture [13]. The results of the developed models can be seen in Fig. 5.6. As

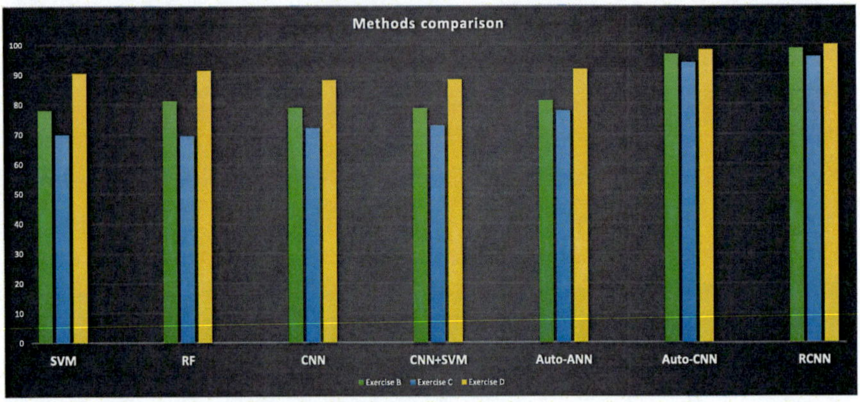

Fig. 5.6 Classification results for different classifiers

we can see in the figure, while the other models had a relatively good performance, the developed CRNN architecture had a better accuracy for exercises B, C, and D.

5.6 Conclusion

In this work, we presented some traditional methods developed for the classification of EMG signals where different feature extraction, feature selection, and feature engineering methods were used. These features were then fed to various machine learning models such as SVM, Random Forest, and kNN among others to perform the gesture recognition. We also presented different deep learning approaches where the feature extraction step was skipped as it is integrated in this new architecture. CNNs have built-in feature extractors, which makes them a powerful tool. Others used method like RNNs and LSTMs, which are able to capture temporal dependencies.

For us, we wanted to merge the strengths of the CNN architecture with their feature extraction capabilities and for capturing spatial hierarchies, and the LSTM model where they help with capturing temporal patterns.

The results achieved in this work are encouraging. Some inconsistencies in the results were found especially for exercise C. To be able to have a better understanding and more unbiased results, testing the data on more EMG datasets can be done. Another way of further proving the reliability of the model is testing it on the other 30 subjects existing in the dataset.

References

1. Gazzoni Marco, Botter Alberto, and Vieira Taian. Surface EMG and muscle fatigue: multi-channel approaches to the study of myoelectric manifestations of muscle fatigue. *Physiological measurement*, 38(5):R27, 2017.
2. Sami Briouza, Hassène Gritli, Nahla Khraief, Safya Belghith, and Dilbag Singh. A brief overview on machine learning in rehabilitation of the human arm via an exoskeleton robot. In *2021 International Conference on Data Analytics for Business and Industry (ICDABI)*, pages 129–134. IEEE, 2021.
3. Fatih Demir, Varun Bajaj, Melih C Ince, Sachin Taran, and Abdulkadir Şengür. Surface EMG signals and deep transfer learning-based physical action classification. *Neural Computing and Applications*, 31(12):8455–8462, 2019.
4. Nurhazimah Nazmi, Mohd Azizi Abdul Rahman, Shin-Ichiroh Yamamoto, Siti Anom Ahmad, Hairi Zamzuri, and Saiful Amri Mazlan. A review of classification techniques of EMG signals during isotonic and isometric contractions. *Sensors*, 16(8):1304, 2016.
5. Mamun Bin Ibne Reaz, M Sazzad Hussain, and Faisal Mohd-Yasin. Techniques of EMG signal analysis: detection, processing, classification and applications. *Biological procedures online*, 8(1):11–35, 2006.
6. Diana C Toledo-Pérez, Juvenal Rodríguez-Reséndiz, Roberto A Gómez-Loenzo, and JC Jauregui-Correa. Support vector machine-based EMG signal classification techniques: A review. *Applied Sciences*, 9(20):4402, 2019.
7. Alvaro David Orjuela-Cañón, Andrés F Ruíz-Olaya, and Leonardo Forero. Deep neural network for EMG signal classification of wrist position: Preliminary results. In *2017 IEEE Latin American Conference on Computational Intelligence (LA-CCI)*, pages 1–5. IEEE, 2017.
8. Anand Kumar Mukhopadhyay and Suman Samui. An experimental study on upper limb position invariant EMG signal classification based on deep neural network. *Biomedical Signal Processing and Control*, 55:101669, 2020.
9. Inderjeet Singh Dhindsa, Ravinder Agarwal, and Hardeep Singh Ryait. Performance evaluation of various classifiers for predicting knee angle from electromyography signals. *Expert Systems*, 36(3):e12381, 2019.
10. Sami Briouza, Hassène Gritli, Nahla Khraief, Safya Belghith, and Dilbag Singh. EMG signal classification for human hand rehabilitation via two machine learning techniques: kNN and SVM. In *2022 5th International Conference on Advanced Systems and Emergent Technologies (IC_ASET)*, pages 412–417. IEEE, 2022.
11. Sami Briouza, Hassène Gritli, Nahla Khraief, Safya Belghith, and Dilbag Singh. Classification of sEMG biomedical signals for upper-limb rehabilitation using the random forest method. In *2022 5th International Conference on Advanced Systems and Emergent Technologies (IC_ASET)*, pages 161–166. IEEE, 2022.
12. Sami Briouza, Hassène Gritli, Nahla Khraief, Safya Belghith, and Dilbag Singh. A convolutional neural network-based architecture for EMG signal classification. In *2021 International Conference on Data Analytics for Business and Industry (ICDABI)*, pages 107–112. IEEE, 2021.
13. Sami Briouza, Hassène Gritli, Nahla Khraief, Safya Belghith, and Dilbag Singh. Classification of sEMG biomedical signals for upper-limb and hand rehabilitation using a hybrid CNN-SVM architecture. In *2022 International Conference on Data Analytics for Business and Industry (ICDABI)*, pages 467–471, 2022.
14. Asim Waris and Ernest Nlandu Kamavuako. Effect of threshold values on the combination of EMG time domain features: Surface versus intramuscular EMG. *Biomedical Signal Processing and Control*, 45:267–273, 2018.
15. Cemil Altın and Orhan Er. Comparison of different time and frequency domain feature extraction methods on elbow gestureâs EMG. *European journal of interdisciplinary studies*, 2(3):25–34, 2016.

16. DE Hernández, L Trujillo, E Z-Flores, OM Villanueva, and O Romo-Fewell. Detecting epilepsy in EEG signals using time, frequency and time-frequency domain features. *Computer science and engineering–theory and applications*, pages 167–182, 2018.
17. Vigneashwara Pandiyan, Rita Drissi-Daoudi, Sergey Shevchik, Giulio Masinelli, Roland Logé, and Kilian Wasmer. Analysis of time, frequency and time-frequency domain features from acoustic emissions during laser powder-bed fusion process. *Procedia CIRP*, 94:392–397, 2020.
18. Mücahid Günay and Ahmet Alkan. Classification of EMG signals by spectral methods and SVM classifier. *Kahramanmaras Sütçü Imam Üniversitesi Mühendislik Bilimleri Dergisi*, 13(2):63–80, 2016.
19. Zhong Li, Xiaorong Guan, Kaifan Zou, and Cheng Xu. Estimation of knee movement from surface EMG using random forest with principal component analysis. *Electronics*, 9(1):43, 2020.
20. Tao Zhou, Olatunji Mumini Omisore, Wenjing Du, Lei Wang, and Yuan Zhang. Adapting random forest classifier based on single and multiple features for surface electromyography signal recognition. In *2019 12th International Congress on Image and Signal Processing, BioMedical Engineering and Informatics (CISP-BMEI)*, pages 1–6. IEEE, 2019.
21. Sami Briouza, Hassène Gritli, Nahla Khraief, Safya Belghith, and Dilbag Singh. An auto-ANN for EMG classification for the motion recognition of the human hand. In *2022 IEEE Information Technologies & Smart Industrial Systems (ITSIS)*, pages 1–6, 2022.
22. Zhen Ding, Chifu Yang, Zhihong Tian, Chunzhi Yi, Yunsheng Fu, and Feng Jiang. sEMG-based gesture recognition with convolution neural networks. *Sustainability*, 10(6):1865, 2018.
23. Linda Resnik, He Huang, Anna Winslow, Dustin L. Crouch, Fan Zhang, and Nancy Wolk. Evaluation of EMG pattern recognition for upper limb prosthesis control: a case study in comparison with direct myoelectric control. *Journal of NeuroEngineering and Rehabilitation*, 15(23):1–13, 2018.
24. Yogesh Paul, Vibha Goyal, and Ram Avtar Jaswal. Comparative analysis between SVM & KNN classifier for EMG signal classification on elementary time domain features. In *2017 4th International Conference on Signal Processing, Computing and Control (ISPCC)*, pages 169–175. IEEE, 2017.
25. Manfredo Atzori, Arjan Gijsberts, Claudio Castellini, Barbara Caputo, Anne-Gabrielle Mittaz Hager, Simone Elsig, Giorgio Giatsidis, Franco Bassetto, and Henning Müller. Electromyography data for non-invasive naturally-controlled robotic hand prostheses. *Scientific data*, 1(1):1–13, 2014.
26. Ercan Gokgoz and Abdulhamit Subasi. Comparison of decision tree algorithms for EMG signal classification using dwt. *Biomedical Signal Processing and Control*, 18:138–144, 2015.
27. Ram Murat Singh, Vivek Ahlawat, S Chatterji, and Amod Kumar. Comparative analysis of SVM and ANN classifier based on surface EMG signals for elbow movement classification. *Journal of Interdisciplinary Mathematics*, 23(1):153–161, 2020.
28. Ismail Mohd Khairuddin, Shahrul Naim Sidek, Anwar PP Abdul Majeed, Mohd Azraai Mohd Razman, Asmarani Ahmad Puzi, and Hazlina Md Yusof. The classification of movement intention through machine learning models: the identification of significant time-domain EMG features. *PeerJ Computer Science*, 7:e379, 2021.
29. Yusuke Yamanoi, Yosuke Ogiri, and Ryu Kato. EMG-based posture classification using a convolutional neural network for a myoelectric hand. *Biomedical Signal Processing and Control*, 55:101574, 2020.
30. Lin Chen, Jianting Fu, Yuheng Wu, Haochen Li, and Bin Zheng. Hand gesture recognition using compact CNN via surface electromyography signals. *Sensors*, 20(3):672, 2020.
31. Jun Li, Lixin Wei, Yintang Wen, Xiaoguang Liu, and Hongrui Wang. Hand gesture recognition based improved multi-channels CNN architecture using EMG sensors. *Journal of Intelligent & Fuzzy Systems*, 43:643–656, 2022.
32. Shu Shen, Kang Gu, Xin-Rong Chen, Ming Yang, and Ru-Chuan Wang. Movements classification of multi-channel sEMG based on CNN and stacking ensemble learning. *Ieee Access*, 7:137489–137500, 2019.

33. Jingwei Too, AR Abdullah, N Mohd Saad, N Mohd Ali, and TNS Tengku Zawawi. Featureless EMG pattern recognition based on convolutional neural network. *Indonesian Journal of Electrical Engineering and Computer Science*, 14(3):1291–1297, 2019.
34. Kaan Bakircioğlu and Nalan Özkurt. Classification of EMG signals using convolution neural network. *International Journal of Applied Mathematics Electronics and Computers*, 8(4):115–119, 2020.
35. Alejandro Toro-Ossaba, Juan Jaramillo-Tigreros, Juan C Tejada, Alejandro Peña, Alexandro López-González, and Rui Alexandre Castanho. LSTM recurrent neural network for hand gesture recognition using EMG signals. *Applied Sciences*, 12(19):9700, 2022.
36. Miguel Simão, Pedro Neto, and Olivier Gibaru. EMG-based online classification of gestures with recurrent neural networks. *Pattern Recognition Letters*, 128:45–51, 2019.
37. Dezhen Xiong, Daohui Zhang, Xingang Zhao, and Yiwen Zhao. Deep learning for EMG-based human-machine interaction: a review. *IEEE/CAA Journal of Automatica Sinica*, 8(3):512–533, 2021.
38. Manfredo Atzori and Henning Müller. The ninapro database: a resource for sEMG naturally controlled robotic hand prosthetics. In *2015 37th Annual International Conference of the IEEE Engineering in Medicine and Biology Society (EMBC)*, pages 7151–7154. IEEE, 2015.

Chapter 6
LMI-Based Design of a Robust Affine Control Law for the Position Control of a Knee Exoskeleton Robot: Comparative Analysis of Stability Conditions

Sahar Jenhani and Hassène Gritli

6.1 Introduction

In the recent decades, the world has witnessed an extraordinary demographic transformation characterized by an unprecedented increase in the aging population. This demographic shift, commonly known as population aging, has profound implications for various aspects of society, including healthcare, social welfare, labor markets, and overall economic productivity. One critical area influenced by population aging is the mobility and functional abilities of older individuals, which often experience natural declines in physical capabilities, including reductions in muscular strength, motor neurons, muscle fibers, and aerobic capacity [1–3]. To address the challenges faced by the aging population, there is a growing need for innovative solutions to enhance their mobility and independence. In recent years, exoskeletons have emerged as promising forms of assistive technology in the field of robotics. These wearable robotic devices are designed to augment or restore human physical capabilities by providing mechanical support, strength amplification, and motion assistance. Exoskeletons have gained significant attention and research interest due to their potential applications in rehabilitation, physical therapy, and assistance for the elderly and individuals with mobility impairments [4–7].

S. Jenhani
Laboratory of Robotics, Informatics and Complex Systems (RISC Lab – LR16ES07), National Engineering School of Tunis, University of Tunis El Manar, Tunis, Tunisia

H. Gritli (✉)
Laboratory of Robotics, Informatics and Complex Systems (RISC Lab – LR16ES07), National Engineering School of Tunis, University of Tunis El Manar, Tunis, Tunisia

Higher Institute of Information and Communication Technologies, University of Carthage, Tunis, Tunisia

E. Campos-Cantón et al. (eds.), *Complex Systems and Their Applications*,
https://doi.org/10.1007/978-3-031-51224-7_6

Within the realm of exoskeletons, a particular area of focus lies in the development of exoskeletons specifically designed for rehabilitation purposes. The primary objective of these exoskeletons is to aid individuals with disabilities and the elderly in walking and improving their mobility [8–12]. Rehabilitation exoskeletons offer a unique opportunity to bridge the gap between therapeutic interventions and assistive technologies, providing both rehabilitative exercise training and support for daily activities. This intersection of rehabilitation and robotics has opened up new possibilities for enhancing the lives of individuals with mobility impairments, allowing them to regain independence and actively participate in daily tasks. One of the critical aspects that determines the effectiveness of rehabilitation exoskeletons is the control approach. The control system plays a pivotal role in ensuring the safe and effective operation of exoskeleton devices, facilitating seamless interaction between the user and the robotic system [8, 11, 13–15]. It is responsible for generating and coordinating the motions of the exoskeleton's actuators to assist, amplify, or restore human movements accurately and efficiently [16]. The success of rehabilitation exoskeletons heavily relies on the control algorithm's ability to mimic and augment the wearer's natural movements while providing adequate support and assistance where needed.

The control of rehabilitation exoskeletons is a challenging and vital research area within the field of robotics. The complexity of the human–robot interaction, coupled with the variability in patients' physical conditions, poses significant challenges for developing robust and adaptive control strategies. Researchers and engineers are continuously exploring various control strategies and techniques to optimize the functionality and performance of rehabilitation exoskeletons [17–23]. The development of personalized, robust and adaptive control strategies has become a crucial focus in the field of robotic exoskeletons. The goal is to provide efficient and precise control, ensuring that the exoskeleton seamlessly integrates with the user's body and enhances their mobility and quality of life [24].

Moreover, advancements in sensing technologies, such as wearable sensors and inertial measurement units, have further enriched the possibilities for developing sophisticated control strategies [17, 25, 26]. These sensors provide valuable information about the wearer's gait, posture, and muscle activities, enabling the control system to adjust and adapt in real time based on the user's needs and intentions [27]. Integrating sensor data into the control algorithm allows for a more seamless and intuitive human–robot interaction, facilitating more natural and coordinated movements [17, 28–30].

Hence, the control of rehabilitation exoskeletons represents a multidisciplinary endeavor that merges the principles of robotics, biomechanics, and rehabilitation science. The quest for robust and adaptive control strategies continues to drive research and innovation in this field [31]. By harnessing the power of advanced control algorithms and sensor technologies, we can pave the way for a new generation of exoskeletons that not only enhance stability during rehabilitation but also empower individuals with mobility impairments to lead more independent and fulfilling lives.

This chapter is dedicated to addressing the crucial task of position control for an actuated lower limb exoskeleton robotic system, with a specific focus on the knee joint [32–36]. The knee joint plays a critical role in essential activities such as walking and weight-bearing, making it of paramount importance in the design of exoskeletons for rehabilitation and assistance. Our main goal is to develop and implement an affine state-feedback control law that effectively controls and stabilizes the knee exoskeleton robotic system. The successful control of the knee joint is essential for providing accurate and smooth movements, enhancing the user's mobility and independence. To achieve our objective, we consider the nonlinear dynamics of the robot, which incorporates factors such as solid and viscous frictions. These frictional forces can considerably impact the exoskeleton's performance, underscoring the importance of incorporating them into the control design process. By adopting an affine state-feedback control approach and leveraging the Lyapunov stability theory, we propose three distinct design methods to establish a set of Linear Matrix Inequality requirements on the gain of the adopted controller. These LMI conditions ensure the stabilization of the knee exoskeleton robot, enabling precise and reliable control even in the effect of uncertainties and frictions. In addition to the theoretical analysis and design, we present comprehensive simulation results to validate the effectiveness of the developed LMI conditions and the adopted affine state-feedback control law. The simulation results provide concrete evidence of the controller's ability to achieve stability and robustness, further reinforcing the reliability of our approach.

The rest of the chapter is organized in the following manner: In Sect. 6.2, we present a description of the nonlinear dynamics of the knee exoskeleton and problematic in this study. In Sect. 6.3, we describe the linearized dynamics and the proposed affine state-feedback controller. The design of LMI conditions is explored in Sect. 6.4 through three distinct methods. In Sect. 6.5, the simulation outcomes are presented. Lastly, Sect. 6.6 summarizes the paper and explores potential future directions.

6.2 Nonlinear Dynamics of the Knee Exoskeleton and Challenges Addressed in this Study

6.2.1 Description of Knee Exoskeleton System

In this part, we present the geometric representation of the adopted knee exoskeleton depicted in Fig. 6.1 [33], showcasing its design and functionality. The knee exoskeleton is conceptualized as a seated structure with a freely moving shank, allowing for seamless extension and flexion of the knee joint. At $\theta = 0°$, the knee joint is completely extended, and it achieves maximum flexion at $\theta = 90°$. As depicted in Fig. 6.1, the exoskeleton comprises two segments linked by a rotating axis at the knee joint level. The upper segment represents the thigh and

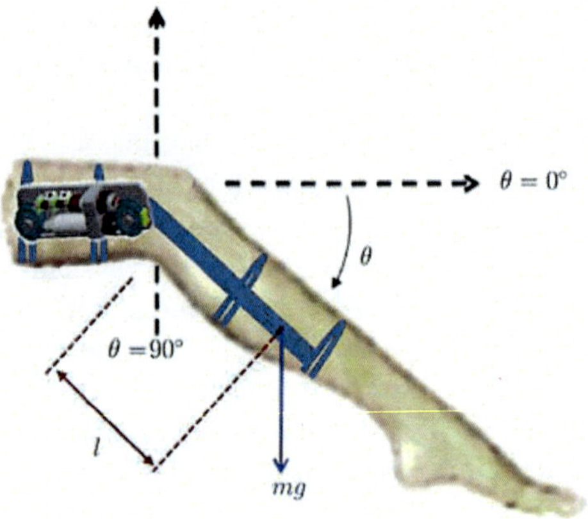

Fig. 6.1 Geometric illustration of the proposed knee exoskeleton design, adapted from [33]

accommodates the actuator, while the lower segment simulates the shank. Both components are securely fastened to the user's lower limb using straps. This design ensures that the exoskeleton mirrors the natural sagittal plane rotational motion of the knee joint.

Selected specifically for this study, the robotic knee joint system with only one degree of freedom (1-DoF) was chosen due to its simplicity and limited range of motion. This feature allows us to focus on developing and testing robust control strategies to enhance stability and precision during lower limb rehabilitation. By precisely regulating the exoskeleton's movement within the constrained degree of freedom, we aim to provide reliable and comfortable assistance to users, promoting their mobility and overall well-being.

6.2.2 Dynamic Model of the Exoskeleton Robot

The equation representing the nonlinear dynamics of the robot incorporates the impact of solid and viscous frictions and can be expressed as follows:

$$\nu\ddot{\theta} = u + a\cos(\theta) - f_v\dot{\theta} - f_s \mathrm{sgn}(\dot{\theta}) - \delta_t. \tag{6.1}$$

In (6.1), $a = mgl$, where m denotes the system's mass, g represents the acceleration brought on by gravity, and l signifies a characteristic length. Moreover, the exoskeleton's inertia is ν, the coefficient of viscous friction is f_v, whereas f_s denotes the solid friction's coefficient. In addition, δ_t denotes the external disturbing signal.

It is crucial to recognize that the parameters of the dynamic model are unable to be precisely determined beforehand, which introduces system uncertainties owing to unaccounted dynamics and the variations of parameter. Thus, the nonlinear dynamics is decomposed into a nominal portion and an indeterminate part, which may be stated mathematically like so:

$$\nu = \nu_\star + \Delta\nu, \tag{6.2a}$$

$$f_v = f_v^\star + \Delta f_v, \tag{6.2b}$$

$$f_s = f_s^\star + \Delta f_s, \tag{6.2c}$$

$$a = a_\star + \Delta a, \tag{6.2d}$$

where the symbol $\star$ is used to denote nominal values of the parameters.

By substituting the expression from Eq. (6.2d) into the nonlinear dynamics (6.1), we get the following result:

$$\nu_\star\ddot{\theta} = u + a_\star\cos(\theta) - f_v^\star\dot{\theta} - f_s^\star\mathrm{sgn}(\dot{\theta}) + \Delta_t, \tag{6.3}$$

where Δ_t denotes here the term encompassing the uncertainties and disturbances and is defined as follows:

$$\Delta_t = -\Delta\nu\ddot{\theta} + \Delta a\cos(\theta) - \Delta f_v\dot{\theta} - \Delta f_s\mathrm{sgn}(\dot{\theta}) - \delta(t). \tag{6.4}$$

In the following, we will assume that Δ_t satisfies the subsequent restriction:

$$|\Delta_t| \leq \bar{\delta}, \tag{6.5}$$

with $\bar{\delta} > 0$.

Let us define θ_d to represent the intended state at which the knee exoskeleton system will be regulated. By introducing the variables $\phi = \theta - \theta_d$, $\dot{\theta} = \dot{\phi}$, and $\ddot{\theta} = \ddot{\phi}$, the equation of motion (6.1) becomes:

$$\nu_\star\ddot{\phi} = u - f_v^\star\dot{\phi} + a_\star\cos(\phi + \theta_d) - f_s^\star\mathrm{sgn}(\dot{\phi}) + \Delta_t. \tag{6.6}$$

6.2.3 *Problem Formulation*

This paper is focused on controlling a lower extremity exoskeleton device designed to support standing and walking tasks for elderly individuals. While exoskeletons have a wide range of applications, our primary focus is on providing assistive mobility to the elderly population. Wearable technologies, such as exoskeletons and orthotics, have shown great promise in helping older and disabled individuals with limb deficiencies regain their physical movement and independence.

Recently, the control of lower limb exoskeletons has garnered increasing attention among robotics researchers, as these robotic systems continue to demonstrate their usefulness and potential. Our primary goal in this study is to control the knee exoskeleton robot, characterized by 1-DoF, as depicted in Fig. 6.1. In this study, at the level of the knee joint, we regard the rod and the foot to be one spinning system.

Our primary goal is to achieve robust control and stabilization of the knee exoskeleton robot, as defined by its equation of motion (6.1). To attain this objective, we will develop Linear Matrix Inequality (LMI) stability conditions, a powerful approach that enables robust control of robotic systems [37]. By employing LMI-based control strategies, we aim to enhance the performance and stability of the knee exoskeleton during rehabilitation exercises. These stability conditions will ensure that the controlled system remains stable at the desired configuration state, allowing for precise and reliable control of the knee exoskeleton robot even when uncertainties and fraction are present and then the signal Δ_t.

6.3 Linearized Dynamics and Utilized Affine State-Feedback Control law

6.3.1 *Intended Control Effort at the Equilibrium State*

The robot requires a special control effort to maintain this equilibrium condition when the controller u has successfully moved it to the intended configuration θ_d with zero velocity $\dot{\theta}_d = 0$. Let u_d represent the necessary control input at the equilibrium. As a result, we arrive at the following equation using the nonlinear dynamics (6.6):

$$0 = u_d + a_\star \cos(\theta_d). \tag{6.7}$$

At the state of balance, the desired control input u_d can be expressed by the following equation:

$$u_d = -a_\star \cos(\theta_d). \tag{6.8}$$

6.3.2 *Linearized Dynamic Model*

In this section, our main goal is to linearize the nonlinear dynamic model (6.6) around the desired position (θ_d, u_d). To achieve this, we make the assumptions that $f_s^\star \text{sgn}(\dot{\theta}) = 0$ and $\Delta_t = 0$. Under these assumptions, we can derive the following expression:

$$\nu_\star \ddot{\phi} = u - f_v^\star \dot{\phi} + a_\star \cos(\theta). \tag{6.9}$$

Moreover, applying the first-order Taylor approximation, we derive the following relation:

$$\cos(\theta) = \cos(\theta_d) - \sin(\theta_d)\phi. \tag{6.10}$$

Thus, the expression (6.9) becomes:

$$\nu_\star \ddot{\phi} = u - f_v^\star \dot{\phi} + a_\star \cos(\theta_d) - a_\star \sin(\theta_d)\phi. \tag{6.11}$$

Hence, by utilizing the obtained desired control law (6.8) and taking into account $\Delta u = u - u_d$, we can represent the linearized approximation of the nonlinear dynamics (6.6) as follows:

$$\nu_\star \ddot{\phi} = \Delta u - f_v^\star \dot{\phi} - a_\star \sin(\theta_d)\phi. \tag{6.12}$$

6.3.3 *Extended Nonlinear Dynamic Model*

We may create a dynamics that illustrates the distinction between their nonlinear dynamics of robotic systems (6.6) and the specified linearized approximation provided by (6.12), as shown below:

$$\begin{aligned}\nu_\star \ddot{\phi} =&\Delta u - f_v^\star \dot{\phi} + a_\star \cos(\phi + \theta_d) - f_s^\star \mathrm{sgn}(\dot{\phi}) + \Delta_t - a_\star \sin(\theta_d)\phi \\ &+ u_d + a_\star \sin(\theta_d)\phi.\end{aligned} \tag{6.13}$$

Furthermore, by utilizing (6.8), we can rephrase the previous expression (6.13) as follows:

$$\begin{aligned}\nu_\star \ddot{\phi} =&\Delta u - f_v^\star \dot{\phi} + a_\star \cos(\phi + \theta_d) - f_s^\star \mathrm{sgn}(\dot{\phi}) + \Delta_t - a_\star \sin(\theta_d)\phi \\ &- a_\star \cos(\theta_d) + a_\star \sin(\theta_d)\phi.\end{aligned} \tag{6.14}$$

6.3.4 *State Space Representation*

Let us define the state vector $\mathcal{X} = \begin{bmatrix} \phi \\ \dot{\phi} \end{bmatrix}$. Based on the nonlinear dynamic model (6.14), we can express it under the following state space representation:

$$\dot{\mathcal{X}} = \mathcal{A}\mathcal{X} + \mathcal{B}\Delta u + \mathcal{D}_1 \Phi_1(\mathcal{X}) + \mathcal{D}_2 \Phi_2(\mathcal{X}) + \mathcal{B}\Delta_t. \tag{6.15}$$

In Eq. (6.15), the functions $\Phi_1(\mathcal{X})$ and $\Phi_2(\mathcal{X})$ have the following descriptions:

$$\Phi_1(\mathcal{X}) = \mathrm{sgn}(\dot{\phi}), \tag{6.16a}$$

$$\Phi_2(\mathcal{X}) = \cos(\phi + \theta_d) + \sin(\theta_d)\phi - \cos(\theta_d), \tag{6.16b}$$

and the matrices $\mathcal{A}$, $\mathcal{B}$, $\mathcal{D}_1$, and $\mathcal{D}_2$ are described by:

$$\mathcal{A} = \begin{bmatrix} 0 & 1 \\ -\frac{a_\star}{\nu_\star}\sin(\theta_d) & -\frac{f_v^\star}{\nu_\star} \end{bmatrix} \quad \mathcal{B} = \begin{bmatrix} 0 \\ \frac{1}{\nu_\star} \end{bmatrix} \quad \mathcal{D}_1 = \begin{bmatrix} 0 \\ -\frac{f_s^\star}{\nu_\star} \end{bmatrix} \quad \mathcal{D}_2 = \begin{bmatrix} 0 \\ \frac{a_\star}{\nu_\star} \end{bmatrix} \tag{6.17}$$

6.3.5 *Proposed Affine State-Feedback Controller*

Our primary goal in this research is to produce precise position control of the knee exoskeleton robot. To attain this objective and effectively control the robot defined by its equation of motion (6.1), we will develop an affine state-feedback control law.

Then, let us consider the following expression of the affine state-feedback controller:

$$u = u_d + \Delta u, \tag{6.18}$$

where Δu is described as follows:

$$\Delta u = \begin{cases} \mathcal{K}X & \text{if } \|X\|^2 \geq \epsilon^2, \\ 0 & \text{if } \|X\|^2 < \epsilon^2, \end{cases} \tag{6.19}$$

with ϵ is a small constant that will be determined and fixed later.

Thus, the following is an expression for the nonlinear dynamics of the robotic exoskeleton system under the suggested control law (6.18):

$$\dot{X} = (\mathcal{A} + \mathcal{B}\mathcal{K})\,X + \mathcal{D}_1\Phi_1(X) + \mathcal{D}_2\Phi_2(X) + \mathcal{B}\Delta_t \quad \text{where} \quad \|X\|^2 \geq \epsilon^2. \tag{6.20}$$

6.4 Design of LMI Stability Conditions

This section is dedicated to establishing an LMI condition for the gain $\mathcal{K}$ of the proposed control law (6.18) to achieve regulation and stability of the knee exoskeleton robot. Our approach involves utilizing the Lyapunov technique to accomplish this objective. Hence, let's adopt the subsequent candidate Lyapunov function:

$$\mathcal{V}(X) = X^{\mathrm{T}}\mathcal{P}X \text{ such that } \|X\|^2 \geq \epsilon^2. \tag{6.21}$$

Then, the following is a description of the candidate Lyapunov function's time-dependent derivative:

$$\dot{\mathcal{V}}(X) = 2X^{\mathrm{T}}\mathcal{P}\dot{X} \text{ such that } \|X\|^2 \geq \epsilon^2. \tag{6.22}$$

According to the nonlinear dynamics under controller (6.20), the expression (6.22) becomes:

$$\begin{aligned}\dot{\mathcal{V}}(\mathcal{X}) =& 2\mathcal{X}^{\mathrm{T}}\mathcal{P}(\mathcal{A}+\mathcal{B}\mathcal{K})\mathcal{X}+2\mathcal{X}^{\mathrm{T}}\mathcal{P}\mathcal{D}_1\Phi_1(\mathcal{X})\\ &+2\mathcal{X}^{\mathrm{T}}\mathcal{P}\mathcal{D}_2\Phi_2(\mathcal{X})+2\mathcal{X}^{\mathrm{T}}\mathcal{P}\mathcal{B}\Delta_t.\end{aligned} \tag{6.23}$$

To more enhance the control results, we consider the subsequent requirement for exponential stabilization:

$$\dot{\mathcal{V}}(\mathcal{X})+\alpha\mathcal{V}(\mathcal{X})<0, \tag{6.24}$$

with $\alpha>0$ denotes a fixed positive scalar.

Next, we will derive three different methods to develop LMI conditions on the gain $\mathcal{K}$ for the stabilization problem based on expression (6.23).

6.4.1 First Design Approach of LMI Condition

In the first development approach, we assume that the function $\Phi_2(\mathcal{X})$ is bounded by a constant value. Let us note first that the constraint on $\mathcal{V}(\mathcal{X})$, as denoted by Eq. (6.21), can be reformulated as follows:

$$\begin{bmatrix}\mathcal{X}\\ \Phi_1(\mathcal{X})\\ \Phi_2(\mathcal{X})\\ \Delta_t\\ 1\end{bmatrix}^{\mathrm{T}}\begin{bmatrix}-\mathcal{I} & \mathcal{O} & \mathcal{O} & \mathcal{O} & \mathcal{O}\\ \star & \mathcal{O} & \mathcal{O} & \mathcal{O} & \mathcal{O}\\ \star & \star & \mathcal{O} & \mathcal{O} & \mathcal{O}\\ \star & \star & \star & \mathcal{O} & \mathcal{O}\\ \star & \star & \star & \star & \epsilon^2\end{bmatrix}\begin{bmatrix}\mathcal{X}\\ \Phi_1(\mathcal{X})\\ \Phi_2(\mathcal{X})\\ \Delta_t\\ 1\end{bmatrix}\leq 0 \tag{6.25}$$

with $\mathcal{I}$ and $\mathcal{O}$ representing the identity matrix and the zero matrix, respectively, both here and in the sequel.

Furthermore, using the expression (6.23), the condition (6.24) may be reformulated like so:

$$\begin{bmatrix}\mathcal{X}\\ \Phi_1(\mathcal{X})\\ \Phi_2(\mathcal{X})\\ \Delta_t\\ 1\end{bmatrix}^{\mathrm{T}}\begin{bmatrix}\Lambda_1 & \mathcal{P}\mathcal{D}_1 & \mathcal{P}\mathcal{D}_2 & \mathcal{P}\mathcal{B} & \mathcal{O}\\ \star & \mathcal{O} & \mathcal{O} & \mathcal{O} & \mathcal{O}\\ \star & \star & \mathcal{O} & \mathcal{O} & \mathcal{O}\\ \star & \star & \star & \mathcal{O} & \mathcal{O}\\ \star & \star & \star & \star & \mathcal{O}\end{bmatrix}\begin{bmatrix}\mathcal{X}\\ \Phi_1(\mathcal{X})\\ \Phi_2(\mathcal{X})\\ \Delta_t\\ 1\end{bmatrix}<0 \tag{6.26}$$

where $\Lambda_1=(\mathcal{P}\mathcal{A}+\mathcal{P}\mathcal{B}\mathcal{K})+(\mathcal{P}\mathcal{A}+\mathcal{P}\mathcal{B}\mathcal{K})^{\mathrm{T}}+\alpha\mathcal{P}$.

Moreover, as $\Phi_1(\mathcal{X}) = \text{sgn}(\dot{\phi}) = \pm 1$, then $\Phi_1(\mathcal{X})^2 = 1$. This expression may be rewritten as follows:

$$\begin{bmatrix} \mathcal{X} \\ \Phi_1(\mathcal{X}) \\ \Phi_2(\mathcal{X}) \\ \Delta_t \\ 1 \end{bmatrix}^{\mathrm{T}} \begin{bmatrix} \mathcal{O} & \mathcal{O} & \mathcal{O} & \mathcal{O} & \mathcal{O} \\ \star & \mathcal{I} & \mathcal{O} & \mathcal{O} & \mathcal{O} \\ \star & \star & \mathcal{O} & \mathcal{O} & \mathcal{O} \\ \star & \star & \star & \mathcal{O} & \mathcal{O} \\ \star & \star & \star & \star & -1 \end{bmatrix} \begin{bmatrix} \mathcal{X} \\ \Phi_1(\mathcal{X}) \\ \Phi_2(\mathcal{X}) \\ \Delta_t \\ 1 \end{bmatrix} = 0. \tag{6.27}$$

Furthermore, as $\Phi_2(\mathcal{X}) = \cos(\phi+\theta_d)+\sin(\theta_d)\phi-\cos(\theta_d) = \cos(\theta)+\sin(\theta_d)\phi-\cos(\theta_d)$ and given that $0^\circ \le \theta \le 90^\circ$ and $0^\circ \le \theta_d \le 90^\circ$, it is evident that $-(\sin(\theta_d)\phi_{\max} + 1) \le \Phi_2(\mathcal{X}) \le \sin(\theta_d)\phi_{\max} + 1$. Thus, we can demonstrate the following inequality:

$$\Phi_2(\mathcal{X})^2 \le \mu_1^2, \tag{6.28}$$

where

$$\mu_1 = \sin(\theta_d)\phi_{\max} + 1, \tag{6.29}$$

with $\phi_{\max} = \max(\phi)$.

Consequently, the inequality (6.28) becomes:

$$\begin{bmatrix} \mathcal{X} \\ \Phi_1(\mathcal{X}) \\ \Phi_2(\mathcal{X}) \\ \Delta_t \\ 1 \end{bmatrix}^{\mathrm{T}} \begin{bmatrix} \mathcal{O} & \mathcal{O} & \mathcal{O} & \mathcal{O} & \mathcal{O} \\ \star & \mathcal{O} & \mathcal{O} & \mathcal{O} & \mathcal{O} \\ \star & \star & \mathcal{I} & \mathcal{O} & \mathcal{O} \\ \star & \star & \star & \mathcal{O} & \mathcal{O} \\ \star & \star & \star & \star & -\mu_1^2 \end{bmatrix} \begin{bmatrix} \mathcal{X} \\ \Phi_1(\mathcal{X}) \\ \Phi_2(\mathcal{X}) \\ \Delta_t \\ 1 \end{bmatrix} \le 0. \tag{6.30}$$

Based on constraint (6.5), it follows that $\Delta_t^2 \le \bar{\delta}^2$. Then, we demonstrate the subsequent expression:

$$\begin{bmatrix} \mathcal{X} \\ \Phi_1(\mathcal{X}) \\ \Phi_2(\mathcal{X}) \\ \Delta_t \\ 1 \end{bmatrix}^{\mathrm{T}} \begin{bmatrix} \mathcal{O} & \mathcal{O} & \mathcal{O} & \mathcal{O} & \mathcal{O} \\ \star & \mathcal{O} & \mathcal{O} & \mathcal{O} & \mathcal{O} \\ \star & \star & \mathcal{O} & \mathcal{O} & \mathcal{O} \\ \star & \star & \star & \mathcal{I} & \mathcal{O} \\ \star & \star & \star & \star & -\bar{\delta}^2 \end{bmatrix} \begin{bmatrix} \mathcal{X} \\ \Phi_1(\mathcal{X}) \\ \Phi_2(\mathcal{X}) \\ \Delta_t \\ 1 \end{bmatrix} \le 0. \tag{6.31}$$

Hence, based on the S-procedure lemma and the Finsler lemma [38–42] and using the previously developed expression (6.25), (6.26), (6.27), (6.30) and (6.31), we demonstrate the subsequent expression:

$$\begin{bmatrix} \Lambda_1+\xi_1\mathcal{I} & \mathcal{P}\mathcal{D}_1 & \mathcal{P}\mathcal{D}_2 & \mathcal{P}\mathcal{B} & O \\ \star & -\tau_1\mathcal{I} & O & O & O \\ \star & \star & -\tau_2\mathcal{I} & O & O \\ \star & \star & \star & -\tau_3\mathcal{I} & O \\ \star & \star & \star & \star & \tau_1+\mu_1^2\tau_2+\bar{\delta}^2\tau_3-\xi_1\epsilon^2 \end{bmatrix} < 0, \qquad (6.32)$$

with $\tau_1 > 0$, $\tau_2 > 0$, $\tau_3 > 0$ and $\xi_1 > 0$.

Therefore, utilizing the Schur complement lemma [38–42] and according to the previous matrix inequality obtained in (6.32), we can establish the following two inequalities:

$$\Lambda_1+\xi_1\mathcal{I}+\tau_1^{-1}\mathcal{P}\mathcal{D}_1\mathcal{D}_1^{\mathrm{T}}\mathcal{P}+\tau_2^{-1}\mathcal{P}\mathcal{D}_2\mathcal{D}_2^{\mathrm{T}}\mathcal{P}+\tau_3^{-1}\mathcal{P}\mathcal{B}\mathcal{B}^{\mathrm{T}}\mathcal{P} < 0 \qquad (6.33a)$$

$$\tau_1+\mu_1^2\tau_2+\bar{\delta}^2\tau_3-\xi_1\epsilon^2 < 0. \qquad (6.33b)$$

Then, after multiplying the first extended condition defined by (6.33a) on the right side and the left side by $\mathcal{S}=\mathcal{P}^{-1}$, we get the subsequent inequality:

$$\Lambda_2+\xi_1\mathcal{S}^2+\tau_1^{-1}\mathcal{D}_1\mathcal{D}_1^{\mathrm{T}}+\tau_2^{-1}\mathcal{D}_2\mathcal{D}_2^{\mathrm{T}}+\tau_3^{-1}\mathcal{B}\mathcal{B}^{\mathrm{T}} < 0. \qquad (6.34)$$

where $\Lambda_2=(\mathcal{A}\mathcal{S}+\mathcal{B}\mathcal{R})+(\mathcal{A}\mathcal{S}+\mathcal{B}\mathcal{R})^{\mathrm{T}}+\alpha\mathcal{S}$ and $\mathcal{R}=\mathcal{K}\mathcal{S}$.

Posing $\alpha_1=\tau_1^{-1}$, $\alpha_2=\tau_2^{-1}$, $\alpha_3=\tau_3^{-1}$ and $\gamma=\xi_1^{-1}$, and utilizing the Schur complement [38–41], we get the subsequent LMI stability requirement:

$$\begin{bmatrix} \Lambda_2+\alpha_1\mathcal{D}_1\mathcal{D}_1^{\mathrm{T}}+\alpha_2\mathcal{D}_2\mathcal{D}_2^{\mathrm{T}}+\alpha_3\mathcal{B}\mathcal{B}^{\mathrm{T}} & \mathcal{S} \\ \mathcal{S} & -\gamma\mathcal{I} \end{bmatrix} < 0. \qquad (6.35)$$

Furthermore, by multiplying the second extended stability condition (6.33b) by γ^2, we obtain the following expression:

$$\gamma^2\tau_1+\mu_1^2\gamma^2\tau_2+\bar{\delta}^2\gamma^2\tau_3-\gamma\epsilon^2 < 0. \qquad (6.36)$$

By utilizing the Schur complement and the expression (6.36), we can derive the subsequent Linear Matrix Inequality (LMI):

$$\begin{bmatrix} \epsilon^2\gamma & \star & \star & \star \\ \bar{\delta}\gamma & \alpha_3 & \star & \star \\ \mu_1\gamma & O & \alpha_2 & \star \\ \gamma & O & O & \alpha_1 \end{bmatrix} > 0. \qquad (6.37)$$

Hence, we have two requirements for stabilization represented by (6.35) and (6.37). These LMIs are expressed in terms of the unknowable factors $\mathcal{S}$, $\mathcal{R}$, α_1, α_2, α_3, and γ.

Theorem 6.1 *Posing that there exist positive constant parameters $\bar{\delta}$ and ϵ and matrices $\mathcal{S}$ and $\mathcal{R}$, as well as scalars α_1, α_2, α_3 and γ, such that:*

LMI (6.35) and LMI (6.37)

are feasible. Therefore, it can be concluded that the proposed state-feedback controller defined in (6.18) stabilizes the nonlinear dynamics (6.1). Moreover, the gain $\mathcal{K}$ of the adopted control law can be calculated using the subsequent expression:

$$\mathcal{K} = \mathcal{R}\mathcal{S}^{-1}. \tag{6.38}$$

6.4.2 Second Design Approach of LMI Condition

In this design approach, we make the assumption that the function $\Phi_2(\mathcal{X})$ remains bounded within the constraint of a linear equation. As highlighted earlier, the constraint applied to the candidate Lyapunov function (6.21) can be redefined using the expression (6.25). Furthermore, we will adhere to the same formulation (6.23), which can be equivalently presented as (6.26). Additionally, let recall that $\Phi_1(\mathcal{X}) = \text{sgn}(\dot{\phi}) = \pm 1$, thus yielding $\Phi_1(\mathcal{X})^2 = 1$, and hence, we can substantiate the expression (6.27).

Furthermore, since $\Phi_2(\mathcal{X}) = \cos(\phi + \theta_d) + \sin(\theta_d)\phi - \cos(\theta_d) = \cos(\theta) + \sin(\theta_d)\phi - \cos(\theta_d)$, therefore $\Phi_2(\mathcal{X})^2 = \cos^2(\theta) + \sin^2(\theta_d)\phi^2 + \cos^2(\theta_d) + 2\cos(\theta)\sin(\theta_d)\phi - 2\cos(\theta_d)\cos(\theta) - 2\sin(\theta_d)\cos(\theta_d)\phi$. Hence, since $\phi = \theta - \theta_d$, we can express it as follows:

$$\begin{aligned}\Phi_2(\mathcal{X})^2 =& \cos^2(\theta) + \sin^2(\theta_d)\phi^2 + \cos^2(\theta_d) + 2\cos(\theta)\sin(\theta_d)\theta \\ & - 2\cos(\theta)\sin(\theta_d)\theta_d - 2\cos(\theta_d)\cos(\theta) - 2\sin(\theta_d)\cos(\theta_d)\phi.\end{aligned} \tag{6.39}$$

It is straightforward to expand the following inequalities:

$$\sin^2(\theta_d) \le 1, \tag{6.40a}$$

$$\cos^2(\theta) \le -\theta^2 + b, \tag{6.40b}$$

$$2\cos(\theta)\theta \le 2\theta, \tag{6.40c}$$

$$-2\cos(\theta) \le 0, \tag{6.40d}$$

with $b = \frac{\pi^2}{4}$.

Given that $\theta = \phi + \theta_d$, we can proceed to deduce the following expressions:

$$\sin^2(\theta_d)\phi^2 \leq \phi^2, \tag{6.41a}$$

$$\cos^2(\theta) \leq -\phi^2 - 2\theta_d\phi - \theta_d^2 + b, \tag{6.41b}$$

$$-2\cos(\theta_d)\cos(\theta) \leq 0, \tag{6.41c}$$

$$-2\sin(\theta_d)\theta_d\cos(\theta) \leq 0, \tag{6.41d}$$

$$2\sin(\theta_d)\cos(\theta)\theta \leq 2\sin(\theta_d)\phi + 2\sin(\theta_d)\theta_d. \tag{6.41e}$$

Thus, based on the above evaluations, we are able to establish the subsequent inequality:

$$\Phi_2(\mathcal{X})^2 \leq 2\eta_1\phi + \eta_2, \tag{6.42}$$

where

$$\eta_1 = -\theta_d + \sin(\theta_d) - \sin(\theta_d)\cos(\theta_d), \tag{6.43a}$$

$$\eta_2 = -\theta_d^2 + b + \cos^2(\theta_d) + 2\sin(\theta_d)\theta_d. \tag{6.43b}$$

It is crucial to emphasize that for all values of θ_d ranging from 0 to $\frac{\pi}{2}$, we observe that $\eta_1 < 0$.

When we take into account $\phi = C_1\mathcal{X}$, with $C_1 = \begin{bmatrix} 1 & 0 \end{bmatrix}$, the Eq. (6.42) becomes:

$$\Phi_2(\mathcal{X})^2 \leq 2\eta_1 C_1\mathcal{X} + \eta_2. \tag{6.44}$$

Therefore, we can reformulate the inequality (6.44) as follows:

$$\begin{bmatrix} \mathcal{X} \\ \Phi_1(\mathcal{X}) \\ \Phi_2(\mathcal{X}) \\ \Delta_t \\ 1 \end{bmatrix}^{\mathrm{T}} \begin{bmatrix} O & O & O & O & -\eta_1 C_1^{\mathrm{T}} \\ \star & O & O & O & O \\ \star & \star & I & O & O \\ \star & \star & \star & O & O \\ \star & \star & \star & \star & -\eta_2 \end{bmatrix} \begin{bmatrix} \mathcal{X} \\ \Phi_1(\mathcal{X}) \\ \Phi_2(\mathcal{X}) \\ \Delta_t \\ 1 \end{bmatrix} \leq 0. \tag{6.45}$$

Furthermore, we will consider that Δ_t adheres to the same constraint as defined by (6.5). This leads us to the resultant expression outlined in (6.31). Consequently, using the S-procedure lemma and utilizing the expressions previously derived in (6.25), (6.26), (6.27), (6.45), and (6.31), we are able to establish the subsequent matrix inequality:

$$\begin{bmatrix} \Lambda_1 + \xi_1 \mathcal{I} & \mathcal{P}\mathcal{D}_1 & \mathcal{P}\mathcal{D}_2 & \mathcal{P}\mathcal{B} & \tau_2\eta_1 C_1^{\mathrm{T}} \\ \star & -\tau_1 \mathcal{I} & \mathcal{O} & \mathcal{O} & \mathcal{O} \\ \star & \star & -\tau_2 \mathcal{I} & \mathcal{O} & \mathcal{O} \\ \star & \star & \star & -\tau_3 \mathcal{I} & \mathcal{O} \\ \star & \star & \star & \star & \tau_1 + \tau_2\eta_2 + \tau_3\bar{\delta}^2 - \xi_1\epsilon^2 \end{bmatrix} < 0, \tag{6.46}$$

with $\tau_1 > 0$, $\tau_2 > 0$, $\tau_3 > 0$ and $\xi_1 > 0$.

Hence, employing the Schur complement lemma and in accordance with the previously derived expression (6.46), we obtain:

$$\Lambda_1 + \xi_1 \mathcal{I} + \tau_1^{-1}\mathcal{P}\mathcal{D}_1\mathcal{D}_1^{\mathrm{T}}\mathcal{P} + \tau_2^{-1}\mathcal{P}\mathcal{D}_2\mathcal{D}_2^{\mathrm{T}}\mathcal{P} + \tau_3^{-1}\mathcal{P}\mathcal{B}\mathcal{B}^{\mathrm{T}}\mathcal{P}$$
$$- \left(\tau_1 + \tau_2\eta_2 + \tau_3\bar{\delta}^2 - \xi_1\epsilon^2\right)^{-1} \tau_2^2\eta_1^2 C_1^{\mathrm{T}} C_1 < 0. \tag{6.47}$$

By multiplying the expression (6.47) on the right and the left by $\mathcal{S} = \mathcal{P}^{-1}$, we arrive at the subsequent inequality:

$$\Lambda_2 + \xi_1 \mathcal{S}^2 + \tau_1^{-1}\mathcal{D}_1\mathcal{D}_1^{\mathrm{T}} + \tau_2^{-1}\mathcal{D}_2\mathcal{D}_2^{\mathrm{T}} + \tau_3^{-1}\mathcal{B}\mathcal{B}^{\mathrm{T}} + \eta_1^2\Gamma_1\mathcal{S}C_1^{\mathrm{T}}C_1\mathcal{S} < 0. \tag{6.48}$$

with $\Gamma_1 = -\left(\tau_1 + \tau_2\eta_2 + \tau_3\bar{\delta}^2 - \xi_1\epsilon^2\right)^{-1} \tau_2^2$.

According to the matrix inequality (6.46), it is clear that $\tau_1 + \tau_2\eta_2 + \tau_3\bar{\delta}^2 - \xi_1\epsilon^2 < 0$, then we have $\Gamma_1 > 0$. Therefore, posing $\alpha_1 = \tau_1^{-1}$, $\alpha_2 = \tau_2^{-1}$, $\alpha_3 = \tau_3^{-1}$, and $\gamma = \xi_1^{-1}$, and by employing the Schur complement lemma, we extend the subsequent expression:

$$\begin{bmatrix} \Lambda_2 + \alpha_1\mathcal{D}_1\mathcal{D}_1^{\mathrm{T}} + \alpha_2\mathcal{D}_2\mathcal{D}_2^{\mathrm{T}} + \alpha_3\mathcal{B}\mathcal{B}^{\mathrm{T}} & \mathcal{S} & \eta_1\mathcal{S}C_1^{\mathrm{T}} \\ \star & -\gamma\mathcal{I} & \mathcal{O} \\ \star & \star & -\Gamma_1^{-1} \end{bmatrix} < 0. \tag{6.49}$$

Moreover, it is straightforward to obtain the following expression:

$$-\Gamma_1^{-1} = \alpha_2^2\alpha_1^{-1} + \eta_2\alpha_2 + \bar{\delta}^2\alpha_2^2\alpha_3^{-1} - \alpha_2^2\gamma^{-1}\epsilon^2. \tag{6.50}$$

As $-\alpha_2^2\gamma^{-1}\epsilon^2 < 0$, therefore, using the Young inequality lemma [38–42], we can write:

$$-\alpha_2^2\gamma^{-1}\epsilon^2 \leq \gamma\epsilon^{-2} - 2\alpha_2. \tag{6.51}$$

Subsequently, we derive the following inequality:

$$-\Gamma_1^{-1} \leq \alpha_2^2\alpha_1^{-1} + (\eta_2 - 2)\alpha_2 + \bar{\delta}^2\alpha_2^2\alpha_3^{-1} + \gamma\epsilon^{-2}. \tag{6.52}$$

Therefore, using expression (6.52), and based on the Schur complement lemma, we demonstrate the subsequent LMI stability condition:

$$\begin{bmatrix} \Lambda_3 & \mathcal{S} & \eta_1 \mathcal{S} \mathcal{C}_1^{\mathrm{T}} & \mathit{0} & \mathit{0} & \mathit{0} \\ \star & -\gamma \mathcal{I} & \mathit{0} & \mathit{0} & \mathit{0} & \mathit{0} \\ \star & \star & (\eta_2 - 2)\alpha_2 & \alpha_2 & \alpha_2 & \gamma \\ \star & \star & \star & -\alpha_1 & \mathit{0} & \mathit{0} \\ \star & \star & \star & \star & -\frac{1}{\bar{\delta}^2}\alpha_3 & \mathit{0} \\ \star & \star & \star & \star & \star & -\gamma\epsilon^2 \end{bmatrix} < 0, \tag{6.53}$$

with $\Lambda_3 = \Lambda_2 + \alpha_1 \mathcal{D}_1 \mathcal{D}_1^{\mathrm{T}} + \alpha_2 \mathcal{D}_2 \mathcal{D}_2^{\mathrm{T}} + \alpha_3 \mathcal{B}\mathcal{B}^{\mathrm{T}}$.

It is clear the the previous LMI condition is not feasible since $(\eta_2 - 2)\alpha_2 > 0$ as $\eta_2 > \pi$. Hence, no solution can obtained from the previous LMI. Then, the use of the Young inequality lemma is much restrictive/conservative like that in (6.51). Thus, the idea is to use such lemma by injecting a free parameter like so:

$$-\alpha_2^2 \gamma^{-1} \epsilon^2 \leq \gamma \epsilon^{-2} \sigma^2 - 2\alpha_2 \sigma, \tag{6.54}$$

where σ is a free parameter.

Therefore, we obtain the following inequality:

$$-\Gamma_1^{-1} \leq \alpha_2^2 \alpha_1^{-1} + (\eta_2 - 2\sigma)\alpha_2 + \bar{\delta}^2 \alpha_2^2 \alpha_3^{-1} + \gamma \epsilon^{-2} \sigma^2. \tag{6.55}$$

Accordingly, using expression (6.55), and relying on the Schur complement, we obtain the following LMI condition:

$$\begin{bmatrix} \Lambda_3 & \mathcal{S} & \eta_1 \mathcal{S} \mathcal{C}_1^{\mathrm{T}} & \mathit{0} & \mathit{0} & \mathit{0} \\ \star & -\gamma \mathcal{I} & \mathit{0} & \mathit{0} & \mathit{0} & \mathit{0} \\ \star & \star & (\eta_2 - 2\sigma)\alpha_2 & \alpha_2 & \alpha_2 & \gamma\sigma \\ \star & \star & \star & -\alpha_1 & \mathit{0} & \mathit{0} \\ \star & \star & \star & \star & -\frac{1}{\bar{\delta}^2}\alpha_3 & \mathit{0} \\ \star & \star & \star & \star & \star & -\gamma\epsilon^2 \end{bmatrix} < 0. \tag{6.56}$$

As a result, we deduce the LMI stability condition outlined in (6.56), which is formulated in terms of the unknown variables $\mathcal{S}$, $\mathcal{R}$, α_1, α_2, α_3, and γ.

It is clear that a sufficient condition for the feasibility of the LMI (6.56) is that $\eta_2 - 2\sigma < 0$, and hence, we should select the free parameter σ to be $\sigma > \frac{\eta_2}{2}$ and then we can select for all θ_d between 0 and $\frac{\pi}{2}$, that $\sigma > \frac{\eta_2^{\max}}{2}$ where $\eta_2^{\max} = \max(\eta_2) = b + 1 = \left(\frac{\pi}{2}\right)^2 + 1 \approx 1.7337$.

Theorem 6.2 *Assuming fixed parameters $\bar{\delta} \geq 0$ and $\epsilon \geq 0$, and a free parameter $\sigma > \frac{\pi^2}{8} + \frac{1}{2}$, and given the existence of matrices $\mathcal{S}$ and $\mathcal{R}$, as well as scalars α_1, α_2, α_3, and γ, the subsequent LMI-based problem*

LMI (6.56)

is feasible. Therefore, the control law (6.18) *that has been adopted leads to the stabilization of the nonlinear dynamic model* (6.1) *of the knee exoskeleton robot.*

6.4.3 Third Design Approach of the LMI Condition

In this developed approach, let's make the assumption that the function $\Phi_2(\mathcal{X})$ is bounded by a second-degree equation. Similar to the previous methods, we will utilize the same concept of reformulating the constraint applied to the candidate Lyapunov function (6.21), which can be recast using the expression (6.25). Additionally, we will continue to employ the same expression (6.23), which can be equivalently represented as (6.26). Furthermore, it is worth recalling that $\Phi_1(\mathcal{X}) = \text{sgn}(\dot{\phi}) = \pm 1$, which leads to $\Phi_1(\mathcal{X})^2 = 1$, and then we can demonstrate the expression (6.27).

Furthermore, as $\Phi_2(\mathcal{X}) = \cos(\phi + \theta_d) + \sin(\theta_d)\phi - \cos(\theta_d) = \cos(\theta) + \sin(\theta_d)\phi - \cos(\theta_d)$, it follows that $\Phi_2(\mathcal{X})^2 = \cos^2(\theta) + \sin^2(\theta_d)\phi^2 + \cos^2(\theta_d) + 2\cos(\theta)\sin(\theta_d)\phi - 2\cos(\theta_d)\cos(\theta) - 2\sin(\theta_d)\cos(\theta_d)\phi$. Consequently, considering $\phi = \theta - \theta_d$, we can express $\Phi_2(\mathcal{X})^2$ as follows:

$$\begin{aligned}\Phi_2(\mathcal{X})^2 =& \cos^2(\theta) + \sin^2(\theta_d)\phi^2 + \cos^2(\theta_d) + 2\cos(\theta)\sin(\theta_d)\theta \\ &- 2\cos(\theta)\sin(\theta_d)\theta_d - 2\cos(\theta_d)\cos(\theta) - 2\sin(\theta_d)\cos(\theta_d)\phi.\end{aligned} \tag{6.57}$$

It is easy to establish the following inequalities:

$$\cos^2(\theta) \leq 1, \tag{6.58a}$$

$$2\cos(\theta)\theta \leq 2\theta, \tag{6.58b}$$

$$-2\cos(\theta) \leq 0. \tag{6.58c}$$

Given $\theta = \phi + \theta_d$, we can derive the following expressions:

$$\cos^2(\theta) \leq 1, \tag{6.59a}$$

$$-2\cos(\theta_d)\cos(\theta) \leq 0, \tag{6.59b}$$

$$-2\sin(\theta_d)\theta_d\cos(\theta) \leq 0, \tag{6.59c}$$

$$2\sin(\theta_d)\cos(\theta)\theta \leq 2\sin(\theta_d)\phi + 2\sin(\theta_d)\theta_d. \tag{6.59d}$$

Therefore, building on the above evaluations, we can demonstrate the following inequality:

$$\Phi_2(\mathcal{X})^2 \leq \mu_1\phi^2 + 2\mu_2\phi + \mu_3, \tag{6.60}$$

where

$$\mu_1 = \sin^2(\theta_d), \tag{6.61a}$$

$$\mu_2 = \sin(\theta_d) - \sin(\theta_d)\cos(\theta_d), \tag{6.61b}$$

$$\mu_3 = 1 + \cos^2(\theta_d) + 2\sin(\theta_d)\theta_d. \tag{6.61c}$$

It is important to note that for all values of θ_d between 0 and $\frac{\pi}{2}$, we get $\mu_1 > 0$ and $\mu_3 > 0$.

Considering $\phi = C_1\mathcal{X}$, with $C_1 = \begin{bmatrix} 1 \; 0 \end{bmatrix}$, we can reformulate the expression (6.60) as follows:

$$\Phi_2(\mathcal{X})^2 \leq \mu_1 \mathcal{X}^{\mathrm{T}} C_1^{\mathrm{T}} C_1 \mathcal{X} + 2\mu_2 C_1 \mathcal{X} + \mu_3. \tag{6.62}$$

Therefore, the inequality (6.62) becomes:

$$\begin{bmatrix} \mathcal{X} \\ \Phi_1(\mathcal{X}) \\ \Phi_2(\mathcal{X}) \\ \Delta_t \\ 1 \end{bmatrix}^{\mathrm{T}} \begin{bmatrix} -\mu_1 C_1^{\mathrm{T}} C_1 & O & O & O & -\mu_2 C_1^{\mathrm{T}} \\ \star & O & O & O & O \\ \star & \star & I & O & O \\ \star & \star & \star & O & O \\ \star & \star & \star & \star & -\mu_3 \end{bmatrix} \begin{bmatrix} \mathcal{X} \\ \Phi_1(\mathcal{X}) \\ \Phi_2(\mathcal{X}) \\ \Delta_t \\ 1 \end{bmatrix} \leq 0. \tag{6.63}$$

Furthermore, we will assume that Δ_t satisfies the same constraint as defined by (6.5). As a result, we arrive at the expression defined by (6.31). Therefore, using the S-procedure lemma [38–42] and the Finsler's lemma [39] and utilizing the expressions previously derived in (6.25), (6.26), (6.27), (6.63), and (6.31), we are able to establish the following matrix inequality:

$$\begin{bmatrix} \Lambda_1 + \mu_1\tau_2 C_1^{\mathrm{T}} C_1 + \xi_1 \mathcal{I} & \mathcal{P}\mathcal{D}_1 & \mathcal{P}\mathcal{D}_2 & \mathcal{P}\mathcal{B} & \tau_2\mu_2 C_1^{\mathrm{T}} \\ \star & -\tau_1 \mathcal{I} & O & O & O \\ \star & \star & -\tau_2 \mathcal{I} & O & O \\ \star & \star & \star & -\tau_3 \mathcal{I} & O \\ \star & \star & \star & \star & \tau_1 + \mu_3\tau_2 + \bar{\delta}^2\tau_3 - \xi_1\epsilon^2 \end{bmatrix} < 0, \tag{6.64}$$

with $\tau_1 > 0$, $\tau_2 > 0$, $\tau_3 > 0$ and $\xi_1 > 0$.

Consequently, employing the Schur complement lemma and in accordance with the previously mentioned expression (6.64), we obtain:

$$\begin{aligned} &\Lambda_1 + \xi_1 \mathcal{I} + \mu_1\tau_2 C_1^{\mathrm{T}} C_1 + \tau_1^{-1}\mathcal{P}\mathcal{D}_1\mathcal{D}_1^{\mathrm{T}}\mathcal{P} + \tau_2^{-1}\mathcal{P}\mathcal{D}_2\mathcal{D}_2^{\mathrm{T}}\mathcal{P} + \tau_3^{-1}\mathcal{P}\mathcal{B}\mathcal{B}^{\mathrm{T}}\mathcal{P} \\ &\quad - \left(\tau_1 + \mu_3\tau_2 + \bar{\delta}^2\tau_3 - \xi_1\epsilon^2\right)^{-1} \mu_2^2\tau_2^2 C_1^{\mathrm{T}} C_1 < 0. \end{aligned} \tag{6.65}$$

By multiplying the expression (6.65) on the right and left sides by $\mathcal{S} = \mathcal{P}^{-1}$, we get:

$$\Lambda_2 + \xi_1 \mathcal{S}^2 + \mu_1 \tau_2 \mathcal{S} \mathcal{C}_1^{\mathrm{T}} \mathcal{C}_1 \mathcal{S} + \tau_1^{-1} \mathcal{D}_1 \mathcal{D}_1^{\mathrm{T}} + \tau_2^{-1} \mathcal{D}_2 \mathcal{D}_2^{\mathrm{T}} + \tau_3^{-1} \mathcal{B} \mathcal{B}^{\mathrm{T}} + \mu_2^2 \Gamma_2 \mathcal{S} \mathcal{C}_1^{\mathrm{T}} \mathcal{C}_1 \mathcal{S} < 0, \tag{6.66}$$

with $\Gamma_2 = -\left(\tau_1 + \mu_3 \tau_2 + \bar{\delta}^2 \tau_3 - \xi_1 \epsilon^2\right)^{-1} \tau_2^2$.

According to matrix inequality (6.64), it is clear that $\tau_1 + \mu_3 \tau_2 + \bar{\delta}^2 \tau_3 - \xi_1 \epsilon^2 < 0$, then we can demonstrate that $\Gamma_2 > 0$. Therefore, posing $\alpha_1 = \tau_1^{-1}$, $\alpha_2 = \tau_2^{-1}$, $\alpha_3 = \tau_3^{-1}$, and $\gamma = \xi_1^{-1}$, and employing the Schur complement lemma, we extend the subsequent expression:

$$\begin{bmatrix} \Lambda_2 + \alpha_1 \mathcal{D}_1 \mathcal{D}_1^{\mathrm{T}} + \alpha_2 \mathcal{D}_2 \mathcal{D}_2^{\mathrm{T}} + \alpha_3 \mathcal{B} \mathcal{B}^{\mathrm{T}} & \mathcal{S} & \mathcal{S} \mathcal{C}_1^{\mathrm{T}} & \mu_2 \mathcal{S} \mathcal{C}_1^{\mathrm{T}} \\ \star & -\gamma \mathcal{I} & \mathcal{O} & \mathcal{O} \\ \star & \star & -\frac{1}{\mu_1} \alpha_2 \mathcal{I} & \mathcal{O} \\ \star & \star & \star & -\Gamma_2^{-1} \end{bmatrix} < 0. \tag{6.67}$$

Moreover, it is straightforward to obtain the following expression:

$$-\Gamma_2^{-1} = \alpha_2^2 \alpha_1^{-1} + \mu_3 \alpha_2 + \bar{\delta}^2 \alpha_2^2 \alpha_3^{-1} - \alpha_2^2 \gamma^{-1} \epsilon^2, \tag{6.68}$$

As $-\alpha_2^2 \gamma^{-1} \epsilon^2 < 0$, therefore, using the Young inequality lemma [38–42], we can write:

$$-\alpha_2^2 \gamma^{-1} \epsilon^2 \leq \gamma \epsilon^{-2} - 2\alpha_2. \tag{6.69}$$

Subsequently, we derive the following inequality:

$$-\Gamma_2^{-1} = \alpha_2^2 \alpha_1^{-1} + (\mu_3 - 2)\alpha_2 + \bar{\delta}^2 \alpha_2^2 \alpha_3^{-1} + \gamma \epsilon^{-2}. \tag{6.70}$$

Therefore, using expression (6.52), and based on the Schur complement lemma, we demonstrate the subsequent LMI stability condition:

$$\begin{bmatrix} \Lambda_3 & \mathcal{S} & \mathcal{S} \mathcal{C}_1^{\mathrm{T}} & \mu_2 \mathcal{S} \mathcal{C}_1^{\mathrm{T}} & \mathcal{O} & \mathcal{O} & \mathcal{O} \\ \star & -\gamma \mathcal{I} & \mathcal{O} & \mathcal{O} & \mathcal{O} & \mathcal{O} & \mathcal{O} \\ \star & \star & -\frac{1}{\mu_1} \alpha_2 \mathcal{I} & \mathcal{O} & \mathcal{O} & \mathcal{O} & \mathcal{O} \\ \star & \star & \star & (\mu_3 - 2)\alpha_2 & \alpha_2 & \alpha_2 & \gamma \\ \star & \star & \star & \star & -\alpha_1 & \mathcal{O} & \mathcal{O} \\ \star & \star & \star & \star & \star & -\frac{1}{\bar{\delta}^2} \alpha_3 & \mathcal{O} \\ \star & \star & \star & \star & \star & \star & -\gamma \epsilon^2 \end{bmatrix} < 0, \tag{6.71}$$

with $\Lambda_3 = \Lambda_2 + \alpha_1 \mathcal{D}_1 \mathcal{D}_1^{\mathrm{T}} + \alpha_2 \mathcal{D}_2 \mathcal{D}_2^{\mathrm{T}} + \alpha_3 \mathcal{B} \mathcal{B}^{\mathrm{T}}$.

It is clear the the previous LMI condition is not feasible since $(\mu_3 - 2)\alpha_2 > 0$ as $\mu_3 > \pi$. Hence, no solution can obtained from the previous LMI. Then, the use of the Young inequality lemma is much restrictive/conservative like that in (6.69). Thus, the idea is to use such lemma by injecting a free parameter like so:

$$-\alpha_2^2\gamma^{-1}\epsilon^2 \leq \gamma\epsilon^{-2}\sigma_1^2 - 2\alpha_2\sigma_1, \tag{6.72}$$

where σ_1 is a free parameter.

Therefore, we obtain the following inequality:

$$-\Gamma_2^{-1} \leq \alpha_2^2\alpha_1^{-1} + (\mu_3 - 2\sigma_1)\alpha_2 + \bar{\delta}^2\alpha_2^2\alpha_3^{-1} + \gamma\epsilon^{-2}\sigma_1^2. \tag{6.73}$$

Accordingly, using expression (6.73), and relying on the Schur complement, we obtain the following LMI condition:

$$\begin{bmatrix} \Lambda_3 & \mathcal{S} & \mathcal{S}C_1^{\mathrm{T}} & \eta_1\mathcal{S}C_1^{\mathrm{T}} & O & O & O \\ \star & -\gamma I & O & O & O & O & O \\ \star & \star & -\frac{1}{\mu_1}\alpha_2 & O & O & O & O \\ \star & \star & \star & (\mu_3 - 2\sigma_1)\alpha_2 & \alpha_2 & \alpha_2 & \gamma\sigma_1 \\ \star & \star & \star & \star & -\alpha_1 & O & O \\ \star & \star & \star & \star & \star & -\frac{1}{\bar{\delta}^2}\alpha_3 & O \\ \star & \star & \star & \star & \star & \star & -\gamma\epsilon^2 \end{bmatrix} < 0. \tag{6.74}$$

As a result, we deduce the LMI stability condition outlined in (6.74), which is formulated in terms of the unknown variables $\mathcal{S}$, $\mathcal{R}$, α_1, α_2, α_3, and γ.

It is clear that a sufficient condition for the feasibility of the LMI (6.74) is that $\mu_3 - 2\sigma_1 < 0$, and hence we should select the free parameter σ_1 to be $\sigma_1 > \frac{\mu_3}{2}$. Using expression of the scalar μ_3 develped before, it is easy to demonstrate that for all θ_d between 0 and $\frac{\pi}{2}$, we have $0 \leq \mu_3 \leq 1+\pi$. Therefore, the previous condition on σ_1 becomes $\sigma_1 > \frac{\mu_3^{\max}}{2}$, where $\mu_3^{\max} = \max(\mu_3)$. Then, we should select σ_1 to be $\sigma_1 > \frac{1+\pi}{2} \approx 2.0708$.

Theorem 6.3 *Assuming fixed parameters $\bar{\delta} \geq 0$ and $\epsilon \geq 0$, and a free parameter $\sigma_1 > \frac{\pi+1}{2}$, and given the existence of matrices $\mathcal{S}$ and $\mathcal{R}$, as well as scalars α_1, α_2, α_3, and γ, the following LMI*

LMI (6.74)

is feasible. Therefore, the proposed controller (6.18) leads to the stability of the nonlinear dynamic model (6.1) of the knee rehabilitation exoskeleton robot.

6.5 Results of Simulations

In order to guarantee the stability of the actuated lower limb orthosis/exoskeleton at the knee joint level, three distinct LMI techniques were developed. The numerical and simulation findings from these three approaches are presented in this part. Therefore, we present the outcomes of our detailed numerical analyzes, highlighting the effectiveness of each designed LMI condition. Table 6.1 provides the values of the different parameters for the actuated knee exoskeleton robotic system that are specified by Eq. (6.1).

Then, we proceed to illustrate the control outcomes of the exoskeleton robotic system's response to the intended state θ_d, specifically set at $\theta_d = 30°$. Additionally, we introduce an external disturbance denoted as Δ_t, specifically chosen to take the form of a sinusoidal input:

$$\Delta_t = \bar{\delta} \times \sin(\omega_f \times t). \tag{6.75}$$

In this equation, ω_f stands for the frequency of the signal. For our purposes, we set ω_f to be 5π. It is important to note that this disturbance input is applied exclusively within the time interval from 2 to 7 seconds. Beyond this timeframe, the value of Δ_t is reset to zero.

6.5.1 *Results Obtained Utilizing the First Design Approach*

Based on the established LMI stability condition outlined in Theorem 6.1, resolving the corresponding LMI-based issue derived from this theorem and setting the values $\alpha = 1$, $\epsilon = 1$, and $\bar{\delta} = 1$, we computed this gain:

$$\mathcal{K} = \begin{bmatrix} -25.6776 & -15.5344 \end{bmatrix}. \tag{6.76}$$

By introducing the feedback gain (6.76) into the proposed state-feedback controller (6.18), we observe the graphical results depicted in Fig. 6.2. The robot has clearly been successfully adjusted to the desired position θ_d, specifically set at

Table 6.1 Parameters' value of the knee exoskeleton robot [36]

Parameter	Value	Unit
$\nu_\star$	0.323	Kg.m^2/rad
$m_\star$	0.000105	Kg
$l_\star$	0.25	m
$f_v^\star$	0.35	N.m.s/rad
$f_s^\star$	0.059	N.m/A
g	9.8	N

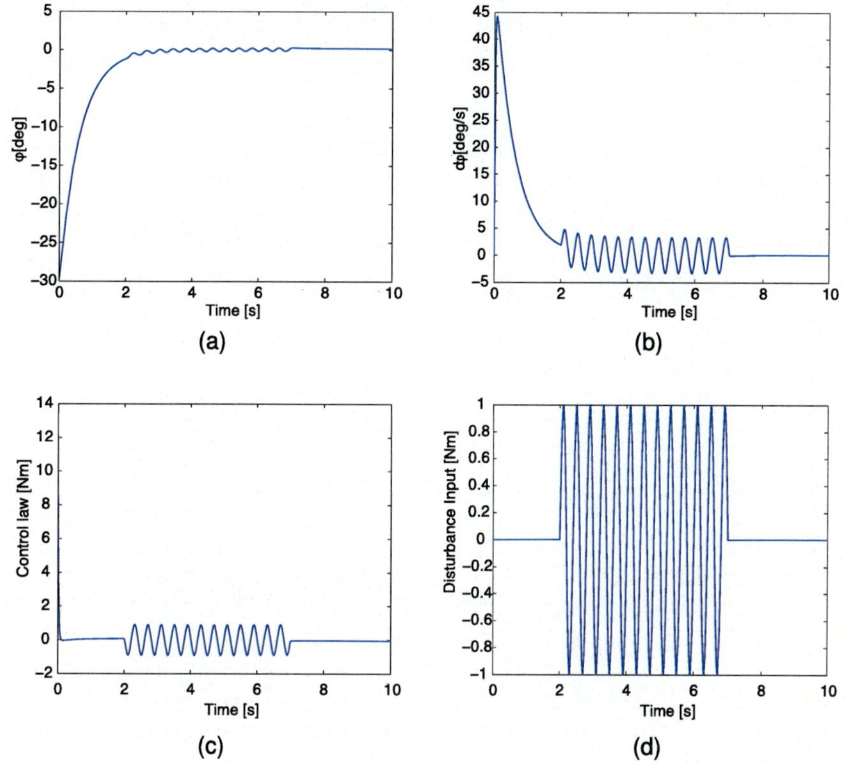

Fig. 6.2 Graphical results of the controlled robotic system using the first development approach. (**a**) Angular position error ϕ, (**b**) Angular velocity error $\dot{\phi}$, (**c**) Adopted Controller u, and (**d**) External disturbance Δ_t

$\theta_d = 30°$. Furthermore, once the exoskeleton robotic system attains stabilization, the control effort u converges to u_d.

6.5.2 *Results Obtained Utilizing the Second Design Approach*

Utilizing the established LMI stability condition presented in Theorem 6.2, we have solved the associated LMI-based problem derived from this theorem. By selecting the values of parameters $\sigma = 3$, $\alpha = 1$, $\epsilon = 1$, and $\bar{\delta} = 1$, we get the following result of the feedback gain $\mathcal{K}$:

$$\mathcal{K} = \begin{bmatrix} -34.2011 & -19.5944 \end{bmatrix}. \tag{6.77}$$

Using the feedback gain (6.77) in the designed state-feedback controller (6.18), we can observe the graphical results illustrated in Fig. 6.3. As the robot precisely

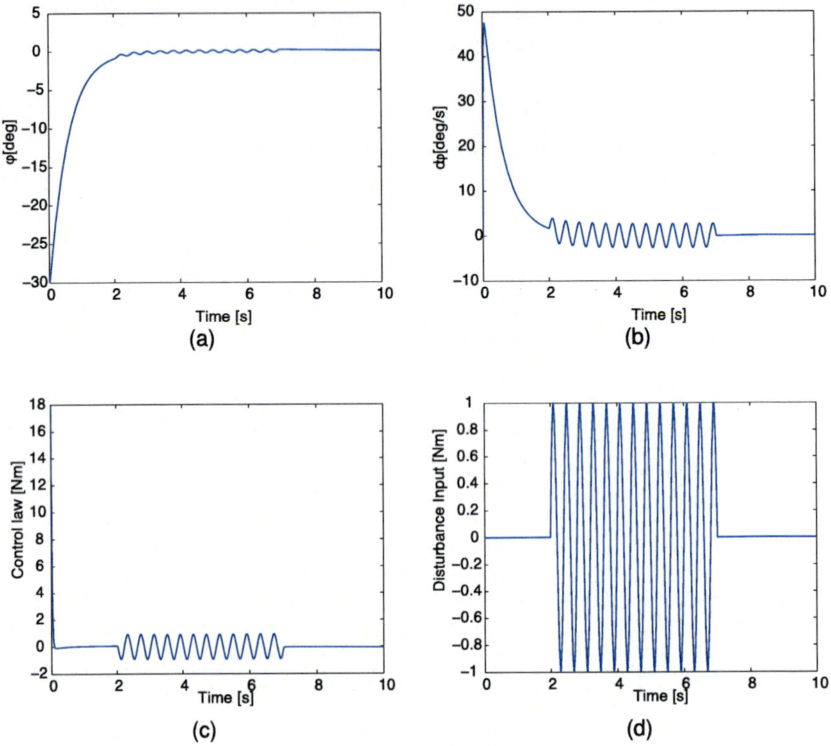

Fig. 6.3 Graphical results of the controlled robotic system using the second development approach. (**a**) Angular position error ϕ, (**b**) Angular velocity error $\dot{\phi}$, (**c**) Adopted Controller u, and (**d**) External disturbance Δ_t

converges to the desired state θ_d, which in this present case $\theta_d = 30°$, the efficacy of the control approach becomes clear. Furthermore, once the robotic system achieves a stable state, the control effort u gradually diminishes, ultimately converging to u_d.

6.5.3 Results Obtained Utilizing the Third Design Approach

Based on the LMI stability condition stated in Theorem 6.3, resolving the related LMI issue derived from this theorem and setting the values $\sigma_1 = 2$, $\alpha = 1$, $\epsilon = 1$, and $\bar{\delta} = 1$, we get:

$$\mathcal{K} = \begin{bmatrix} -29.0460 & -17.5350 \end{bmatrix}. \tag{6.78}$$

Utilizing the feedback gain (6.78) into the adopted control law (6.18), we demonstrate the graphical results shown in Fig. 6.4. Thus, the exoskeleton robot

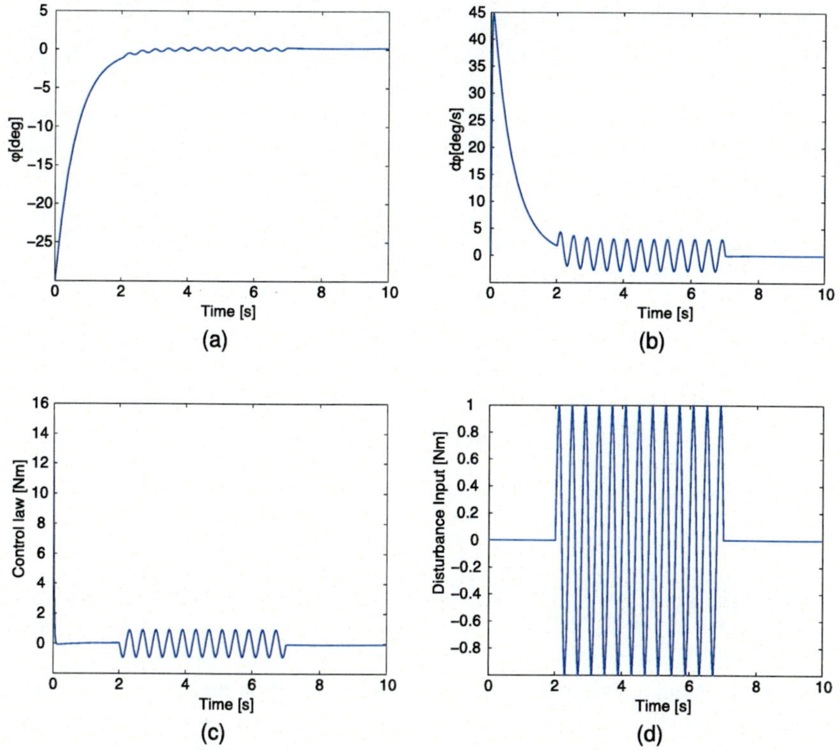

Fig. 6.4 Graphical results of the controlled robotic system using the third development approach. (**a**) Angular position error ϕ, (**b**) Angular velocity error $\dot{\phi}$, (**c**) Adopted Controller u, and (**d**) External disturbance Δ_t

is successfully controlled to the intended position. Furthermore, when the system is stabilized in the intended state, the control effort u converges to the desired level u_d.

6.5.4 A Brief Comparison Between the Three Design Approaches

In order to provide a brief comparison of the three distinct design approaches in the robust position control of the knee exoskeleton robot, Table 6.2 provides a comparison of various values, including angular position, angular velocities, control gains, and the desired state. In this part, we will setting the values of $\alpha = 1$, $\epsilon = 1$, and $\bar{\delta} = 1$. Moreover, we will take different values of the desired position θ_d.

It is evident from Table 6.2 that for $\theta_d = 10°$, the third approach provides the most accurate value for θ and the lowest value for $\dot{\theta}$. However, the second approach exhibits the lowest control gains. Hence, when $\theta_d = 30°$, the first approach

Table 6.2 Simulation results for different values of θ_d

Method	θ (°)	$\dot{\theta}$ (°/s)	$\mathcal{K}$	θ_d (°)
First approach	9.902°	0.0036	$\begin{bmatrix}-23.0703 & -14.3446\end{bmatrix}$	10°
Second approach	9.817°	0.0016	$\begin{bmatrix}-9.3441 & -6.4600\end{bmatrix}$	10°
Third approach	10°	0.001	$\begin{bmatrix}-218.2545 & -7.2907\end{bmatrix}$	10°
First approach	29.91°	0.0038	$\begin{bmatrix}-25.6776 & -15.5344\end{bmatrix}$	30°
Second approach	29.88°	0.005	$\begin{bmatrix}-18.8696 & -11.9492\end{bmatrix}$	30°
Third approach	29.88°	1.037×10^{-8}	$\begin{bmatrix}-29.1122 & -6.2522\end{bmatrix}$	30°
First approach	49.88°	0.0036	$\begin{bmatrix}-19.3018 & -11.8620\end{bmatrix}$	50°
Second approach	49.87°	0.0036	$\begin{bmatrix}-17.7526 & -10.9461\end{bmatrix}$	50°
Third approach	49.82°	0.006	$\begin{bmatrix}-15.8495 & -6.3208\end{bmatrix}$	50°

delivers the most accurate value for θ value and the third design method provides the lowest value for $\dot{\theta}$. Nonetheless, the second approach still shows the lowest control gains. However, for $\theta_d = 50°$, the results indicate that the third approach is the most accurate in terms of θ. Additionally, the first and second approaches display the lowest values for $\dot{\theta}$. However, the third approach offers the lowest control gains. Therefore, these comparisons highlight the strengths and weaknesses of each approach depending on the value of the desired state θ_d. In conclusion, for $\theta_d = 30°$ and $\theta_d = 50°$, the first approach is generally more accurate in terms of position. However, for $\theta_d = 10°$ and $\theta_d = 30°$, the second approach stands out with lower control gains, while the third approach provides the lowest value in angular.

6.5.5 *Discussion*

The proposed state-feedback control approach plays a significant role in effectively managing the challenging task of controlling the position of the actuated knee exoskeleton robotic system. This strategy employs extended feedback gain from three different methods, enabling precise control of the robot's movement. By guiding the robot to the intended state θ_d, the control method ensures that the angular position error ϕ converges to zero. These results are illustrated in Figs. 6.2a, 6.3a, and 6.4a.

Furthermore, using these three methods not only helps bring angular position error ϕ to zero but also effectively manages angular velocity error $\dot{\phi}$, as seen in Figs. 6.2b, 6.3b, and 6.4b. The slight fluctuations in these plots are due to the influence of an external torque Δ_t, showing that the control method can handle disturbances effectively.

Figures 6.2d, 6.3d, and 6.4d provide a graphical representation of the external disturbance acting on the knee exoskeleton robot, referred to as the applied load Δ_t. It is essential to note that this external torque is only active between the 2 and 7 seconds, and its values typically vary between $-\bar{\delta}$ and $\bar{\delta}$. Moreover, Figs. 6.2c, 6.3c, and 6.4c depict the variation of the proposed controller u. As the knee exoskeleton robotic system attains stability under the influence of the external load, the control input u gradually converges to the desired control effort u_d.

Moreover, the time required for the robotic system to achieve stability varies among the three approaches. The first approach attains stability in approximately 3 seconds, the second approach accomplishes this in under 2.7 seconds, and the third approach stabilizes the actuated knee exoskeleton in less than 2.6 second.

Additionally, across all three methods, we observe that the control effort at the beginning of the simulation is quite similar. In the first approach, it starts at about 13.44, in the second, it is 12.68, and in the third, it is roughly 12.3. Notably, the exerted control effort u on the joint at the start of simulation for all three approaches is remarkably low, as indicated by the plots in Figs. 6.2c, 6.3c and 6.4c. Hence, the outcomes from the three distinct methods underscore the successful regulate of the robot, effectively achieving the intended state with asymptotic stability.

6.6 Conclusion and Future Directions

This study utilizes an affine state-feedback controller to effectively control and stabilize the position of the knee exoskeleton robotic system, accounting for its nonlinear dynamic model with frictions, parameter uncertainties, and external disturbances. The adopted control law was designed using the Lyapunov methodology, and we introduced three distinct methods to establish stability conditions on the feedback gain, employing specific technical lemmas. These conditions were then translated into LMIs. After that, through a comparative analysis, we evaluated the efficiency of each method. Therefore, we used the LMI approach to derive gain conditions, ensuring the stability of the controlled knee exoskeleton. Moreover, we presented simulation results that demonstrated that the designed LMI-based controller achieved asymptotic stability of the knee exoskeleton system.

Moving forward, we have identified several promising avenues for future research. First, we aim to evaluate the robustness of the implemented control law by subjecting it to various scenarios involving external disturbances and uncertainties in the parameters. Secondly, we plan to explore the integration of a filtering mechanism to potentially reduce the amplitude or duration of oscillations in the controlled system. Additionally, our future endeavors will focus on designing a robust static output feedback control law utilizing the LMI approach [43]. Moreover, we are interested in incorporating movement constraints of different lower limb articulations into the controller design. By doing so, we aim to minimize control effort and reduce energy consumption, ultimately enhancing the overall efficiency and usability of the knee exoskeleton robotic system. Finally, an essential aspect of our future work will be the implementation and testing of the control system on an FPGA.

References

1. Anthony A Vandervoort. Aging of the human neuromuscular system. *Muscle & Nerve: Official Journal of the American Association of Electrodiagnostic Medicine*, 25(1):17–25, 2002.
2. Michael Tieland, Inez Trouwborst, and Brian C Clark. Skeletal muscle performance and ageing. *Journal of cachexia, sarcopenia and muscle*, 9(1):3–19, 2018.
3. Maren S Fragala, Eduardo L Cadore, Sandor Dorgo, Mikel Izquierdo, William J Kraemer, Mark D Peterson, and Eric D Ryan. Resistance training for older adults: position statement from the national strength and conditioning association. *The Journal of Strength & Conditioning Research*, 33(8), 2019.
4. Massimo Bergamasco and Hugh Herr. Human–robot augmentation. *Springer handbook of robotics*, pages 1875–1906, 2016.
5. Reem Sulaiman Baragash, Hanan Aldowah, and Samar Ghazal. Virtual and augmented reality applications to improve older adultsâ quality of life: A systematic mapping review and future directions. *Digital health*, 8:20552076221132099, 2022.
6. Aaron J Young and Daniel P Ferris. State of the art and future directions for lower limb robotic exoskeletons. *IEEE Transactions on Neural Systems and Rehabilitation Engineering*, 25(2):171–182, 2016.
7. Olfa Boubaker. *Medical and Healthcare Robotics. New Paradigms and Recent Advances.* Medical Robots and Devices: New Developments and Advances. Elsevier, Academic Press, 1 edition, July 2023.
8. Jyotindra Narayan and Santosha Kumar Dwivedy. Preliminary design and development of a low-cost lower-limb exoskeleton system for paediatric rehabilitation. *Proceedings of the Institution of Mechanical Engineers, Part H: Journal of Engineering in Medicine*, 235(5):530–545, 2021.
9. Baltej Singh Rupal, Sajid Rafique, Ashish Singla, Ekta Singla, Magnus Isaksson, and Gurvinder Singh Virk. Lower-limb exoskeletons: Research trends and regulatory guidelines in medical and non-medical applications. *International Journal of Advanced Robotic Systems*, 14(6):1729881417743554, 2017.
10. Y Wei Hong, Y King, W Yeo, C Ting, Y Chuah, J Lee, and Eu-Tjin Chok. Lower extremity exoskeleton: review and challenges surrounding the technology and its role in rehabilitation of lower limbs. *Australian Journal of Basic and Applied Sciences*, 7(7):520–524, 2013.
11. Walid Hassani, Samer Mohammed, Hala Rifaï, and Yacine Amirat. Powered orthosis for lower limb movements assistance and rehabilitation. *Control Engineering Practice*, 26:245–253, 2014.
12. Aaron M Dollar and Hugh Herr. Lower extremity exoskeletons and active orthoses: Challenges and state-of-the-art. *IEEE Transactions on robotics*, 24(1):144–158, 2008.
13. Yanan Li, Aran Sena, Ziwei Wang, Xueyan Xing, Jan Babic, Edwin HF van Asseldonk, and Etienne Burdet. A review on interaction control for contact robots through intent detection. *Progress in Biomedical Engineering*, 2022.
14. T-J Yeh, Meng-Je Wu, Ting-Jiang Lu, Feng-Kuang Wu, and Chih-Ren Huang. Control of mckibben pneumatic muscles for a power-assist, lower-limb orthosis. *Mechatronics*, 20(6):686–697, 2010.

15. Yasuhisa Hasegawa, Junho Jang, and Yoshiyuki Sankai. Cooperative walk control of paraplegia patient and assistive system. In *2009 IEEE/RSJ International Conference on Intelligent Robots and Systems*, pages 4481–4486. IEEE, 2009.
16. Kyoungchul Kong and Masayoshi Tomizuka. Control of exoskeletons inspired by fictitious gain in human model. *IEEE/ASME Transactions on Mechatronics*, 14(6):689–698, 2009.
17. Jyotindra Narayan and Santosha K. Dwivedy. Lower limb exoskeletons for pediatric gait rehabilitation: A brief review of design, actuation, and control schemes. In *2023 IEEE International Conference on Advanced Systems and Emergent Technologies (IC_ASET)*, pages 1–6, 2023.
18. Richa Sharma, Prerna Gaur, Shaurya Bhatt, and Deepak Joshi. Optimal fuzzy logic-based control strategy for lower limb rehabilitation exoskeleton. *Applied Soft Computing*, 105:107226, 2021.
19. Alberto Topini, William Sansom, Nicola Secciani, Lorenzo Bartalucci, Alessandro Ridolfi, and Benedetto Allotta. Variable admittance control of a hand exoskeleton for virtual reality-based rehabilitation tasks. *Frontiers in neurorobotics*, 15:789743, 2022.
20. Jiaqi Liu, Hongbin Fang, and Jian Xu. Online adaptive pid control for a multi-joint lower extremity exoskeleton system using improved particle swarm optimization. *Machines*, 10(1):21, 2021.
21. S J Gambhire, D Ravi Kishore, P S Londhe, and S N Pawar. Review of sliding mode based control techniques for control system applications. *International Journal of dynamics and control*, 9(1):363–378, 2021.
22. Ghasaq Al Rezage and M.O. Tokhi. Fuzzy pid control of lower limb exoskeleton for elderly mobility. In *2016 IEEE International Conference on Automation, Quality and Testing, Robotics (AQTR)*, pages 1–6, 2016.
23. Xiaofeng Wang, Xing Li, Jianhui Wang, Xiaoke Fang, and Xuefeng Zhu. Data-driven model-free adaptive sliding mode control for the multi degree-of-freedom robotic exoskeleton. *Information Sciences*, 327:246–257, 2016.
24. Yujian Zhou, Jinhua She, Zhen-Tao Liu, Chi Xu, and Zhaohui Yang. Implementation of impedance control for lower-limb rehabilitation robots. In *2021 4th IEEE International Conference on Industrial Cyber-Physical Systems (ICPS)*, pages 700–704. IEEE, 2021.
25. Qi Wang, Panos Markopoulos, Bin Yu, Wei Chen, and Annick Timmermans. Interactive wearable systems for upper body rehabilitation: a systematic review. *Journal of neuroengineering and rehabilitation*, 14(1):1–21, 2017.
26. Zhipeng Wang, Chifu Yang, Shengping Zhang, Shihao Zhang, Chunzhi Yi, Zhen Ding, Baichun Wei, and Feng Jiang. Knee flexion-assisted method for human-exoskeleton system. *IEEE Transactions on Neural Systems and Rehabilitation Engineering*, 31:2800–2808, 2023.
27. Romain Baud, Ali Reza Manzoori, Auke Ijspeert, and Mohamed Bouri. Review of control strategies for lower-limb exoskeletons to assist gait. *Journal of NeuroEngineering and Rehabilitation*, 18(1):119, 2021.
28. Shufei Li, Pai Zheng, Sichao Liu, Zuoxu Wang, Xi Vincent Wang, Lianyu Zheng, and Lihui Wang. Proactive human–robot collaboration: Mutual-cognitive, predictable, and self-organising perspectives. *Robotics and Computer-Integrated Manufacturing*, 81:102510, 2023.
29. Jyotindra Narayan, Bhavik M. Patel, Mohamed Abbas, Ghanshyam Shivhare, and Santosha K. Dwivedy. Cooperative control of a pediatric exoskeleton system for active-assist gait rehabilitation. In *2022 IEEE International Conference on Electronics, Computing and Communication Technologies (CONECCT)*, pages 1–6, 2022.
30. Lukas Bergmann, Daniel Voss, Steffen Leonhardt, and Chuong Ngo. Lower-limb exoskeleton with compliant actuators: Human cooperative control. *IEEE Transactions on Medical Robotics and Bionics*, 5(3):717–729, 2023.
31. Jyotindra Narayan and Santosha Kumar Dwivedy. Towards neuro-fuzzy compensated PID control of lower extremity exoskeleton system for passive gait rehabilitation. *IETE Journal of Research*, 69(2):778–795, 2023.

32. Sahar Jenhani and Hassène Gritli. Lmi-based design of an affine PD controller for the robust stabilization of the knee joint of a lower-limb rehabilitation exoskeleton. In *2023 IEEE International Conference on Advanced Systems and Emergent Technologies (IC_ASET)*, pages 1–7. IEEE, 2023.
33. Hala Rifaï, Samer Mohammed, Walid Hassani, and Yacine Amirat. Nested saturation based control of an actuated knee joint orthosis. *Mechatronics*, 23(8):1141–1149, 2013.
34. Candy M Jansen, Jeffrey E Windau, Peter M Bonutti, and Mark V Brillhart. Treatment of a knee contracture using a knee orthosis incorporating stress-relaxation techniques. *Physical therapy*, 76(2):182–186, 1996.
35. Jack Lysholm, Marketta Nordin, Jan Ekstrand, and Jan Gillquist. The effect of a patella brace on performance in a knee extension strength test in patients with patellar pain. *The American Journal of Sports Medicine*, 12(2):110–112, 1984.
36. Anwer S. Aljuboury, Akram Hashim Hameed, Ahmed R. Ajel, Amjad J. Humaidi, Ahmed Alkhayyat, and Ammar K. Al Mhdawi. Robust adaptive control of knee exoskeleton-assistant system based on nonlinear disturbance observer. *Actuators*, 11(3), 2022.
37. Sahar Jenhani, Hassène Gritli, and Giuseppe Carbone. Comparison between some nonlinear controllers for the position control of Lagrangian-type robotic systems. *Chaos Theory and Applications*, 4(4):179–196, 2022.
38. Hassène Gritli. LMI-based robust stabilization of a class of input-constrained uncertain nonlinear systems with application to a helicopter model. *Complexity*, 2020:7025761, 2020.
39. Firas Turki, Hassène Gritli, and Safya Belghith. An LMI-based design of a robust state-feedback control for the master-slave tracking of an impact mechanical oscillator with double-side rigid constraints and subject to bounded-parametric uncertainty. *Communications in Nonlinear Science and Numerical Simulation*, 82:105020, 2020.
40. Hassène Gritli and Safya Belghith. Robust feedback control of the underactuated Inertia Wheel Inverted Pendulum under parametric uncertainties and subject to external disturbances: LMI formulation. *Journal of The Franklin Institute*, 355(18):9150–9191, 2018.
41. Hassène Gritli. Robust master-slave synchronization of chaos in a one-sided 1-DoF impact mechanical oscillator subject to parametric uncertainties and disturbances. *Mechanism and Machine Theory*, 142:103610, 2019.
42. Hassène Gritli and Safya Belghith. LMI-based synthesis of a robust saturated controller for an underactuated mechanical system subject to motion constraints. *European Journal of Control*, 57:179–193, 2021.
43. Hassène Gritli, Ali Zemouche, and Safya Belghith. On LMI conditions to design robust static output feedback controller for continuous-time linear systems subject to norm-bounded uncertainties. *International Journal of Systems Science*, 52(1):12–46, 2021.

Chapter 7
Position Control of Robotic Systems via Linear Controllers with Application to a Lower Limb Rehabilitation Exoskeleton Robot: Design and Comparative Analysis

Hassène Gritli and Sahar Jenhani

7.1 Introduction

In recent years, robots have been widely used worldwide in various application areas to perform dangerous, arduous, or repetitive tasks [1–4]. To achieve this objective, each robotic system needs to be controlled through a mathematical model that describes its dynamic movement. This mathematical model can be either the transformation model between the operational space and the joint space or the dynamic model that defines the motion equations of the robot in question.

Moreover, each mechanical robotic system must be controlled by an appropriate and simple controller to accomplish such desired objective [5–10]. To achieve its goals, the robot requires the capabilities of movement or manipulation or a combination of movement and manipulation. Moreover, in order to control robotic systems, many control methods have been developed in the literature, such as PID, LQR, sliding mode, and fuzzy logic [11–13], among others, as in [14–18].

The control of robot manipulators is one of the most important tasks in the research field today. Several control approaches have been developed and presented to address the stability problem of manipulator robots. This stabilization problem can be subdivided into two main categories: the position control and the trajectory control [19]. Linear controllers, such as Affine PD-based control law, PID controller,

H. Gritli (✉)
Higher Institute of Information and Communication Technologies, University of Carthage, Tunis, Tunisia

Laboratory of Robotics, Informatics and Complex Systems (RISC Lab—LR16ES07), National Engineering School of Tunis, University of Tunis El Manar, Tunis, Tunisia

S. Jenhani
Laboratory of Robotics, Informatics and Complex Systems (RISC Lab—LR16ES07), National Engineering School of Tunis, University of Tunis El Manar, Tunis, Tunisia

E. Campos-Cantón et al. (eds.), *Complex Systems and Their Applications*,
https://doi.org/10.1007/978-3-031-51224-7_7

and LQR controller, are among the control laws used, as in [20]. Nonlinear controllers, such as PD control with Gravity Compensation, PD control with desired Gravity Compensation, and the CTC controller, are other control laws used. Additionally, some control approaches have been designed based on the linearized dynamics model of the robotic system, as presented in [10, 14–18, 21, 22].

Older people's way of moving might be very different depending on how old they are. Losing muscle strength because of fewer motor neurons, muscle fibers, and aerobic capacity will make walking less efficient [23–25]. Exoskeletons are now widely used in the robotics industry as a type of helpful technology [24]. Many scientists have worked on creating helpful technology to support rehabilitation or to give power to individuals. The aim is to replace a human limb and help the joint move better while also supporting the body's weight [26–30]. These exoskeletons have been considered like manipulator robots to control, and then several control techniques designed for manipulators can be efficiently applied to exoskeletons for the rehabilitation of lower and upper limbs.

More and more exoskeletons are being made to help people with disabilities and older people walk better [30–32]. The control approach is a very important technology in the exoskeleton field. Different kinds of controllers have been used in studies for robotic systems [10]. Many ways to control lower/upper limb braces and exoskeleton devices have been created, such as those mentioned in [15, 31–34].

This chapter focuses on how to control the movement and more particularly the position of a robot that helps people move their legs. In this study, we will look at the human lower limb for its rehabilitation purpose by means of an exoskeleton robot. The main objective is to solve the position control problem for such robotic system [5, 6, 10, 19]. To achieve this goal, we will use the first-order Taylor approximation to develop an approximated linear model of the nonlinear dynamics of robotic systems around the desired position [20, 35–40]. Furthermore, in order to control the robotic systems and ensure stability at the desired state, we will adopt some linear control laws, such as the PD-based controllers, the PID controller [40], the LQR controller, and the Lyapunov-based controller, based on the developed approximated linear dynamic model. Additionally, we will design stability conditions that aid in choosing the appropriate feedback gain values for these controllers, using LMI techniques [41, 42]. Finally, we will use a two-degree-of-freedom (2-DoF) rehabilitation exoskeleton robot as an illustrative example to demonstrate the effectiveness and validity of the designed conditions and the adopted controllers in solving the position control problem.

The structure of this chapter is as follows: Sect. 7.2 provides a brief overview on exoskeletons for rehabilitation assistance. Section 7.3 provides an introduction to the dynamic model of robotic systems and outlines the problem addressed in this work. In Sect. 7.4, we present the development of an approximated linear model. Section 7.5 presents the design of feedback gains for the affine PD-based control law, using first a decoupled-model-based method in Sect. 7.5.1, using secondly a Lyapunov-based method in Sect. 7.5.2, and using thirdly a state feedback controller in Sect. 7.5.3. The design of feedback gains for the affine PID controller is discussed in Sect. 7.6, and for the LQR controller is discussed in Sect. 7.7. The simulation results are presented in Sect. 7.8. Finally, we conclude the chapter in Sect. 7.9.

7.2 On Rehabilitation Assistance via Exoskeletons

7.2.1 *On the Rehabilitation*

These are two technologies of rehabilitation of the lower/upper limbs. First, there is the passive method that promotes the rehabilitation of joint and ligament diseases. These movements are made by a therapist or by an external support so as to force the defective joint to regain its natural movement by time.

The second rehabilitation technique is called the active method. It aims to strengthen muscles and improve motor coordination through patient-induced movements with the help of a medical therapist. Other mechatronic devices, called exoskeletons, have emerged since the 20th century. They are used in various fields of application for the purpose of increasing, assisting or restoring the movements of the wearer.

In the following, we look at the main work on exoskeletons as the mobility and rehabilitation device. A functional knee orthosis corrects a functional deficiency in the knee. An example could be an unstable knee after an anterior cruciate ligament injury, which requires a period of rehabilitation for specific purposes of immobilization or recovery of amplitude. Combined with a treadmill, an exoskeleton may be used to facilitate mobility and rehabilitation of dependents. This technique has a number of advantages, for example:

1. Reproduction of joint movements corresponding to a normal gait to facilitate hip extension, flexion-knee extension alternation, and proper foot placement.
2. Change the walking speed to gradually approach normal walking speed.
3. Maintain a correct trunk extension.
4. Synchronization and coordination between the lower limbs during the two phases of walking.
5. An attempt is made to repeat a large number of (walking) rehabilitation cycles.

7.2.2 *Rehabilitation Techniques*

These are mainly physiotherapy, occupational therapy, and physiotherapy. In terms of functions, rehabilitation is active or passive, passive rehabilitation applies to pathologies of the joints and ligaments, while active rehabilitation aims to develop muscles and improve motor coordination.

7.2.2.1 Passive Rehabilitation

In passive rehabilitation, the patient does not acquire any muscular activity [43]. This rehabilitation requires passive mobilization, postural work, pull-ups, and massages. It mainly aims to restore mobility in terms of amplitude of the affected limb.

Passive mobilization applies cyclical movements to the touch joint as extension flexion movements. The latter are performed by a medical therapist. Passive movement is intended to restore or maintain joint flexibility and counteract muscular, tendon, intraarticular, and capsular stiffness. Passive amplitude mobility is interesting in case of stiffness due to osteoarthritis, after prolonged joint immobilization due to tendonitis, after joint surgery, in case of muscle spasticity due to neurological deficit. On the other hand, working in position consists of doing slow and progressive stretches with maximum amplitude to combat stiffness [44].

7.2.2.2 Active Rehabilitation

Unlike passive rehabilitation, in active rehabilitation, muscle work is either active or actively supported and is accomplished by movements performed by the patient and assisted by a therapist or a specific device called altimeter [45, 46]. Muscle building can be done by isotonic or isodynamic methods. Isobaric work involves muscle contraction without movement. Isobaric work, on the other hand, is done with constant force. It can be concentric or eccentric and can be used in case of tendonitis, for example. Isokinetic work, on the other hand, is done at a constant pace and aims to increase muscle strength through resistance. Active recovery mainly aims to: strengthen muscles, increase muscle mass and reprogram movements, eliminate fat by promoting an increase in muscle mass, and prevent stiffness caused by connective tissues by reducing their spasticity. Active rehabilitation can also be used to restore initiation to include functions such as maintaining balance, maintaining and inducing motor cortical plasticity through the initiation of spontaneous movements.

7.2.3 *A Brief Description of Exoskeletons*

The first scientific definition was given in the USA at the University of Cornelle in 1969 "*the exoskeleton is an external mechanical structure having the shape of the human body, having less degree of human freedom, and which can perform most of the desired tasks.*" Usually, functional requirements on exoskeletons must be necessary transportable (natural appearance), comfortable, have significant autonomous energy and the ability to improve human performance such as speed and endurance.

General, an exoskeleton for the rehabilitation of either the upper limb of the the lower extremity of the disabled humans should be generally composed of [24, 27, 30, 32, 34]:

- Mechanical construction/platform: made of light materials, must be strong enough to support the weight of the wearer.

- Sensors: placed on the mount as well as on the wearer. They ensure the best possible support functionality. The sensors can be: electrical, bioelectric, or a combination of devices (remote control, ...).
- Control unit: it performs the acquisition and processing of information provided by sensors and controls actuators according to the law of control to ensure the efficiency of assistance movement and the stability of the human skeletal system.
- Actuators: they play the role of muscle. They can be electric, hydraulic, or pneumatic. Electric actuators are often used for reasons of ease of energy incorporation.
- Batteries: They provide the exoskeleton with the electrical energy necessary for operation. The requirements are essential and should not burden the system.

7.3 Dynamic Model of Robotic Systems and Problem Formulation

7.3.1 *Dynamic Model of Robotic Systems*

The dynamic model of robotic systems can be derived from the Lagrangian mechanical system, as described in the literature by the following expression:

$$\frac{d}{dt}\left(\frac{\partial \mathcal{L}(q,\dot{q})}{\partial \dot{q}}\right) - \frac{\partial \mathcal{L}(q,\dot{q})}{\partial q} + \frac{\partial \mathcal{R}(\dot{q})}{\partial \dot{q}} = \mathcal{D}(q)\mathcal{U}, \tag{7.1}$$

where $\mathcal{L}(q,\dot{q})$ denotes the Lagrangian and given by:

$$\mathcal{L}(q,\dot{q}) = E_k(q,\dot{q}) - E_p(q), \tag{7.2}$$

where E_k is the kinetic energy of the robotic system, E_p is the potential energy of the system.

In this dynamic model, the following terms are used:

- $q \in \mathbb{R}^n$ represents the configuration/position vector,
- $\dot{q} \in \mathbb{R}^n$ denotes the velocity vector associated with the configuration vector,
- $\mathcal{U} \subset \mathbb{R}^m$ is the input vector of the actuators,
- $\mathcal{R}(\dot{q})$ denotes the dissipation term caused by frictions, and
- $\mathcal{D}(q) \in \mathbb{R}^{n\times m}$ is a non-square matrix that represents external forces applied to the generalized coordinates of the robotic system.

By ignored the presence of friction, elasticity, and external forces, the dynamic model (7.1) can be simplified and reformulated as follows:

$$\mathcal{M}(q)\ddot{q} + \mathcal{H}(q,\dot{q})\dot{q} + \mathcal{G}(q) = \mathcal{D}(q)\mathcal{U}, \tag{7.3}$$

where

- the matrix $\mathcal{M}(q) \in \mathbb{R}^{n \times n}$ denotes the inertia matrix,
- the matrix/vector $\mathcal{H}(q, \dot{q})\dot{q}$ contains two kinds of terms using $\dot{q}_i \dot{q}_j$ known as Centrifugal terms ($i = j$), and Coriolis terms ($i \neq j$),
- the matrix/vector $\mathcal{G}(q)$ denotes the gravity matrix.

Remark 7.1 It is important to note that the classification of a robotic system depends on the values of its configuration/position vector q, its associated velocity vector $\dot{q}$, the input vector of the actuators $\mathcal{U}$, and the external forces distribution matrix $\mathcal{D}(q)$. Based on these parameters, we can distinguish between three main classes of robotic systems.

The first class is called fully actuated, which applies to systems where $n = m$, meaning the number of actuators is equal to the number of degrees of freedom. The second class is under-actuated, where $n > m$, indicating that the system has fewer actuators than degrees of freedom. Finally, the third class is over-actuated, where $n < m$, implying that the system has more actuators than degrees of freedom. Various studies have investigated the dynamics and control of these different types of robotic systems [6, 11, 12, 42, 47, 48].

7.3.2 *Problem Formulation*

Controlling robotic systems can be broadly categorized into two areas: position control and motion control. The position control problem of robots has been studied in the literature, and various control approaches have been developed to address it.

This work focuses on solving the position control problem of robotic systems and with particular application to the lower-limb rehabilitation exoskeletons. The main objective is to design a linear function $\mathcal{U}$ that ensures the robotic system's actual position q approaches a desired state q_d with accuracy. To achieve this, the first step is to linearize the dynamic model (7.3) around the desired position q_d. With this linearized model, four different linear controllers will be adopted to stabilize the robotic system. Additionally, conditions will be developed to aid in choosing the feedback gain values for these controllers. Further details on these topics will be discussed in the following sections.

7.4 Development of the Approximated Linear Dynamic Model of Robotic Systems

This section will go over the process of creating linearized dynamics for robotic systems. This linearized model approximates the nonlinear model (7.3) around the desired state q_d. Our approach for achieving this is to employ the approximation of

Taylor. This technique involves computing the gradient of the nonlinear model at q_d and using it to construct a linear model that approximates the behavior of the system in the vicinity of q_d. By linearizing the model, we can simplify the analysis of the system's behavior and design controllers that can stabilize it. The linearized model will be used as the basis for the development of the four different linear controllers we will adopt to stabilize the robotic system in question.

7.4.1 The Control Reference at the Desired Equilibrium

A desired control input $\mathcal{U}_d$ is required to keep the robotic system in the appropriate location q_d. This is necessary to keep the system at equilibrium. Thus, the nonlinear dynamics defined in (7.3) can be expressed as:

$$\mathcal{M}(q_d)\ddot{q}_d + \mathcal{H}(q_d, \dot{q}_d)\dot{q}_d + \mathcal{G}(q_d) = \mathcal{D}(q_d)\mathcal{U}_d. \tag{7.4}$$

Note that when the robotic system is at the desired position q_d, the velocity and acceleration of the system are zero, that is, $\dot{q}_d = \ddot{q}_d = 0$. Therefore, the nonlinear dynamics equation (7.4) can be simplified as follows:

$$\mathcal{D}(q_d)\mathcal{U}_d = \mathcal{G}(q_d). \tag{7.5}$$

We will investigate the general case where n can be equal to or different from m. Thus, based on Eq. (7.5), we define the control reference level U_d as follows:

$$\mathcal{U}_d = \mathcal{D}^{+}(q_d)\mathcal{G}(q_d). \tag{7.6}$$

We can define the pseudo-inverse of $\mathcal{D}(q_d)$, denoted by $\mathcal{D}^{+}(q_d)$, as follows:

$$\mathcal{D}^{+}(q_d) = \begin{cases} \left(\mathcal{D}(q_d)^{\mathrm{T}}\mathcal{D}(q_d)\right)^{-1}\mathcal{D}(q_d)^{\mathrm{T}} & \text{if } n \geq m, \\ \mathcal{D}(q_d)^{\mathrm{T}}\left(\mathcal{D}(q_d)\mathcal{D}(q_d)^{\mathrm{T}}\right)^{-1} & \text{if } n \leq m. \end{cases} \tag{7.7}$$

7.4.2 The Approximated Linear Model

In [35, 49], an approximate linear dynamic model was developed for the nonlinear dynamics (7.3) around the desired position q_d. The model is obtained by using the first-order Taylor series to approximate the different terms in (7.3) at the desired point $(q_d, \mathcal{U}_d)$. This allows for an easier analysis and design of control strategies for the robotic system.

Let us consider the first term $\mathcal{M}(q)\ddot{q}$, which depends on q and $\ddot{q}$. Its linearized version, developed in [20, 35], is expressed as follows:

$$\mathcal{M}(q)\ddot{q} \approx \mathcal{M}(q_d)\ddot{q}. \tag{7.8}$$

We will now focus on the quantity $\mathcal{H}(q, \dot{q})\dot{q}$, which depends on q and $\dot{q}$. Following the same mathematical manipulation as demonstrated in [20, 35], we can obtain its linear part, given by the following expression:

$$\mathcal{H}(q, \dot{q})\dot{q} \approx 0. \tag{7.9}$$

Let us now consider the gravity term $\mathcal{G}(q)$, which depends only on q. Its linear part is obtained by using the first-order Taylor approximation and is given by:

$$\mathcal{G}(q) \approx \mathcal{G}(q_d) + \mathcal{G}_q(q_d)\,(q - q_d)\,, \tag{7.10}$$

with $\mathcal{G}_q(q_d) = \frac{\partial(\mathcal{G}(q))}{\partial q}|_{(q_d)}$.

Next, we focus on the fourth term in expression (7.3), $\mathcal{D}(q)\mathcal{U}$, which depends on q and $\mathcal{U}$. The linear part of this term, derived in [20, 35], is expressed as:

$$\begin{aligned}\mathcal{D}(q)\mathcal{U} \approx& \mathcal{D}(q_d)\mathcal{U}_d + \mathcal{D}_q(q_d)\widehat{\mathcal{U}_d}\,(q - q_d)\\ &+ \mathcal{D}(q_d)\,(\mathcal{U} - \mathcal{U}_d)\,,\end{aligned} \tag{7.11}$$

where the expression for $\widehat{\mathcal{U}}$ is given as follows:

$$\widehat{\mathcal{U}} = \text{diag}\left(\underbrace{\mathcal{U}, \mathcal{U}, \cdots, \mathcal{U}}_{n \text{ times}}\right). \tag{7.12}$$

Then, by combining the obtained expressions defined by (7.8), (7.9), (7.10) and (7.11), and using the relation (7.5), we can derive the approximated linear model of the nonlinear dynamics (7.3) of the robotic system like so:

$$\begin{aligned}\mathcal{M}(q_d)\ddot{q} + \left(\mathcal{G}_q(q_d) - \mathcal{D}_q(q_d)\widehat{\mathcal{U}_d}\right) q =& \mathcal{D}(q_d)\mathcal{U}\\ &+ \left(\mathcal{G}_q(q_d) - \mathcal{D}_q(q_d)\widehat{\mathcal{U}_d}\right) q_d - \mathcal{G}(q_d).\end{aligned} \tag{7.13}$$

For a simplification of writing, we adopt the subsequent notations:

$$\mathcal{M}_d = \mathcal{M}(q_d), \qquad \mathcal{N}_d = \mathcal{G}_q(q_d) - \mathcal{D}_q(q_d)\widehat{\mathcal{U}_d}, \tag{7.14a}$$

$$\mathcal{D}_d = \mathcal{D}(q_d), \qquad \mathcal{R}_d = \mathcal{N}_d q_d - \mathcal{G}(q_d). \tag{7.14b}$$

Therefore, the expression (7.13) is given like so:

$$\mathcal{M}_d\ddot{q} + \mathcal{N}_d q = \mathcal{D}_d\mathcal{U} + \mathcal{R}_d. \tag{7.15}$$

Hence, the dynamical system represented by Eq. (7.15) characterizes the linear part of the nonlinear dynamics described in Eq. (7.3) for robotic systems. Upon examination of the system in Eq. (7.15) and the expressions presented in (7.14), we get:

$$\mathcal{D}_d \mathcal{U}_d = \mathcal{N}_d q_d - \mathcal{R}_d. \tag{7.16}$$

As mentioned earlier, the objective of this study is to develop a control law $\mathcal{U}$ to stabilize the linear robotic system (7.15) at the desired position q_d. To achieve this goal, we need to rewrite the approximate linear model (7.15) in a different form.

Therefore, we introduce new variables:

$$\phi = q - q_d, \tag{7.17a}$$

$$\Delta\mathcal{U} = \mathcal{U} - \mathcal{U}_d. \tag{7.17b}$$

By defining new variables ϕ as $\dot{\phi} = \dot{q}$ and $\ddot{\phi} = \ddot{q}$, the linear dynamics (7.15) or (7.13) can be expressed in a new form. Since the matrix $\mathcal{M}(q)$ is regular and positive, the linear dynamics can be reformulated as:

$$\ddot{\phi} = -\mathcal{M}_d^{-1} \mathcal{N}_d \phi + \mathcal{M}_d^{-1} \mathcal{D}_d \Delta\mathcal{U}. \tag{7.18}$$

Therefore, we will use the previous linear model (7.18) to design the controller $\Delta\mathcal{U}$ ensuring the stabilization of the position error ϕ at the zero state.

7.5 Design of an Affine PD-Based Control Law

In this section, our objective is to develop a control law ΔU that can stabilize the simplified linear system (7.18) at the zero state $\phi = 0$. To achieve this, we will employ a PD-based controller to ensure the system's stabilization.

$$\Delta\mathcal{U} = \mathcal{K}_p \phi + \mathcal{K}_v \dot{\phi}, \tag{7.19}$$

where $\mathcal{K}_p \in \mathbb{R}^{m \times n}$ denotes the position feedback gain and $\mathcal{K}_v \in \mathbb{R}^{m \times n}$ denotes the velocity feedback gain.

The proposed PD control law defined in (7.19) is introduced to the linear system (7.18) to achieve stabilization at the zero state $\phi = 0$. With this controller, the closed-loop dynamic model can be expressed as follows:

$$\ddot{\phi} - \mathcal{M}_d^{-1} \mathcal{D}_d \mathcal{K}_v \dot{\phi} + \mathcal{M}_d^{-1} \left(\mathcal{N}_d - \mathcal{D}_d \mathcal{K}_p\right) \phi = 0. \tag{7.20}$$

Therefore, the control law to stabilize the nonlinear dynamics (7.3) of the robotic system at the desired position q_d is defined as follows:

$$\mathcal{U} = \mathcal{U}_d + \mathcal{K}_p (q - q_d) + \mathcal{K}_v \dot{q}. \tag{7.21}$$

It is obvious that the expression (7.21) has an affine structure including the PD part. Thus, the controller (7.21) is an affine PD-based controller.

We will now concentrate on designing the two feedback gains, $\mathcal{K}_p$ and $\mathcal{K}_v$, of the affine PD-based control law (7.21) or its simplified version (7.19) to ensure the stability of the obtained closed-loop linear model (7.20). To achieve this objective, we will use three design methodologies.

7.5.1 *First Design Methodology: A Decoupled-Model-Based Method*

The first approach to designing the feedback gains $\mathcal{K}_p$ and $\mathcal{K}_v$ involves imposing some decoupled controlled dynamics and closed loop's desired poles. The purpose is to create feedback gains that stabilize the robotic system by enforcing a stable decoupled linear system's desired behavior. It is assumed that all state variables, q_i, of the robotic system are decoupled for all $i = 1, \cdots, n$, and the intended dynamics associated with each variable q_i are as follows:

$$\ddot{\phi}_i - (\eta_{1i} + \eta_{2i})\, \dot{\phi}_i + (\eta_{1i} \times \eta_{2i})\, \phi_i = 0, \tag{7.22}$$

where $\phi_i = q_i - (q_d)_i$, and η_{1i} and η_{2i} denotes the desired poles of the closed-loop dynamics.

Note that to ensure the stability of the linear model (7.22), the two desired poles η_{1i} and η_{2i} should be with a negative real part.

Based on the linearized model of the linear system (7.22), the entire desired dynamics of the robotic system is described by the subsequent decoupled reference dynamics:

$$\ddot{\phi} + \Omega_v \dot{\phi} + \Omega_p \phi = 0, \tag{7.23}$$

where the two matrices Ω_v and Ω_p are described like so:

$$\Omega_v = -\text{diag}\,(\eta_{11} + \eta_{21}, \eta_{12} + \eta_{22}, \cdots, \eta_{1n} + \eta_{2n}), \tag{7.24a}$$

$$\Omega_p = \text{diag}\,(\eta_{11} \times \eta_{21}, \eta_{12} \times \eta_{22}, \cdots, \eta_{1n} \times \eta_{2n}). \tag{7.24b}$$

Note that $\Omega_v > 0$ and $\Omega_p > 0$. Moreover, it is possible to select the two poles η_{1i} and η_{2i} of the linear model (7.22) to be such that $\eta_{1i} = \eta_{2i} = -w_i$. Thus, by comparing (7.20) and (7.23), the following relations can be obtained:

$$\Omega_v = -\mathcal{M}_d^{-1} \mathcal{D}_d \mathcal{K}_v, \tag{7.25a}$$

$$\Omega_p = \mathcal{M}_d^{-1} \left(\mathcal{N}_d - \mathcal{D}_d \mathcal{K}_p\right), \tag{7.25b}$$

From these two Eqs. (7.25a) and (7.25b), we can get the subsequent expressions of the two gains:

$$\mathcal{K}_p = \mathcal{D}_d^{+}\left(\mathcal{N}_d - \mathcal{M}_d\Omega_p\right), \tag{7.26a}$$

$$\mathcal{K}_v = -\mathcal{D}_d^{+}\mathcal{M}_d\Omega_v. \tag{7.26b}$$

7.5.2 Second Design Methodology: A Lyapunov-Based Method

In this second design methodology, we will employ a Lyapunov-based approach to determine the conditions on the feedback gains $\mathcal{K}_p$ and $\mathcal{K}_v$ that ensure the stability of the closed-loop linear system (7.18). Therefore, let us first choose the following Lyapunov function candidate:

$$\mathcal{V}\left(\phi, \dot{\phi}\right) = \frac{\alpha_1}{2}\dot{\phi}^{\mathrm{T}}\dot{\phi} + \frac{\alpha_2}{2}\phi^{\mathrm{T}}\phi, \tag{7.27}$$

where α_1 and α_2 are two positive constants.

It is easy to show from (7.27) that $\mathcal{V}\left(\phi, \dot{\phi}\right) > 0$. Thus, its derivative with respect to time is described by:

$$\dot{\mathcal{V}}\left(\phi, \dot{\phi}\right) = \alpha_1\dot{\phi}^{\mathrm{T}}\ddot{\phi} + \alpha_2\phi^{\mathrm{T}}\dot{\phi}. \tag{7.28}$$

Relying on the obtained closed-loop dynamic model (7.20), the expression (7.28) becomes:

$$\dot{\mathcal{V}}\left(\phi, \dot{\phi}\right) = \alpha_1\dot{\phi}^{\mathrm{T}}\mathcal{M}_d^{-1}\mathcal{D}_d\mathcal{K}_v\dot{\phi} + \dot{\phi}^{\mathrm{T}}\left(\alpha_1\mathcal{M}_d^{-1}\left(\mathcal{D}_d\mathcal{K}_p - \mathcal{N}_d\right) + \alpha_2\mathcal{I}\right)\phi, \tag{7.29}$$

where $\mathcal{I}$ denotes the identity matrix of appropriate dimension.

To ensure that $\dot{\mathcal{V}}\left(\phi, \dot{\phi}\right) < 0$, the following requirement must be met:

$$\alpha_1\mathcal{M}_d^{-1}\left(\mathcal{D}_d\mathcal{K}_p - \mathcal{N}_d\right) + \alpha_2\mathcal{I} = 0, \tag{7.30a}$$

$$\alpha_1\mathcal{M}_d^{-1}\mathcal{D}_d\mathcal{K}_v < 0. \tag{7.30b}$$

As $\alpha_1 > 0$ and $\mathcal{M}_d > 0$, then the two conditions (7.30a) and (7.30b) can be recast as follows:

$$\mathcal{D}_d\mathcal{K}_p - \mathcal{N}_d + \mu\mathcal{M}_d = 0, \tag{7.31a}$$

$$\mathcal{D}_d\mathcal{K}_v < 0, \tag{7.31b}$$

where $\mu = \frac{\alpha_2}{\alpha_1} > 0$.

Therefore, the two gains $\mathcal{K}_p$ and $\mathcal{K}_v$ of the proposed control law should satisfy to the constraints specified in (7.31). Then, employing (7.31a), the gain $\mathcal{K}_p$ is expressed like so:

$$\mathcal{K}_p = \mathcal{D}_d^{+} \left(\mathcal{N}_d - \mu \mathcal{M}_d\right). \tag{7.32}$$

It is worth noting that the derived expression (7.32) of the feedback gain $\mathcal{K}_p$ is nearly identical to that obtained in the first method. In fact, upon comparing expressions (7.32) and (7.26a), we observe that $\Omega_p = \mu \mathcal{I}$.

Additionally, the inequality (7.31b) can be reformulated like so:

$$\mathcal{D}_d \mathcal{K}_v = -\mathcal{Q}, \tag{7.33}$$

for any positive-definite matrix $\mathcal{Q}$.

Therefore, utilizing the relationship (7.33) and choosing a positive-definite matrix $\mathcal{Q}$, the feedback gain $\mathcal{K}_v$ can be computed using the subsequent equation:

$$\mathcal{K}_v = -\mathcal{D}_d^{+} \mathcal{Q}. \tag{7.34}$$

It is important to note that upon comparing the previous relation (7.34) with (7.26b), we find that $\mathcal{Q} = \mathcal{M}_d \Omega_v$.

7.5.3 *Third Design Methodology: Design of a State-Feedback Controller*

In this section, we will employ a distinct design method compared to the two previous methodologies. This method involves designing a state-feedback controller to stabilize the given linear dynamic model (7.18). Let us begin by considering this state vector $X = \begin{bmatrix} q \\ \dot{q} \end{bmatrix}$ and the desired state vector $X_d = \begin{bmatrix} q_d \\ 0 \end{bmatrix}$. Moreover, recall that $\phi = q - q_d$. Then, let us pose $Z = X - X_d = \begin{bmatrix} \phi \\ \dot{\phi} \end{bmatrix}$.

Hence, the linear system (7.18) can be reformulated as the following state-space model:

$$\dot{Z} = \mathcal{A} Z + \mathcal{B} \Delta \mathcal{U}, \tag{7.35}$$

where

$$\mathcal{A} = \begin{bmatrix} \mathcal{O}_n & \mathcal{I}_n \\ -\mathcal{M}_d^{-1} \mathcal{N}_d & \mathcal{O}_n \end{bmatrix}, \tag{7.36a}$$

$$\mathcal{B} = \begin{bmatrix} O_{n\times m} \\ \mathcal{M}_d^{-1}\mathcal{D}_d \end{bmatrix}. \tag{7.36b}$$

7.5.3.1 The State-Feedback Control Law

To ensure the stability of the linear model (7.35), we will utilize a state-feedback controller denoted by $\Delta\mathcal{U}$. This controller $\Delta\mathcal{U}$ is defined as follows:

$$\Delta\mathcal{U} = \mathcal{K}Z. \tag{7.37}$$

Here, $\mathcal{K}$ represents the state-feedback gain that will be designed using the Lyapunov approach.

Notice that from expression (7.19) and by comparing with the law (7.37), it follows that:

$$\mathcal{K} = \begin{bmatrix} \mathcal{K}_p \ \mathcal{K}_v \end{bmatrix}. \tag{7.38}$$

7.5.3.2 Condition on the Feedback Gains for Stability

By introducing the state-feedback controller defined in (7.37) into the linear dynamics (7.35), we obtain the following expression:

$$\dot{Z} = (\mathcal{A} + \mathcal{B}\mathcal{K})Z. \tag{7.39}$$

Let's take into consideration the subsequent candidate Lyapunov function:

$$\mathcal{V}(Z) = Z^{\mathrm{T}}\mathcal{P}Z, \tag{7.40}$$

with

$$\mathcal{P} = \mathcal{P}^{\mathrm{T}} > 0. \tag{7.41}$$

Therefore, its time derivative can be expressed as follows:

$$\dot{\mathcal{V}}(Z) = \dot{Z}\mathcal{P}Z + Z^{\mathrm{T}}\mathcal{P}\dot{Z}. \tag{7.42}$$

Moreover, we will adopt the following condition of the exponential stability of the linear system (7.35):

$$\dot{\mathcal{V}}(Z) < -\alpha\mathcal{V}(Z), \tag{7.43}$$

where α is some fixed positive parameter, which is the decay rate of the Lyapunov function.

Utilizing (7.43) and (7.35), we can establish the subsequent expression:

$$(\mathcal{A}+\mathcal{B}\mathcal{K})^{\mathrm{T}}\mathcal{P}+\mathcal{P}(\mathcal{A}+\mathcal{B}\mathcal{K})+\alpha\mathcal{P}<0. \tag{7.44}$$

By multiplying the expression (7.44) on the left and right by the matrix $\mathcal{S}=\mathcal{P}^{-1}$, we arrive at the subsequent condition expressed as a Linear Matrix Inequality (LMI):

$$(\mathcal{A}\mathcal{S}+\mathcal{B}\mathcal{R})+(\mathcal{A}\mathcal{S}+\mathcal{B}\mathcal{R})^{\mathrm{T}}+\alpha\mathcal{S}<0, \tag{7.45}$$

where $\mathcal{R}=\mathcal{K}\mathcal{S}$.

Then, by solving the derived LMI condition (7.45), the state-feedback gain $\mathcal{K}$ is determined and computed using the subsequent equation:

$$\mathcal{K}=\mathcal{R}\mathcal{S}^{-1}. \tag{7.46}$$

Therefore, the gains $\mathcal{K}_p$ and $\mathcal{K}_v$ of the PD-based controller (7.19) will be expanded using the expression (7.38).

7.6 Design of an Affine PID-Based Control Law

As previously, the primary goal of this section is to design the controller $\Delta\mathcal{U}$ to ensure stability of the linear model (7.18) at the desired position q_d. Our second idea is to adopt the PID controller.

7.6.1 *The PID-Based Control Law*

Let us propose the subsequent PID-based controller:

$$\begin{cases} \Delta\mathcal{U}=\mathcal{K}_p\phi+\mathcal{K}_v\dot{\phi}+\mathcal{K}_i z, \\ \dot{z}=\phi, \end{cases} \tag{7.47}$$

where the gains $\mathcal{K}_p$, $\mathcal{K}_v$ and $\mathcal{K}_i$ denotes the position gain, the velocity gain, and the integral position gain, respectively. Moreover, in (7.47), notice that

$$z=\int\phi\,dt. \tag{7.48}$$

Therefore, the term $\mathcal{K}_i z$ corresponds to the integral action in the controller $\Delta\mathcal{U}$.

The affine PID-based controller, utilized for the nonlinear dynamic model (7.3) of the robotic system to control and stabilize it at the desired state q_d, is formulated as follows:

$$\begin{cases} \mathcal{U} = \mathcal{U}_d + \mathcal{K}_p (q - q_d) + \mathcal{K}_v \dot{q} + \mathcal{K}_i z, \\ \dot{z} = q - q_d. \end{cases} \tag{7.49}$$

By introducing the adopted PID controller defined in (7.47) in the linear system (7.18), we get the subsequent expression:

$$\begin{cases} \ddot{\phi} = \mathcal{M}_d^{-1} (\mathcal{D}_d \mathcal{K}_p - \mathcal{N}_d) \phi + \mathcal{M}_d^{-1} \mathcal{D}_d \mathcal{K}_v \dot{\phi} + \mathcal{M}_d^{-1} \mathcal{D}_d \mathcal{K}_i z, \\ \dot{z} = q - q_d. \end{cases} \tag{7.50}$$

Next, we aim to determine the selection of the gains $\mathcal{K}_p$, $\mathcal{K}_v$, and $\mathcal{K}_i$ for the PID-based controller (7.47), ensuring the stability of the controlled dynamics (7.50).

Assuming $Y = \begin{bmatrix} \phi \\ \dot{\phi} \\ z \end{bmatrix}$ as the state vector, the resulting closed-loop model (7.50) can be reformulated in the subsequent state-space representation:

$$\dot{Y} = (\mathcal{A}_1 + \mathcal{B}_1 \mathcal{K}) Y, \tag{7.51}$$

with

$$\mathcal{A}_1 = \begin{bmatrix} O_n & \mathcal{I}_n & O_n \\ -\mathcal{M}_d^{-1} \mathcal{N}_d & O_n & O_n \\ \mathcal{I}_n & O_n & O_n \end{bmatrix}, \tag{7.52a}$$

$$\mathcal{B}_1 = \begin{bmatrix} O_{n\times m} \\ \mathcal{M}_d^{-1} \mathcal{D}_d \\ O_{n\times m} \end{bmatrix}, \tag{7.52b}$$

$$\mathcal{K} = \left[\mathcal{K}_p \; \mathcal{K}_v \; \mathcal{K}_i \right]. \tag{7.52c}$$

7.6.2 Condition on the Feedback Gains for Stability

To determine the condition on the gain $\mathcal{K}$ and subsequently the gain matrices $\mathcal{K}_p$, $\mathcal{K}_v$, and $\mathcal{K}_i$, we will employ the subsequent candidate Lyapunov function:

$$\mathcal{V}(Y) = Y^{\mathrm{T}} \mathcal{P} Y \quad \text{with} \quad \mathcal{P} = \mathcal{P}^{\mathrm{T}} > 0. \tag{7.53}$$

Its time derivative is defined like so:

$$\dot{\mathcal{V}}(Y) = \dot{Y} \mathcal{P} Y + Y^{\mathrm{T}} \mathcal{P} \dot{Y}. \tag{7.54}$$

As previously, we will consider the following condition for exponential stability:

$$\dot{\mathcal{V}}(Y) < -\alpha\mathcal{V}(Y), \tag{7.55}$$

where $\alpha > 0$ is some fixed decay rate.

By utilizing (7.55) and (7.51), we derive the subsequent inequality:

$$(\mathcal{A}_1 + \mathcal{B}_1\mathcal{K})^{\mathrm{T}}\mathcal{P} + \mathcal{P}(\mathcal{A}_1 + \mathcal{B}_1\mathcal{K}) + \alpha\mathcal{P} < 0. \tag{7.56}$$

By multiplying (7.56) on the left and right by the matrix $\mathcal{S} = \mathcal{P}^{-1}$, we establish the subsequent stability condition:

$$\mathcal{A}_1\mathcal{S} + \mathcal{B}_1\mathcal{R} + (\mathcal{A}_1\mathcal{S} + \mathcal{B}_1\mathcal{R})^{\mathrm{T}} + \alpha\mathcal{S} < 0, \tag{7.57}$$

where $\mathcal{R} = \mathcal{K}\mathcal{S}$.

By solving the formulated LMI condition (7.57), the state-feedback gain $\mathcal{K}$ can be calculated using expression (7.46). Consequently, the gains $\mathcal{K}_p$, $\mathcal{K}_v$, and $\mathcal{K}_i$ for the employed controller (7.47) can be obtained from (7.52c).

7.7 Design of the LQR Control Law

In this section, our focus shifts to the adoption of an alternative control law known as the Linear Quadratic Regulator (LQR), which is an optimal control strategy. To apply the LQR controller and regulate the robotic system effectively, we will utilize the linearized dynamic model described by Eq. (7.35).

7.7.1 The LQR controller

The suggested LQR control law is a state-feedback controller, given like so:

$$\Delta\mathcal{U} = -\mathcal{K}Z. \tag{7.58}$$

Next, we will present the expression of the gain $\mathcal{K}$ of the optimal LQR controller (7.58).

It should be noted that the controller to be implemented on the robotic system is described like so:

$$\mathcal{U} = \mathcal{U}_d - \mathcal{K}Z, \tag{7.59}$$

$$= \mathcal{U}_d - \mathcal{K}_p\,(q - q_d) - \mathcal{K}_v\dot{q}. \tag{7.60}$$

7.7.2 *Condition on the Feedback Gain for Stability*

Under the adopted LQR controller defined by (7.58), the controlled linear dynamics (7.35) is given by:

$$\dot{Y} = (\mathcal{A} - \mathcal{B}\mathcal{K})Y. \tag{7.61}$$

The matrices $\mathcal{A}$ and $\mathcal{B}$ in the control law (7.61) are determined by Eqs. (7.36a) and (7.36b), respectively.

The simplified problem of the LQR controller consists of finding the matrix $\mathcal{K}$ that minimizes the following cost function:

$$\mathcal{J} = \int \left(Z^{\mathrm{T}} Q Z + \Delta\mathcal{U}^{\mathrm{T}} \mathcal{R} \Delta\mathcal{U} \right) dt. \tag{7.62}$$

Here, Q and $\mathcal{R}$ are constant, positive definite, and symmetric matrices.

The gain vector $\mathcal{K}$ of the LQR control law can be obtained using the following equation:

$$\mathcal{K} = -\mathcal{R}^{-1}\mathcal{B}^{\mathrm{T}}\mathcal{P}. \tag{7.63}$$

Where, the equation below provides the solution for a positive definite, symmetric, and constant matrix $\mathcal{P}$:

$$\mathcal{A}^{\mathrm{T}}\mathcal{P} + \mathcal{P}\mathcal{A} - \mathcal{P}\mathcal{B}\mathcal{R}^{-1}\mathcal{B}^{\mathrm{T}}\mathcal{P} + Q = 0. \tag{7.64}$$

7.8 Simulation Results

In this section, we will showcase the simulation results of the various controllers implemented to demonstrate the effectiveness and validity of the stability conditions designed to address the position control challenge in a Lagrangian-type robotic system. To illustrate this, a 2-DoF exoskeleton robot will be employed as an illustrative example.

7.8.1 *The Adopted 2-DoF Exoskeleton Robot*

In Fig. 7.1, we can observe the double inverted pendulum, which serves as the adopted 2-DoF exoskeleton robot. The relative angular positions are represented by the vector $\theta = \left[\, \theta_1 \; \theta_2 \,\right]$, while the absolute angular positions are denoted by the

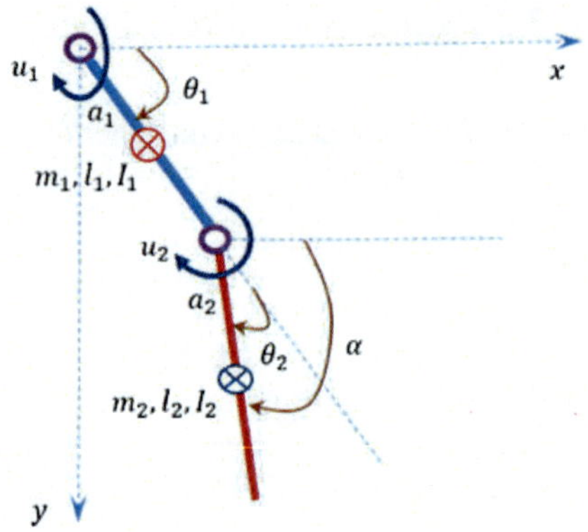

Fig. 7.1 The shcematic model of the adopted 2-DoF exoskeleton robot for the rehabilitation of the human lower limb

vector $q = \begin{bmatrix} \theta_1 & \alpha \end{bmatrix}$, with $\alpha = \theta_1 + \theta_2$. Additionally, the control inputs are represented by the vector $\mathcal{U} = \begin{bmatrix} u_1 \\ u_2 \end{bmatrix}$.

To simplify the expression of the matrices in the nonlinear dynamic model (7.3), we adopt generalized absolute coordinates represented by the vector q. Hence, the dynamic model of the 2-DoF exoskeleton robot can be described by the nonlinear dynamics (7.3), with the matrices defined as follows:

$$\mathcal{M}(\theta) = \begin{bmatrix} m_1 a_1^2 + m_2 l_1^2 + I_1 & m_2 l_1 a_2 \cos(\theta_1 - \alpha) \\ m_2 l_1 a_2 \cos(\theta_1 - \alpha) & m_2 a_2^2 + I_2 \end{bmatrix}, \tag{7.65a}$$

$$\mathcal{H}(\theta, \dot{\theta}) = m_2 l_1 a_2 \sin(\theta_1 - \alpha) \begin{bmatrix} 0 & \dot{\alpha} \\ -\dot{\theta}_1 & 0 \end{bmatrix}, \tag{7.65b}$$

$$\mathcal{G}(\theta) = g \begin{bmatrix} (m_1 a_1 + m_2 l_1) \cos\theta_1 \\ m_2 a_2 \cos\alpha \end{bmatrix}, \tag{7.65c}$$

$$\mathcal{D}(\theta) = \begin{bmatrix} 1 & -1 \\ 0 & 1 \end{bmatrix}. \tag{7.65d}$$

In Table 7.1, we can find the description and corresponding values of the parameters associated with the matrices of the robotic system mentioned earlier.

It is worth mentioning that the generalized relative position θ can be calculated via the following relation:

$$\theta = \begin{bmatrix} 1 & 0 \\ -1 & 1 \end{bmatrix} q. \tag{7.66}$$

We will choose the desired position like so:

Table 7.1 Details on the different parameters used for the simulation of the proposed 2-DoF exoskeleton rehabilitation robot

Parameter	Description	Value
m_1	First segment's Mass	2 Kg
m_2	Second segment's Mass	1 Kg
l_1	First segment's Length	0.5 m
l_2	Second segment's Length	0.4 m
a_1	Distance from the first articulation to the center of mass of the first segment	0.375 m
a_2	Distance from the second articulation to the center of mass of the second segment	0.25 m
I_1	Coefficient of the rotational inertia of the first segment	0.02 kg.m^2
I_2	Coefficient of the rotational inertia of the second segment	0.01 kg.m^2
g	Gravity constant	9.81 m/s^2

$$\theta_d = \begin{bmatrix} 90^\circ \\ -90^\circ \end{bmatrix}, \tag{7.67}$$

and then via relation (7.66):

$$q_d = \begin{bmatrix} 90^\circ \\ 0^\circ \end{bmatrix}. \tag{7.68}$$

In the following, we will show the temporal evolution of the relative angular positions θ_1 and θ_2, instead of the absolute angular positions $q_1 = \theta_1$ and $q_2 = \alpha$.

7.8.2 *Numerical Results Under the Affine PD-Based Controller*

Using this desired state q_d in the linear dynamics (7.18), we obtain then:

$$\mathcal{M}_d^{-1}\mathcal{N}_d = \begin{bmatrix} -22.2449 & 0 \\ 0 & 0 \end{bmatrix}, \tag{7.69a}$$

$$\mathcal{M}_d^{-1}\mathcal{D}_d = \begin{bmatrix} 1.8141 & -1.8141 \\ 0 & 13.7931 \end{bmatrix}. \tag{7.69b}$$

7.8.2.1 Results Obtained for the First Design Approach

Based on the discussion in Sect. 7.5.1, we make the assumption that all poles of the reference model (7.23) are equal, denoted as $\eta_{1i} = \eta_{2i} = -w_i = -5$. Consequently, the matrices Ω_v and Ω_p in Eq. (7.24) can be expressed as follows:

$$\Omega_v = \text{diag}\,(10, 10)\,, \tag{7.70a}$$

$$\Omega_p = \text{diag}\,(25, 25)\,. \tag{7.70b}$$

By utilizing the matrices Ω_v and Ω_p obtained earlier, we can compute the feedback gains $\mathcal{K}_p$ and $\mathcal{K}_v$ for the PD-based controller $\Delta\mathcal{U}$ as defined in Eq. (7.19). These gains can be calculated using the equations given in (7.26), resulting in the following values:

$$\mathcal{K}_p = \begin{bmatrix} -26.0438 & -1.8125 \\ 0 & -1.8125 \end{bmatrix}, \tag{7.71a}$$

$$\mathcal{K}_v = \begin{bmatrix} -5.5125 & -0.7250 \\ 0 & -0.7250 \end{bmatrix}. \tag{7.71b}$$

By incorporating the obtained gains $\mathcal{K}_p$ and $\mathcal{K}_v$ into the affine PD-based controller (7.21), the nonlinear dynamics (7.3) of the robotic system are effectively regulated toward the desired position state q_d.

The simulation results of the controlled robotic system are presented in Fig. 7.2. Figure 7.2a illustrates the temporal variation of the angular positions θ_1 and θ_2, demonstrating their convergence toward the desired position. Figure 7.2b displays

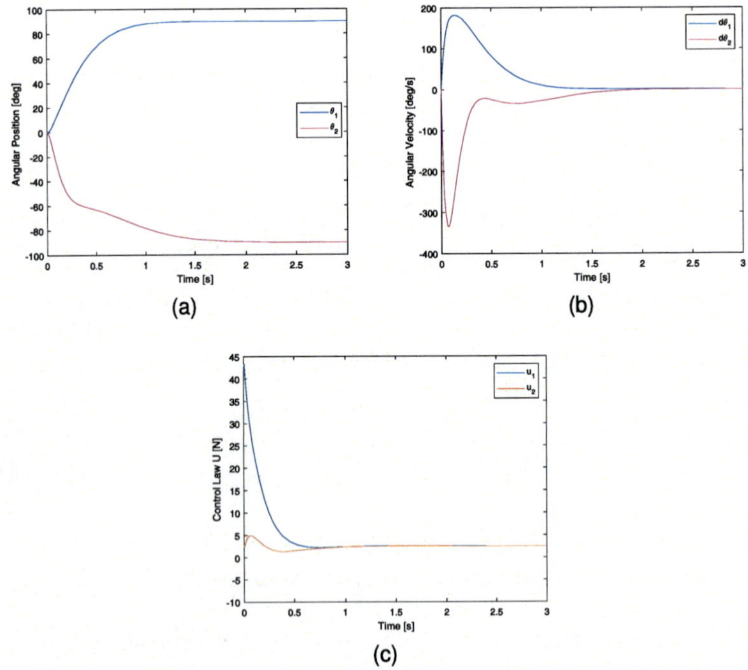

Fig. 7.2 Temporal evolution of (**a**) the two angular positions θ_1 and θ_2, (**b**) the two angular velocities $\dot{\theta}_1$ and $\dot{\theta}_2$ of the robotic system, and (**c**) the affine PD-based control law $\mathcal{U}$

the temporal variation of the angular velocities $\dot{\theta}1$ and $\dot{\theta}2$, indicating their convergence to zero. Additionally, Fig. 7.2c illustrates the temporal evolution of the adopted controller $\mathcal{U}$. Clearly, once the robot stabilizes at the desired state q_d, the control inputs u_1 and u_2 remain constant at the value of 2.4526.

7.8.2.2 Results Obtained for the Second Design Approach

The second approach for determining the feedback gains $\mathcal{K}_p$ and $\mathcal{K}_v$ was presented in Sect. 7.5.2. By utilizing the expressions (7.32) and (7.34), along with a predefined positive definite matrix Q, we can compute the values of $\mathcal{K}_p$ and $\mathcal{K}_v$. Interestingly, it can be shown that by selecting $Q = \mathcal{M}_d\Omega_v$, as described in (7.70), and setting $\mu = 25$ according to (7.70), we get the same values for $\mathcal{K}_p$ and $\mathcal{K}_v$ as given in (7.71).

7.8.2.3 Results Obtained for the Third Design Approach

We compute the matrices $\mathcal{A}$ and $\mathcal{B}$ of the linear system (7.35) using the desired state (7.68) as follows:

$$\mathcal{A} = \begin{bmatrix} 0 & 0\,1\,0 \\ 0 & 0\,0\,1 \\ 22.2449 & 0\,0\,0 \\ 0 & 0\,0\,0 \end{bmatrix}, \tag{7.72a}$$

$$\mathcal{B} = \begin{bmatrix} 0 & 0 \\ 0 & 0 \\ 1.8141 & -1.8141 \\ 0 & 13.7931 \end{bmatrix}. \tag{7.72b}$$

Based on the computed matrices $\mathcal{A}$ and $\mathcal{B}$, with a fixed value of $\alpha = 10$, we solve the Linear Matrix Inequality (LMI) (7.45) derived in Sect. 7.6. Subsequently, utilizing (7.26), we derived the state-feedback gain $\mathcal{K}$ as follows:

$$\mathcal{K} = \begin{bmatrix} -68.7364 & 77.2257 & -9.0951 & 12.5715 \\ -10.9442 & -5.9880 & -1.7799 & -0.9621 \end{bmatrix}. \tag{7.73}$$

Subsequently, by incorporating the obtained feedback gain $\mathcal{K}$ into the employed affine PD-based controller (7.37), within the nonlinear dynamics (7.3) of the exoskeleton robot, the system is effectively guided toward the desired position q_d.

The simulation results of the controlled robotic system's response are depicted in Fig. 7.3. Figure 7.3a illustrates the temporal evolution of the angular positions θ_1 and θ_2. It is evident that both angular positions converge toward their desired

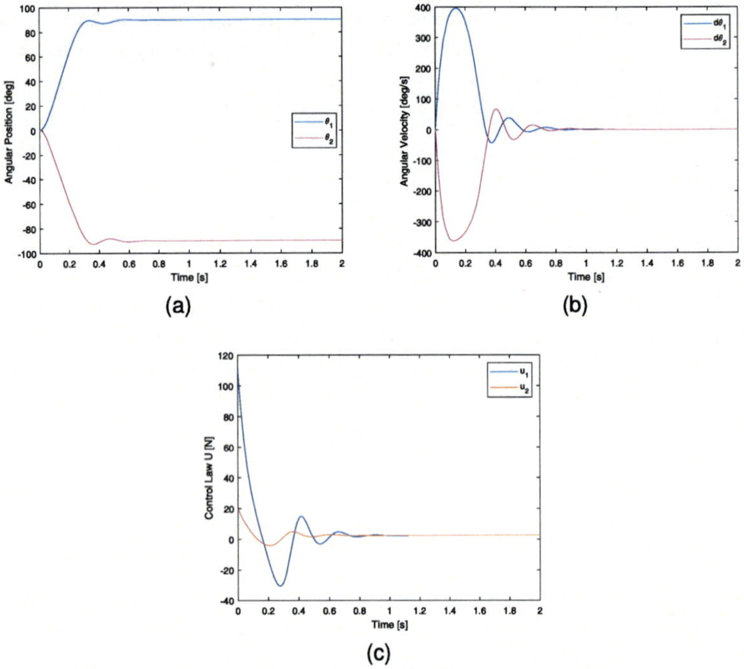

Fig. 7.3 Temporal evolution of (**a**) the two angular positions θ_1 and θ_2, (**b**) the two angular velocities $\dot{\theta}_1$ and $\dot{\theta}_2$ of the robotic system, and (**c**) the state-feedback controller $\mathcal{U}$

values. Moreover, the variation of the associated angular velocities $\dot{\theta}_1$ and $\dot{\theta}_2$, which tend toward zero, is also shown in Fig. 7.3b. The temporal evolution of the used controller $\mathcal{U}$ is also shown in Fig. 7.3c, where the two control inputs u_1 and u_2 of the controlled robotic system tend to a constant value of 2.4525 at equilibrium, q_d.

7.8.3 *Numerical Results Under the Affine PID-Based Controller*

The two matrices $\mathcal{A}_1$ and $\mathcal{B}_1$ presented in the linear model (7.51), and described by (7.52a) and (7.52b), are calculated as follows using the selected desired state (7.68) in the matrices presented in the nonlinear dynamics (7.3):

$$\mathcal{A} = \begin{bmatrix} 0 & 0 & 1.0000 & 0 & 0 & 0 \\ 0 & 0 & 0 & 1.0000 & 0 & 0 \\ 22.2449 & 0 & 0 & 0 & 0 & 0 \\ -0.0000 & 0 & 0 & 0 & 0 & 0 \\ 1.0000 & 0 & 0 & 0 & 0 & 0 \\ 0 & 1.0000 & 0 & 0 & 0 & 0 \end{bmatrix}, \tag{7.74a}$$

$$\mathcal{B} = \begin{bmatrix} 0 & 0 \\ 0 & 0 \\ 1.8141 & -1.8141 \\ 0 & 13.7931 \\ 0 & 0 \\ 0 & 0 \end{bmatrix}. \tag{7.74b}$$

Using these two matrices $\mathcal{A}_1$ and $\mathcal{B}_1$, and by setting $\alpha = 10$, we solved the LMI condition (7.57). Then, we get the following feedback gains:

$$\mathcal{K}_p = \begin{bmatrix} -156.5403 & -11.5723 \\ -0.9571 & -18.8494 \end{bmatrix}, \tag{7.75a}$$

$$\mathcal{K}_v = \begin{bmatrix} -12.2747 & -0.9834 \\ -0.0816 & -1.6036 \end{bmatrix}, \tag{7.75b}$$

$$\mathcal{K}_i = \begin{bmatrix} -502.4965 & -40.3005 \\ -3.3339 & -65.6495 \end{bmatrix}. \tag{7.75c}$$

By incorporating the obtained gains $\mathcal{K}_p$, $\mathcal{K}_v$, and $\mathcal{K}_i$ into the employed affine PID-based controller $\mathcal{U}$, as defined in Eq. (7.47), the nonlinear dynamics (7.3) is utilized to achieve control and stabilization of the robotic system at the desired state q_d.

Figure 7.4 illustrates the simulation results of the response of the controlled robotic system. Thus, the temporal evolution of θ_1 and θ_2, which defines the angular positions, is presented in Fig. 7.4a, where it is obvious that both of the angular positions θ_1 and θ_2 tend toward to the desired states. Figure 7.4b represents the variation of $\dot{\theta}_1$ and $\dot{\theta}_2$, where the two the angular velocities converge toward zero. In addition, Fig. 7.4c illustrates the temporal variation of the employed controller $\mathcal{U}$. Once the robotic system has reached the desired state q_d, both of the control inputs u_1 and u_2 tend to a constant value 2.452.

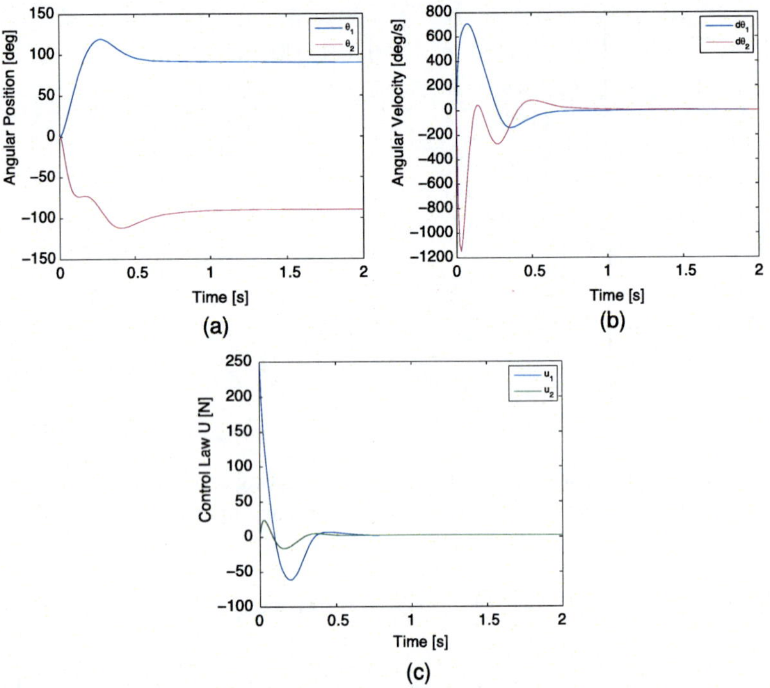

Fig. 7.4 Temporal variation of (**a**) the two angular positions θ_1 and θ_2, (**b**) the two angular velocities $\dot{\theta}_1$ and $\dot{\theta}_2$ of the robotic system, and (**c**) the PID controller $\mathcal{U}$

7.8.4 Numerical Results Under the LQR Control Law

Considering the linear system (7.51), let's examine the matrices $\mathcal{A}$ and $\mathcal{B}$ computed using Eqs. (7.72a) and (7.72b). Furthermore, we can choose the matrices $\mathcal{Q}$ and $\mathcal{R}$ as follows:

$$\mathcal{Q} = \begin{bmatrix} 20 & 0 & 0 & 0 \\ 0 & 20 & 0 & 0 \\ 0 & 0 & 20 & 0 \\ 0 & 0 & 0 & 20 \end{bmatrix}, \tag{7.76a}$$

$$\mathcal{R} = \begin{bmatrix} 10 & 0 \\ 0 & 10 \end{bmatrix}. \tag{7.76b}$$

Based on the matrices $\mathcal{A}$, $\mathcal{B}$, $\mathcal{Q}$, and $\mathcal{R}$, we calculated the feedback gain $\mathcal{K}$ as follows:

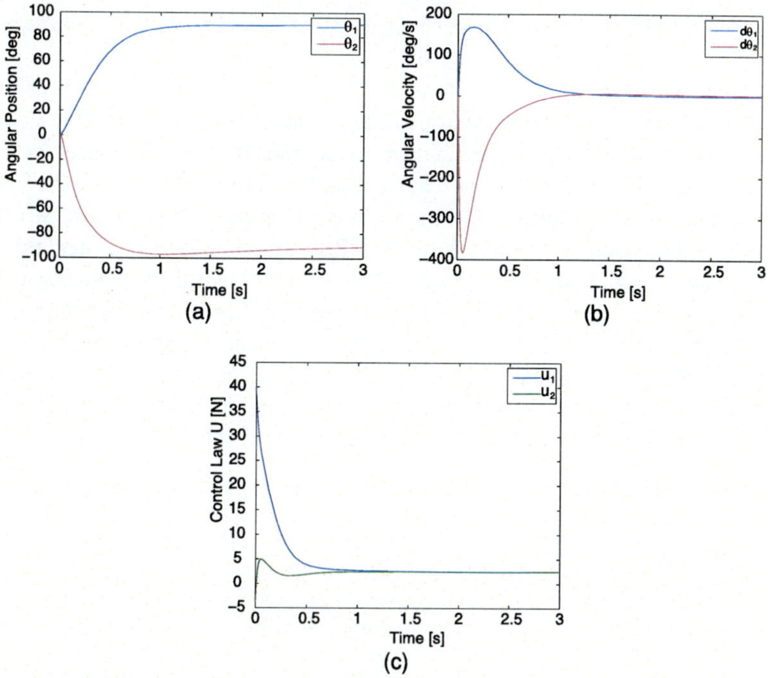

Fig. 7.5 Temporal variation of (**a**) the two angular positions θ_1 and θ_2, (**b**) the two angular velocities $\dot{\theta}_1$ and $\dot{\theta}_2$ of the robotic system, and (**c**) the LQR controller $\mathcal{U}$

$$\mathcal{K}=\begin{bmatrix} 23.9544 & 0.4535 & 5.2493 & 0.5688 \\ -3.9583 & 1.3395 & -0.9243 & 1.3916 \end{bmatrix}. \tag{7.77}$$

By applying the obtained feedback gain $\mathcal{K}$ in the LQR control law (7.60), the robotic system is steered toward the desired state q_d.

The simulation results of the controlled robotic system are depicted in Fig. 7.5. Figure 7.5a illustrates the temporal evolution of the angular positions θ_1 and θ_2, demonstrating the convergence of both angles towards their desired values. Additionally, Fig. 7.5b displays the temporal variation of the angular velocities $\dot{\theta}1$ and $\dot{\theta}2$, which converge to zero. Moreover, Fig. 7.5c illustrates the temporal evolution of the applied control law $\mathcal{U}$. It is evident that at the equilibrium state q_d of the controlled robotic system, the two control inputs u_1 and u_2 tend to a constant value of 2.438.

7.9 Conclusion

In this study, our focus was on addressing the position control problem of Lagrangian-type robots using various linear control laws. The primary goal was to control the robot by approximating its nonlinear dynamic model around a desired state q_d. To achieve this objective and drive the robotic system from its current position to the desired position q_d, we employed three control strategies: the affine PD-based controller, the affine PID-based control law, and the LQR control law. Additionally, we derived stability conditions for the feedback gains in the approximate linear dynamics, ensuring the stability of the controlled robotic system at the desired position q_d. To demonstrate the effectiveness of the proposed linear controllers, we utilized a 2-DoF exoskeleton robot as an illustrative example. Simulation results were presented, and a comparison among the different control laws was performed to evaluate their performance.

References

1. Juan Angel Gonzalez-Aguirre, Ricardo Osorio-Oliveros, Karen L. Rodriguez-Hernandez, Javier Lizarraga-Iturralde, Rubén Morales Menendez, Ricardo A. Ramirez-Mendoza, Mauricio Adolfo Ramirez-Moreno, and Jorge de Jesus Lozoya-Santos. Service robots: Trends and technology. *Applied Sciences*, 11(22):10702, 2021.
2. Bence Tipary and Gabor Erdos. Generic development methodology for flexible robotic pick-and-place workcells based on digital twin. *Robotics and Computer-Integrated Manufacturing*, 71:102140, 2021.
3. Jiankun Wang, Weinan Chen, Xiao Xiao, Yangxin Xu, Chenming Li, Xiao Jia, and Max Q.-H. Meng. A survey of the development of biomimetic intelligence and robotics. *Biomimetic Intelligence and Robotics*, 1:100001, 2021.
4. Italo Renan da Costa Barros and Tiago Pereira Nascimento. Robotic mobile fulfillment systems: A survey on recent developments and research opportunities. *Robotics and Autonomous Systems*, 137:103729, 2021.
5. Mark W. Spong, Seth Hutchinson, and M. Vidyasagar. *Robot Modeling and Control*. Robotics. John Wiley & Sons Inc, 2 edition, 2020.
6. Amal Choukchou-Braham, Brahim Cherki, Mohamed Djemai, and Krishna Busawon.
7. E. C. Orozco-Magdaleno, F. Gomez-Bravo, E. Castillo-Castaneda, and G. Carbone. Evaluation of locomotion performances for a mecanum-wheeled hybrid hexapod robot. *IEEE/ASME Transactions on Mechatronics*, 26(3):1657–1667, 2021.
8. Luca Gualtieri, Erwin Rauch, and Renato Vidoni. Development and validation of guidelines for safety in human-robot collaborative assembly systems. *Computers & Industrial Engineering*, page 107801, 2021.
9. Raouf Fareh, Sofiane Khadraoui, Mahmoud Y. Abdallah, Mohammed Baziyad, and Maamar Bettayeb. Active disturbance rejection control for robotic systems: A review. *Mechatronics*, 80:102671, 2021.
10. Sahar Jenhani, Hassène Gritli, and Giuseppe Carbone. Comparison between some nonlinear controllers for the position control of Lagrangian-type robotic systems. *Chaos Theory and Applications*, 4(4):179 – 196, 2022.
11. Soukaina Krafes, Zakaria Chalh, and Abdelmjid Saka. A review on the control of second order underactuated mechanical systems. *Complexity*, 2018:9573514, 2018.

12. Yang Liu and Hongnian Yu. A survey of underactuated mechanical systems. *IET Control Theory Applications*, 7(7):921–935, 2013.
13. Saleh Mobayen, Fairouz Tchier, and Lakhdar Ragoub. Design of an adaptive tracker for n-link rigid robotic manipulators based on super-twisting global nonlinear sliding mode control. *International Journal of Systems Science*, 48(9):1990–2002, 2017.
14. Ashish Singla and Gurminder Singh. Real-time swing-up and stabilization control of a cart-pendulum system with constrained cart movement. *International Journal of Nonlinear Sciences and Numerical Simulation*, 18(6):525–539, 2017.
15. Jyotindra Narayan and Santosha K. Dwivedy. Robust LQR-based neural-fuzzy tracking control for a lower limb exoskeleton system with parametric uncertainties and external disturbances. *Applied Bionics and Biomechanics*, 2021:5573041, 2021.
16. Ishan Chawla and Ashish Singla. Real-time stabilization control of a rotary inverted pendulum using LQR-based sliding mode controller. *Arabian Journal for Science and Engineering*, 46(3):2589–2596, 2021.
17. Mohamed Abbas, Sami Al Issa, and Santosha K. Dwivedy. Event-triggered adaptive hybrid position-force control for robot-assisted ultrasonic examination system. *Journal of Intelligent & Robotic Systems*, 102(4):84, 2021.
18. Mohamed Abbas, Jyotindra Narayan, and Santosha K. Dwivedy. A systematic review on cooperative dual-arm manipulators: modeling, planning, control, and vision strategies. *International Journal of Intelligent Robotics and Applications*, 2023.
19. Rafael Kelly, Victor Santibáñez Davila, and Antonio Lora. *Control of Robot Manipulators in Joint Space*. Advanced Textbooks in Control and Signal Processing. Springer-Verlag, London, 2005.
20. Sahar Jenhani, Hassène Gritli, and Giuseppe Carbone. Design of an affine control law for the position control problem of robotic systems based on the development of a linear dynamic model. In *2022 5th International Conference on Advanced Systems and Emergent Technologies (IC_ASET)*, pages 403–411, 2022.
21. Jonathan S. Terry, Levi Rupert, and Marc D. Killpack. Comparison of linearized dynamic robot manipulator models for model predictive control. In *2017 IEEE-RAS 17th International Conference on Humanoid Robotics (Humanoids)*, pages 205–212, 2017.
22. Amit Kumar, Shrey Kasera, and L. B. Prasad. Optimal control of 2-link underactuated robot manipulator. In *2017 International Conference on Innovations in Information, Embedded and Communication Systems (ICIIECS)*, pages 1–6, 2017.
23. François Prince, Hélène Corriveau, Réjean Hébert, and David A Winter. Gait in the elderly. *Gait & posture*, 5(2):128–135, 1997.
24. Olfa Boubaker. *Medical and Healthcare Robotics. New Paradigms and Recent Advances.* Medical Robots and Devices: New Developments and Advances. Elsevier, Academic Press, 1 edition, July 2023.
25. Jyotindra Narayan and Santosha Kumar Dwivedy. Preliminary design and development of a low-cost lower-limb exoskeleton system for paediatric rehabilitation. *Proceedings of the Institution of Mechanical Engineers, Part H: Journal of Engineering in Medicine*, 235(5):530–545, 2021.
26. José L Pons. *Wearable robots: biomechatronic exoskeletons*. John Wiley & Sons, 2008.
27. Y Wei Hong, Y King, W Yeo, C Ting, Y Chuah, J Lee, and Eu-Tjin Chok. Lower extremity exoskeleton: review and challenges surrounding the technology and its role in rehabilitation of lower limbs. *Australian Journal of Basic and Applied Sciences*, 7(7):520–524, 2013.
28. Aaron M Dollar and Hugh Herr. Lower extremity exoskeletons and active orthoses: Challenges and state-of-the-art. *IEEE Transactions on robotics*, 24(1):144–158, 2008.
29. Hugh Herr. Exoskeletons and orthoses: classification, design challenges and future directions. *Journal of neuroengineering and rehabilitation*, 6(1):1–9, 2009.
30. Md Rasedul Islam, Brahim Brahmi, Tanvir Ahmed, Md. Assad-Uz-Zaman, and Mohammad Habibur Rahman. Chapter 9 - exoskeletons in upper limb rehabilitation: A review to find key challenges to improve functionality. In Olfa Boubaker, editor, *Control Theory in Biomedical Engineering*, pages 235–265. Academic Press, 2020.

31. Brahim Brahmi, Maarouf Saad, Cristobal Ochoa Luna, Philippe S. Archambault, and Mohammad H. Rahman. Passive and active rehabilitation control of human upper-limb exoskeleton robot with dynamic uncertainties. *Robotica*, 36(11):1757–1779, 2018.
32. Di Shi, Wuxiang Zhang, Wei Zhang, and Xilun Ding. A review on lower limb rehabilitation exoskeleton robots. *Chinese Journal of Mechanical Engineering*, 32(1):74, 2019.
33. Jyotindra Narayan and Santosha K. Dwivedy. Lower limb exoskeletons for pediatric gait rehabilitation: A brief review of design, actuation, and control schemes. In *2023 IEEE International Conference on Advanced Systems and Emergent Technologies (IC_ASET)*, pages 1–6, 2023.
34. Jyotindra Narayan and Santosha Kumar Dwivedy. Towards neuro-fuzzy compensated PID control of lower extremity exoskeleton system for passive gait rehabilitation. *IETE Journal of Research*, 69(2):778–795, 2023.
35. Hassène Gritli, Sahar Jenhani, and Giuseppe Carbone. Position control of robotic systems via an affine PD-based controller: Comparison between two design approaches. In *2022 5th International Conference on Advanced Systems and Emergent Technologies (IC_ASET)*, pages 424–432, 2022.
36. Sahar Jenhani, Hassène Gritli, and Giuseppe Carbone. Determination of conditions on feedback gains for the position control of robotic systems under an affine PD-based control law. In *2022 5th International Conference on Advanced Systems and Emergent Technologies (IC_ASET)*, pages 518–526, 2022.
37. Sahar Jenhani, Hassène Gritli, and Giuseppe Carbone. Design and computation aid of command gains for the position control of manipulator robots. In *2022 International Conference on Decision Aid Sciences and Applications (DASA)*, pages 1558–1564, 2022.
38. Sahar Jenhani, Hassène Gritli, and Giuseppe Carbone. Position feedback control of Lagrangian robotic systems via an affine PD-based control law. Part 1: Design of LMI conditions. In *2022 IEEE 2nd International Maghreb Meeting of the Conference on Sciences and Techniques of Automatic Control and Computer Engineering (MI-STA)*, pages 171–176, 2022.
39. Sahar Jenhani, Hassène Gritli, and Giuseppe Carbone. Position feedback control of Lagrangian robotic systems via an affine PD-based control law. Part 2: Improved results. In *2022 IEEE 2nd International Maghreb Meeting of the Conference on Sciences and Techniques of Automatic Control and Computer Engineering (MI-STA)*, pages 177–182, 2022.
40. Sahar Jenhani, Hassène Gritli, and Giuseppe Carbone. Position control of Lagrangian robotic systems via an affine PID-based controller and using the LMI approach. In Vincenzo Niola, Alessandro Gasparetto, Giuseppe Quaglia, and Giuseppe Carbone, editors, *Advances in Italian Mechanism Science*, pages 727–737. Springer International Publishing, Cham, 2022.
41. Stephen Boyd, Laurent El-Ghaoui, Eric Feron, and Venkataramanan Balakrishnan. *Linear matrix inequalities in system and control theory*, volume 15 of *Studies in Applied and Numerical Mathematics*. Society for Industrial and Applied Mathematics (SIAM), Philadelphia, PA., 1 edition, 1994.
42. Hassène Gritli and Safya Belghith. Robust feedback control of the underactuated Inertia Wheel Inverted Pendulum under parametric uncertainties and subject to external disturbances: LMI formulation. *Journal of The Franklin Institute*, 355(18):9150–9191, 2018.
43. Tanvir Ahmed, Emily Longwell-Grice, Md Rasedul Islam, Inga Wang, and Mohammad Rahman. Robot-aided rehabilitation with SREx: A smart robotic exoskeleton for reducing therapists' physical stress. *Archives of Physical Medicine and Rehabilitation*, 103(12):e184–e185, 2022.
44. Ruben Fuentes-Alvarez, Joel Hernandez Hernandez, Ivan Matehuala-Moran, Mariel Alfaro-Ponce, Ricardo Lopez-Gutierrez, Sergio Salazar, and Rogelio Lozano. Assistive robotic exoskeleton using recurrent neural networks for decision taking for the robust trajectory tracking. *Expert Systems with Applications*, 193:116482, 2022.
45. Brahim Brahmi, Maarouf Saad, M.H. Rahman, Cristobal Ochoa-Luna, and Islam Rasedul. Chapter 2 - development and control of an upper extremity exoskeleton robot for rehabilitation. In Jacob Rosen and Peter Walker Ferguson, editors, *Wearable Robotics*, pages 23–44. Academic Press, 2020.

46. Amir Molaei, Nima Amooye Foomany, Mahsa Parsapour, and Javad Dargahi. A portable low-cost 3d-printed wrist rehabilitation robot: Design and development. *Mechanism and Machine Theory*, 171:104719, 2022.
47. Pengcheng Liu, M. Nazmul Huda, Li Sun, and Hongnian Yu. A survey on underactuated robotic systems: Bio-inspiration, trajectory planning and control. *Mechatronics*, 72:102443, 2020.
48. Hassène Gritli and Safya Belghith. LMI-based synthesis of a robust saturated controller for an underactuated mechanical system subject to motion constraints. *European Journal of Control*, 57:179–193, 2021.
49. Sahar Jenhani, Hassène Gritli, and Giuseppe Carbone. Design of an affine control law for the position control problem of robotic systems based on the development of a linear dynamic model. In *2022 5th International Conference on Advanced Systems and Emergent Technologies (IC_ASET)*, pages 403–411. IEEE, 2022.

Part III
Controlling Dynamical and Complex Systems

Chapter 8
Synchronization of Memristive Hindmarsh-Rose Neurons Connected by Memristive Synapses

A. Anzo-Hernández, I. Carro-Pérez, B. Bonilla-Capilla, and J. G. Barajas-Ramírez

8.1 Introduction

Understanding the emergent behaviors of a set of connected biological neurons and how it is related to brain function specialization as memory, thinking or learning is one of the fundamental problems in neuroscience [1]. In particular, the synchronization phenomenon of neural electrical activities is a central problem in some brain diseases such as Alzheimer, Parkinson, and epilepsy among others [2, 3]. For the above reasons, a special interest has awakened on scientific researchers which have addressed this problem from distinct points of view, including nonlinear dynamical systems and chaos [4, 5]. Some relevant mathematical models of neurons that have been proposed and analyzed from the perspective of dynamical systems are Hodgkin-Huxley (HH), FitzHugh-Nagumo (FHN), Morris-Lecar (ML), and Hindmarsh-Rose (HR) neuron model, as well as the Hopfield and cellular neural network (HNN and CNN, respectively) models, among others [6]. One of the main successes of these models is that by varying parameters associated with potassium and sodium ion channels, these models can reproduce some of the typical dynamical behaviors of neurons such as spiking, bursting, quiescent, or chaotic behaviors.

On the other hand, neuromorphic systems have been recently highlighted, as a consequence of their potential new applications in advanced technologies associated with artificial intelligence. Neuromorphic systems are circuit, or computing

A. Anzo-Hernández · B. Bonilla-Capilla
Investigadoras e Investigadores por México, CONACYT, CEMMAC-FCFM-BUAP, Puebla, Mexico
e-mail: beatriz.bonillac@correo.buap.mx

I. Carro-Pérez · J. G. Barajas-Ramírez (✉)
Instituto Potosino de Investigación Científica y Tecnológica, División de Control y Sistemas Dinámicos, San Luis Potosi, Mexico
e-mail: iliani.carro@ipicyt.edu.mx; jgbarajas@ipicyt.edu.mx

E. Campos-Cantón et al. (eds.), *Complex Systems and Their Applications*,
https://doi.org/10.1007/978-3-031-51224-7_8

systems, focused on emulating the behavior of neural systems [7]; in this system, the memristor plays a relevant role in current research works [8]. Conceptually, the memristor was originally proposed by L. 0. Chua in 1971 [9], as the fourth basic circuit, referred to as an *ideal* memristor, i.e., resistor with memory, and it was introduced as a relation between magnetic flux and electric charge in a two-terminal device. In recent years, these ideas have been extended to mean any two-terminal device that satisfies a state-dependent Ohm's law where the nature of the internal state is generalized. As such, generalized memristors have been used to describe the relationships of ion currents and channel gates in the aforementioned neuron models. That is, memristors are used to represent the effect of potassium and sodium conductance channels, as well as magnetic fluxes generated for polarization and depolarization effects [10]. To incorporate the effect of electromagnetic induction, some extensions of neurons models have been proposed where the ideal memristor is used to realize the induced current. A relevant example for this work is the Hindmarsh-Rose (HR) neuron model with a memristor, that is, the so-called memristive Hindmarsh-Rose (mHR) neuron model [1, 11, 12].

An open research problem in modeling biological neuron networks is to represent adequately their connections. The most commonly used assumption about their connection is to assume that it is diffusive, that is, they are connected by the difference between their state variables [13]. In the context of a network of neurons, this form of coupling is called electrical coupling. But chemical coupling has been also considered [14], where the exchange of neurotransmitters between neurons is modeled throughout a sigmoidal function. Furthermore, in recent research works, the effect of electromagnetic induction during the fluctuation of neurotransmitters from synapse has been considered, where the signal exchange between a pair of neurons is modeled via a memristor synapse [15–17].

In this paper, we analyze the dynamical behavior of the mHR neuron model; we study the equilibrium points and their stability features in order to produce action potentials. Further, we select pairs of model parameters to generate different action potential patterns. Next, two mHR neurons are connected via a memristive synapse. Specifically, we use a hyperbolic tangent as the memductance function of an ideal flux-controlled memristor synapses. Using this arrangement, we study the synchronization phenomena using a statistical measure called normalized average synchronization error, which when it is zero means that identical synchronization is achieved. We explore two possible scenarios: when both mHR neurons have the same external excitation, therefore the same type of action potential patterns, and when the neurons have different external excitation, that is, different action potential patterns. We conclude that in both scenarios, there is a large set of coupling values such that synchronization is achieved even when the original trajectories of the neurons are markedly different.

This contribution is organized as follows: In Sect. 8.2, we introduce the mHR neuron model, and in Sect. 8.3, we analyze its stability and the generation of action potentials. In Sect. 8.4, two mHR neurons are connected via memristor synapse and analyze its synchronized behavior, while in Sect. 8.5, we present some concluding remarks.

8.2 The *m*HR Neuron Model

The original neuron model proposed by Hindmarsh and Rose (HR) is a three-dimensional dynamical system given by Hindmarsh and Rose [18]

$$\begin{aligned}\frac{dx(t)}{dt} &= y(t) - x(t)^3 + bx(t)^2 - z(t) + I(t),\\ \frac{dy(t)}{dt} &= 1 - dx(t)^2 - y(t),\\ \frac{dz(t)}{dt} &= r(s(x(t) - x_o) - z(t)),\end{aligned} \tag{8.1}$$

where $x(t)$ represents the membrane potential; $y(t)$ is the recovery current or spiking variable, which is associated with fast ion currents; and $z(t)$ is the slow adaption current or bursting variable, associated with slow ion concentrations. The variable $I(t)$ represents the externally applied current into the neuron; b is a parameter that controls the spiking frequency and allows to change the neuron dynamics from bursting to spiking; the slow activation of the variable $z(t)$ is modulated through the small parameter $0 < r << 1$, which is treated as a rate of current activation; while x_o is viewed as a control parameter that inhibits or induces the activation of the slow current.

Depending on the value of its parameters, system (8.1) is able to reproduce behaviors of real neurons, for example [19], (I) *quiescence*, in which the neuron reaches a stationary regime; (II) *spiking*, in that the neuron membrane voltage is composed of equally spaced spikes; (III) *bursting*, in which the neuron membrane voltage is made up of groups of two or more spikes (called bursts) separated by periods of inactivity; and (IV) *irregular spiking or bursting*, at which nonperiodic series of spikes at chaotic burst constitute the neuron dynamics.

A typical choice of parameter values for the HR model (8.1) is $d = 4$, $s = 4$, $x_o = -1.6$, and $r = 0.006$. For these parameter values, bifurcation analysis reveals the importance of the parameters b and I, such that when $I = 4$, the square wave-type bursting is observed at $b = 2.7$, while at $b = 2.52$, the neuron dynamics switch to a plateau-like bursting [20, 21].

Based on the three-dimensional HR neuron model (8.1), a variation was proposed by Thottil and Ignatius [22], Bao et al. [11], Usha and Subha [12], and Wouapi et al. [1], where the time-varying electromagnetic field across the neuron membrane is considered to include the effect of the repeated exchange of charged ions. In this sense, the magnetic flux generated by such ions is introduced in the three-dimensional HR model by introducing a flux-controlled ideal memristor into the membrane potential, so that the equations of the model are modified as follows:

$$
\begin{aligned}
\frac{dx(t)}{dt} &= y(t) - x(t)^3 + bx(t)^2 - z(t) - k_1 W(\phi(t))x(t) + I(t),\\
\frac{dy(t)}{dt} &= 1 - dx(t)^2 - y(t),\\
\frac{dz(t)}{dt} &= r(s(x(t) - x_o) - z(t)),\\
\frac{d\phi(t)}{dt} &= x(t) - k_2\phi(t),
\end{aligned}
\tag{8.2}
$$

where $\phi(t)$ is the magnetic flux across the neuron's membrane. The additional term $k_1 W(\phi(t))x(t)$, at the right hand of the membrane potential $x(t)$ equation, is seen as the current originated by externally applied electromagnetic induction, where k_1 becomes a relevant parameter for neuron dynamics since it defines the modulation gain of induced current on the membrane potential. The parameter k_2, on the other hand, represents the modulation of magnetic flux $\phi(t)$ induced by the membrane potential of the neuron. In what follows, we call system (8.2) the memristive HR model (mHR).

It is worth mentioning that the term $W(\phi(t))$ corresponds to the memductance function of the ideal flux-controlled memristor, that is, it is the derivative of its electric charge vs magnetic flux characteristic function. An alternative interpretation of this term is as a state-dependent Ohm's law between voltage and current in the memristor's terminals, with this relationship changing in accordance with characteristic function of the memristor. For the ideal current ($i(t) = \frac{dq(\phi(t))}{dt}$) and voltage ($v(t) = \frac{d\phi(t)}{dt}$) conceptualizations, one obtains [12]:

$$
i(t) = \frac{dq(\phi(t))}{d\phi}\frac{d\phi(t)}{dt} = W(\phi(t))v(t), \tag{8.3}
$$

where $q(t)$ is the electric charge and $\phi(t)$ the magnetic flux across the memristor, with the memductance function $W(\phi(t))$ representing the internal state ($\phi(t)$)-dependent component of its Ohm's law description [11].

One of the most frequent used descriptions of ideal memristors is to assume that the flux-dependent charge is described by a smooth continuous cubic function of the form $q(\phi(t)) = \alpha\phi(t) + \beta\phi(t)^3$, with the corresponding memductance function [22]:

$$
W(\phi(t)) = \alpha + 3\beta\phi(t)^2, \tag{8.4}
$$

where α and β are memresistor parameters. With the above memductance function, the ideal memristor exhibits the three main characteristic fingerprints of memristors: pinched hysteresis loop, passivity, and nonvolatile memory [6, 11].

In Fig. 8.1, we observe the dynamical behavior of the mHR (8.2) and (8.4) with $k_1 = 1$, $k_2 = 0.5$, $\alpha = 0.1$, $\beta = 0.02$, and initial condition $(-0.9282, -3.2539, 3.6881, -1.6033)$.

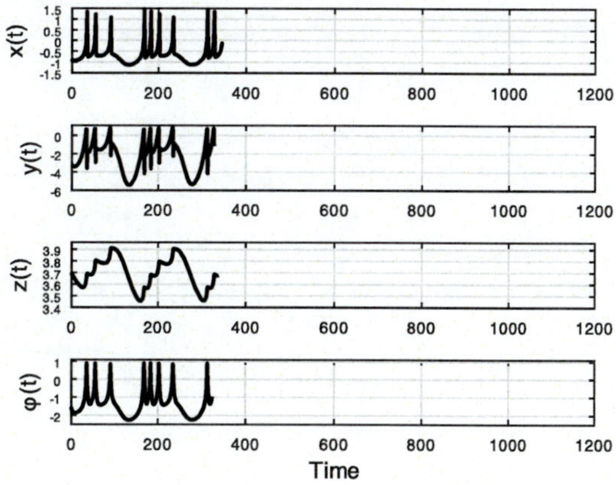

Fig. 8.1 Dynamical behavior of the mHR neuron model (8.2) with the memductance function (8.4), $b = 3$, $k_1 = 1$, $k_2 = 0.5$, $\alpha = 0.1$, $\beta = 0.02$, and initial condition $(-0.9282, -3.2539, 3.6881, -1.6033)$

8.3 Analysis of the mHR Model

By setting to zero the left-hand side of the mHR neuron model (8.2), it is straightforward to demonstrate that the equilibrium points are located at

$$E_i = (\bar{x}_i, c - d\bar{x}_i^2, s(\bar{x}_i - x_o), \bar{x}_i/k_2),$$

where $\bar{x}_i$, with $i = 1, 2, 3$, are the solution of the following cubic equation:

$$\bar{x}^3 + m\bar{x}^2 + n\bar{x} + I_0 = 0, \tag{8.5}$$

with $m = \dfrac{k_2^2(d-b)}{ak_2^2 + 3\beta k_1}$; $n = \dfrac{k_2^2(s + k_1\alpha)}{ak_2^2 + 3\beta k_1}$; $I_0 = -k_2^2 \dfrac{I + c + sx_o}{ak_2^2 + 3\beta k_1}$.

According to Cardano's formula, the solution of (8.5) depends on the value of the discriminant given by

$$\Delta = \left(\frac{q}{2}\right)^2 + \left(\frac{p}{3}\right)^3, \tag{8.6}$$

where $p = (3n - m^2)/3$ and $q = (2m^3 - 9mn + 27I_0)/27$. Then, the solution of (8.5) has the following features [23]:

- If $\Delta < 0$, then there are three real roots in (8.5). The above implies that the mHR (8.2) has three equilibrium points.
- If $\Delta > 0$, then there exist in (8.5) one real root and two complex roots. In this case, the equilibrium point of the mHR neuron model (8.2) is one.
- If $\Delta = 0$, then all the roots of (8.5) are real, and at least two of them are equal; which implies that the mHR neuron model (8.2) has two equilibrium points.

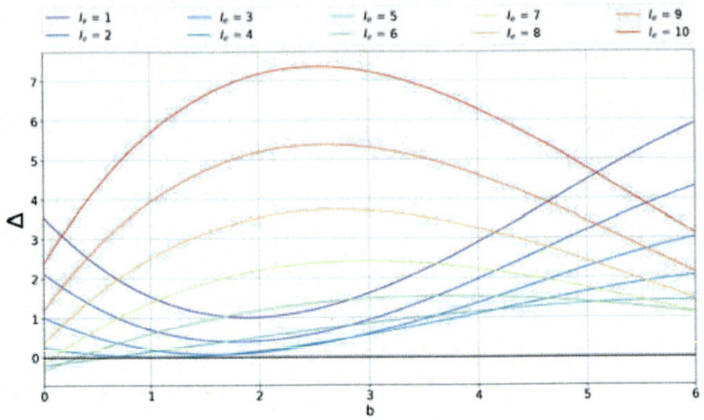

Fig. 8.2 Values of Cardano's discriminant Δ (8.6) with $b \in [0, 6]$, $I \in [1, 10]$ and $k_1 = 1$, $k_2 = 0.5$, $\alpha = 0.5$, and $\beta = 0.02$

For the case $\Delta > 0$, some authors consider just one possible equilibrium point under the argument that these cannot be complex [1]. In this paper, we hold the same reasoning. In Fig. 8.2, we observe the values of the Cardano's discriminant for $b \in [0, 6]$ and $I \in [1, 10]$, keeping fixed the parameters: $k_1 = 1$, $k_2 = 0.5$, $\alpha = 0.5$, and $\beta = 0.02$. We observe that $\Delta > 0$ except for small values of b and $I_e \in [5, 7]$.

It is worth remarking that Cardano's discriminant defines the number of possible equilibrium points of the mHR neuron model (8.2). However, it does not specify if such equilibrium points are nod-stable, saddle, spiral-stable, etc. In order to state the stability of the equilibrium points, we analyze the eigenvalues of the Jacobian matrix (8.2), which is given by

$$J(\bar{x}) = \begin{bmatrix} \Gamma(\bar{x}) & 1 & -1 & -\dfrac{6k_1\beta\bar{x}^2}{k_2} \\ -2d\bar{x} & -1 & 0 & 0 \\ rs & 0 & -r & 0 \\ 1 & 0 & 0 & -k_2 \end{bmatrix}, \tag{8.7}$$

with $\Gamma(\bar{x}) = -3a\bar{x}^2 + 2b\bar{x} - k_1\alpha - 3k_1\beta\bar{x}^2/k_2^2$. Note that the Jacobian matrix (8.7) depends on the solution of (8.5), which, at the same time, depends on the mHR biophysical parameters throughout the coefficients m, n, and I_0. Then, for a given combination of parameter's values, we solve the cubic equation (8.5) and evaluate the corresponding solution $\bar{x}$ in matrix (8.7), and then the eigenvalues are evaluated with the objective of classifying the unique equilibrium point E_1 of (8.2) (since $\Delta > 0$, the complex solution of (8.5) is discriminated).

In Fig. 8.3, we observe the classification of the mHR equilibrium point by varying the values of parameters: (a)I_e vs k_1, (b) I_e vs b, and (c) k_1 vs k_2. In blue color are the regions, in the parameter space, where the equilibrium point E_1 is node-stable (the eigenvalues of J are real and negative), in green color the regions where E_1 is

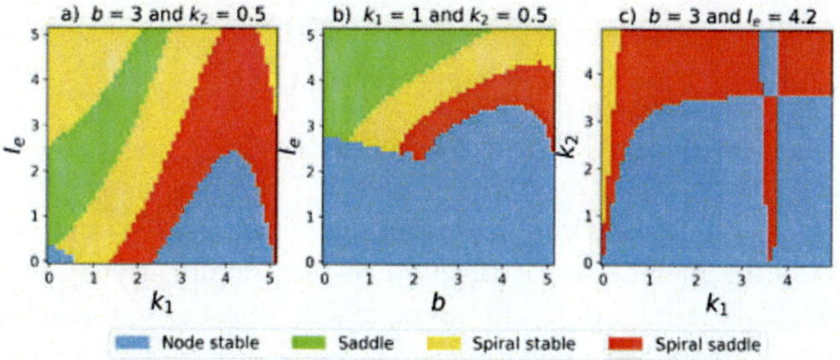

Fig. 8.3 Classification of the mHR equilibrium point by varying the values of parameters **(a)** I_e vs k_1, **(b)** I_e vs b, and **(c)** k_1 vs k_2

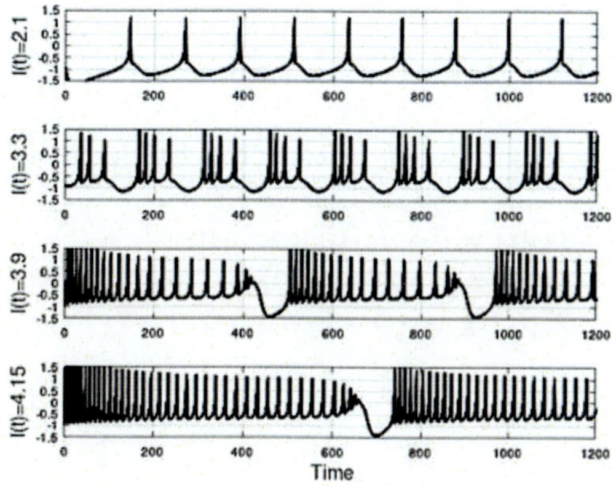

Fig. 8.4 Spiking patterns of variable $x(t)$ in the mHR neuron model (8.2) with the memductance function (8.4) and parameters $b = 3$, $k_1 = 1, k_2 = 0.5, \alpha = 0.1$, $\beta = 0.02$, and initial condition $(-0.9282, -3.2539, 3.6881, -1.6033)$

a saddle point (at least one, of the three real eigenvalues of J, has different sign), in yellow color the regions where E_1 is spiral-stable (i.e., the eigenvalues of J are complex and its real part is negative), and in red color are the regions where E_1 is a spiral saddle (i.e., the eigenvalues of J are complex and with different sign).

Using the parameter set such that equilibrium point E_1 is a saddle equilibrium point (green regions of Fig. 8.3) and changing the external current parameter $I(t)$ in the range [2, 5], different action potential patterns are generated. Four different spiking patterns are shown in Fig. 8.4 resulting from different external currents $I_{(t)}$.

8.4 Two mHR Connected via Memristor Synapse

Consider two mHR neuron models (8.2) with memductance function (8.4) connected via a memristive synapse. Further, let $x_{1i}(t)$, with $i = 1, 2, 3, 4$, be the four state variables of neuron 1 and $x_{2i}(t)$ the corresponding variables for neuron 2. The membrane potential difference ($\Delta X_{12}(t) = x_{11}(t) - x_{21}(t)$) between neurons generates a current derived from the state-dependent Ohm's law of the memristive synapses as follows:

$$\begin{aligned} I_{syn}(t) &= \mathbf{W}(u_{12}(t))(x_{11}(t) - x_{21}(t)), \\ I_{syn}(t) &= \tanh(u_{12}(t))\Delta X_{12}(t), \\ \frac{du_{12}(t)}{dt} &= \Delta X_{12}(t), \end{aligned} \tag{8.8}$$

where $u_{12}(t)$ is an interneuron magnetic flux-like variable and $\mathbf{W}(u_{12}(t)) = \tanh(u_{12}(t))$ is the memductance function of the coupling synapses between the two neurons. It is straightforward to note that this current combines a usual diffusive coupling between chaotic systems and a chemical coupling, described by a hyperbolic function.

From the above, the equations for the two neurons coupled through a memristive synapse are given by the following differential equations:

Neuron 1

$$\begin{aligned} \dot{x}_{11}(t) &= x_{12}(t) - ax_{11}(t)^3 + bx_{11}(t)^2 - x_{13}(t) - k_1 W(x_{14}(t))x_{11}(t) + I_1(t) \\ &\quad + \epsilon \mathbf{W}(u_{21}(t))\Delta X_{21}(t), \\ \dot{x}_{12}(t) &= c - dx_{11}(t)^2 - x_{12}(t), \\ \dot{x}_{13}(t) &= r[s(x_{11}(t) - x_0) - x_{13}(t)], \\ \dot{x}_{14}(t) &= x_{11}(t) - k_2 x_{14}(t). \end{aligned} \tag{8.9}$$

Neuron 2

$$\begin{aligned} \dot{x}_{21}(t) &= x_{22}(t) - ax_{21}(t)^3 + bx_{21}(t)^2 - x_{23}(t) - k_1 W(x_{24}(t))x_{21}(t) + I_2(t) \\ &\quad + \epsilon \mathbf{W}(u_{12}(t))\Delta X_{12}(t), \\ \dot{x}_{22}(t) &= c - dx_{21}(t)^2 - x_{22}(t), \\ \dot{x}_{23}(t) &= r[s(x_{21}(t) - x_0) - x_{23}(t)], \\ \dot{x}_{24}(t) &= x_{21}(t) - k_2 x_{24}(t). \end{aligned} \tag{8.10}$$

States of the Memristive Synapses

$$\begin{aligned} \dot{u}_{12}(t) &= \Delta X_{12}(t), \\ \dot{u}_{21}(t) &= \Delta X_{21}(t). \end{aligned} \tag{8.11}$$

We say that the mHR neurons, in the system described by (8.9), (8.10), and (8.11), achieve, asymptotically, identical synchronization if their membrane potentials $x_{11}(t)$ and $x_{21}(t)$ satisfy

$$x_{1i}(t) \approx x_{2i}(t) \quad \text{as} \quad t \to \infty \quad \text{for} \quad i = 1, 2, 3, 4. \tag{8.12}$$

8.4.1 Normalized Average Synchronization Error

We propose a numerical approach to detect the emergence of synchronization between the mHR neurons (Eqs. (8.9) and (8.10)) coupled via memristive synapse (Eqs. (8.11)). Therefore, in order to quantify the degree of synchronization between neurons, we calculate their normalized average synchronization error E as a statistical measure of the strength of their synchrony in the following way [24]:

$$E = \frac{1}{M}\sum_{h=1}^{M} \frac{\sqrt{\sum_{i=1}^{4}(x_{1i}(h) - x_{2i}(h))^2}}{\sqrt{\sum_{k=1}^{2}\sum_{i=1}^{4} x_{ki}^2(h)}}. \tag{8.13}$$

where h is an index that moves between one and M, which is the number of evaluation points on a simulation from an initial time t_0 to a final time t_f with a fixed time step of τ.

From (8.12), we have that $E \approx 0$ indicates the highest correlation between neurons, that is, identical synchronization; $E = 1$ implies that the two neurons are completely uncorrelated, while $E > 1$ indicates a form of anticorrelation [25].

Letting the coupling strength ϵ be the tuning variable to achieve synchronization, we investigate the resulting E value with all other parameters of (8.9), (8.10), and (8.11) fixed at their usual values, $d = 5$, $s = 4$, $x_o = -\frac{8}{5}$, $r = 0.006$, $b = 3$, $\alpha = 0.1$, and $\beta = 0.02$, with $k_2 = 0.5$. Initially, the external current for both neurons set as identical and k_1 is changed between different values. As a second case, we consider that the external currents are nonidentical.

We implement the numerical simulation of the coupled mHR neurons with conventional integration libraries (e.g., lsoda or odeint as implemented in scientific python *SciPy*) with the following initial conditions: neuron 1 (−1.45, −1.76, −4.3, −6.81) and neuron 2 (2.91, 1.88, 3.25, 4.31) for $t_0 = 0$ and $t_f = 6000$ time units with a time step of $\tau = 0.001$ resulting in $M = 6 \times 10^6$ evaluation points. Then, we search for an appropriate value of $\epsilon \in (0, 5]$ such that the E is near zero. For the case of $I_1(t) = I_2(t) = 4.2$ and $k_1(t) \in \{1, 1.2, 2.3\}$, the corresponding coupling strengths such that synchronization is achieved are $\epsilon \in \{0.5, 1.0, 1.5\}$. In Fig. 8.5 we observe the dynamics of the two mHR neuron membrane potential variables $x_{11}(t)$ and $x_{21}(t)$ for the corresponding k_1 and ϵ values. In the numerical solution of the coupled mHR model for the first 3000 time units are for $\epsilon = 0$, then for the last 3000 time units (the transition is marked as a dashed red line in Fig. 8.5), are simulated with the corresponding coupling strength ϵ to achieve identical synchronization (Figs. 8.6 and 8.7).

The results shown in Fig. 8.8 for $k_1 = 0.5, 1.0, 1.5, 2.0, 2.5, 3.0.$ are the average evaluated value of E for their corresponding coupling strength ϵ. It is worth noting that the synchronization enhances as we increase the coupling strength, but as the modulation gain k_1 increases, the neurons are marginally more difficult to synchronize, requiring larger coupling strength.

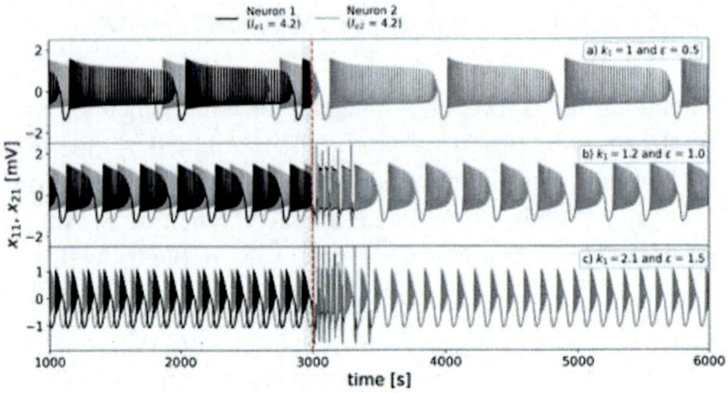

Fig. 8.5 Dynamics of two identical mHR neuron membrane potentials $x_{11}(t)$ and $x_{21}(t)$ coupled through a memristive synapse (8.8), with $I_1 = I_2 = 4.2$ and different values of k_1 and ϵ

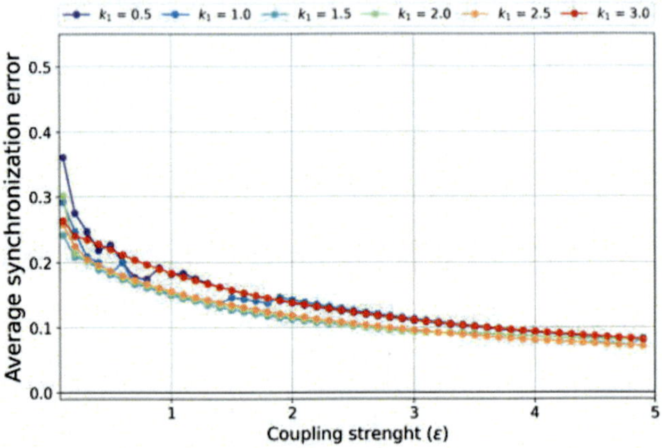

Fig. 8.6 Normalized average synchronization error for two mHR neurons with the same external current $I_1(t) = I_2(t) = 4.2$ and different values of coupling strength ϵ and modulation gain k_1

As a second case, we consider that the external currents of the mHR neurons are nonidentical. In Fig. 8.8, we show the dynamical behavior of the membrane voltage of two mHR neurons coupled with a memristive synapses with different external input currents $I_1 = 3.2$ and $I_2 = 2.4$. As one can corroborate from Fig. 8.1, for these different external currents, the neurons present periodic spiking and bursting behavior, respectively. The results in Fig. 8.8 are for mHR neurons coupled via a memristor synapse (Eq. (8.8)), with $k_1 \in \{1, 1.2, , 2.1\}$, respectively. The appropriate coupling strength ϵ such that identical synchronization is achieved is found to be $\epsilon \in \{0.5, 1.0, , 1.5\}$. Please note that although synchronization is achieved, the resulting emergent behavior has different characteristics than the original activation patterns. The synchronized behavior shows periodic bursting trains. Moreover, for this case of $k_1 = 2.1$ and $\epsilon = 1.5$, we observe that neuron

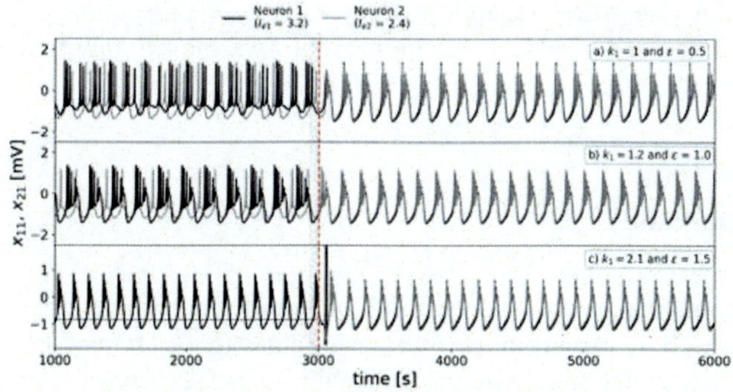

Fig. 8.7 Dynamics of membrane potentials $x_{11}(t)$ and $x_{21}(t)$ for mHR neurons with different activation potential patterns coupled through a memristive synapse (8.8), with $I_1(t) = 3.2$ and $I_2(t) = 2.3$ and different values of k_1 and ϵ

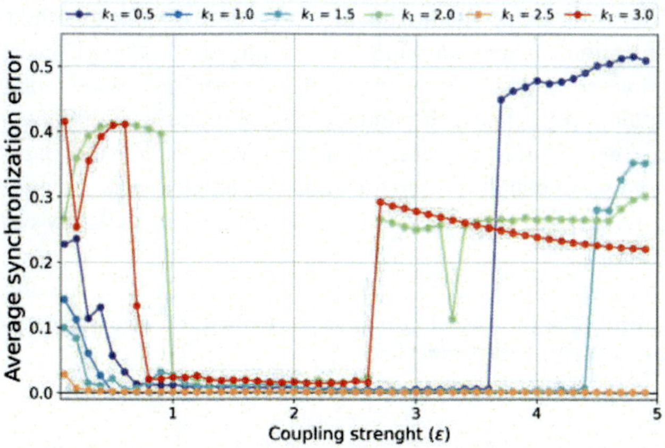

Fig. 8.8 Normalized average synchronization error for two mHR neurons with different external current $I_1(t) = 3.2$, $I_2(t) = 2.4$ and different values of modulation gain k_1 and their corresponding coupling strength ϵ

2 has a stationary behavior before connection and its dynamics become activated to follow the synchronized behavior of neuron 1.

On the other hand, in Fig. 8.8, the normalized average synchronization error E is calculated for nonidentical mHR neurons, i.e., with different external currents $I_1(t) = 3.2$ for neuron 1 and $I_2(t) = 2.4$ for neurons 2. We consider different values of coupling strength $\epsilon \in (0, 5]$.

Again, we solve the mHR neuron models (8.9)–(8.10) with different values of modulation gain k_1. Unlike the first numerical example (for neurons with identical external current), we observe a minimum and maximum value of the coupling strength ϵ such that E is almost zero. Furthermore, as the coupling strength increase, the value of E also grows, regardless of the value of k_1. The above implies that for nonidentical neurons, identical synchronization is achieved only for some chosen values of ϵ.

8.5 Conclusion

We investigate the nature and stability of equilibrium points for a version of mHR model. Then, varying some parameter values (e.g., $I(t)$ and k_1), we identify regions, in parameter space, where the stability changes from stable node, stable spiral, and saddle local behaviors. We observe that the most relevant pair of mHR parameters are b and the external current $I(t)$ to generate different activation patterns in the model. Then, we propose that two mHR neurons are connected via memristive synapse. We use the hyperbolic tangent as the memductance function of an ideal flux-controlled memristor in the synapses. With this model, we study the synchronization phenomena using the normalized average synchronization error. We explore two possible scenarios: when the two neurons have identical activation potential patterns (i.e., the same parameter values) and when the two neurons are nonidentical (they differ in the external current). We conclude that in both scenarios, there is a large set of coupling strength values such that synchronization is achieved even when the original trajectories of the neurons have markedly different active potential patterns.

Acknowledgments The first author acknowledges CONACYT, through the program 278 of Investigadoras e Investigadores por México.

References

1. M.K. Wouapi, B.H. Fotsin, E.B.M. Ngouonkadi, F.F. Kemwoue, and Z.T. Njitacke. Complex bifurcation analysis and synchronization optimal control for hindmarshârose neuron model under magnetic flow effect. *Cognitive Neurodynamics*, 15(2):315–347, 2020.
2. T. Wennekers and F. Pasemann. Generalized types of synchronization in networks of spiking neurons. *Neurocomputing*, 38–40:1037–1042, 2001.
3. R.M.G. Reinhart. Synchronizing neural rhythms. *Science*, 377(6606):588–589, 2022.
4. M. A. García-Vellisca, R. Jaimes-Reátegui, and A. N. Pisarchik. Chaos in neural oscillators induced by unidirectional electrical coupling. *Mathematical Modelling of Natural Phenomena*, 12(4):43–52, 2017.
5. A. N. Pisarchik, R. Jaimes-Reátegui, and M. A. Garcí-a-Vellisca. Asymmetry in electrical coupling between neurons alters multistable firing behavior. *Chaos: An Interdisciplinary Journal of Nonlinear Science*, 28(3):033605, 2018.
6. H. Lin, C. Wang, Q. Deng, C. Xu, Z. Deng, and C. Zhou. Review on chaotic dynamics of memristive neuron and neural network. *Nonlinear Dynamics*, 106(1):959–973, 2021.
7. D. Ivanov, A. Chezhegov, M. Kiselev, A. Grunin, and D. Larionov. Neuromorphic artificial intelligence systems. *Frontiers in Neuroscience*, 16:1–20, 2022.

8. J. Faridi and M. Kafeel. Memristor-a promising candidate for neural circuits in neuromorphic computing systems. *International Journal of Electrical and Computer Engineering*, 13(3):174–177, 2019.
9. L. O. Chua. Memristor-the missing circuit element. *IEEE Transactions on Circuit Theory*, 18(5):507–519, 1971.
10. Y. Wang, J. Ma, Y. Xu, F. Wu, and P. Zhou. The electrical activity of neurons subject to electromagnetic induction and Gaussian white noise. *International Journal of Bifurcation and Chaos*, 27(2):1750030, 2017.
11. B. Bao, A. Hu, H. Bao, Q. Xu, M. Chen, and H.Luo Wu. Three-dimensional memristive hindmarshârose neuron model with hidden coexisting asymmetric behaviors. *Complexity*, 2018:1–11, 2018.
12. K. Usha and P.A. Subha. Hindmarsh–rose neuron model with memristors. *Biosystems*, 178:1–9, 2019.
13. X.F. Wang and G. Chen. Synchronization in scale-free dynamical networks: robustness and fragility. *IEEE Transactions on Circuits and Systems I: Fundamental and Applications*, 49(1):54–62, 2002.
14. S. Mostaghimi, F. Nazarimehr, S. Jafari, and J. Ma. Chemical and electrical synapse-modulated dynamical properties of coupled neurons under magnetic flow. *Applied Mathematics and Computation*, 348:42–56, 2019.
15. J. Ma, L. Mi, P. Zhou, Y. Xu, and T. Hayat. Phase synchronization between two neurons induced by coupling of electromagnetic field. *Applied Mathematics and Computation*, 307:321–328, 2017.
16. Y. Xu, Y. Jia, J. Ma, A. Alsaedi, and B. Ahmad. Synchronization between neurons coupled by memristor. *Chaos, Solitons & Fractals*, 104:435–442, 2017.
17. H. Bao, Y. Zhang, W. Liu, and B. Bao. Memristor synapse-coupled memristive neuron network: synchronization transition and occurrence of chimera. *Nonlinear Dynamics*, 100(1):937–950, 2020.
18. J. L. Hindmarsh and R. M. Rose. A model of neuronal bursting using three coupled first order differential equations. *Proceedings of the Royal Society of London. Series B. Biological Sciences*, 221(1222):87–102, 1984.
19. M. Storace, D. Linaro, and E. de Lange. The hindmarsh–rose neuron model: Bifurcation analysis and piecewise-linear approximations. *Chaos: An Interdisciplinary Journal of Nonlinear Science*, 18(3):033128, 2008.
20. R. Barrio and A. Shilnikov. Parameter-sweeping techniques for temporal dynamics of neuronal systems: case study of hindmarsh-rose model. *The Journal of Mathematical Neuroscience*, 1(1):6, 2008.
21. R. Barrio, S. Ibáñez, and L. Pérez. Homoclinic organization in the hindmarsh–rose model: A three parameter study. *Chaos: An Interdisciplinary Journal of Nonlinear Science*, 30(5):053132, 2020.
22. S.K. Thottil and R.P. Ignatius. Nonlinear feedback coupling in hindmarsh–rose neurons. *Nonlinear Dynamics*, 87(3):1879–1899, 2016.
23. J.V. Uspensky. *Theory of equations*. McGraw-Hill, New York, 1976.
24. K. Yadav, A. Sharma, and M.D. Shrimali. Dynamics of nonlinear oscillators with time-varying conjugate coupling. *Indian Academy of Sciences - Conference Series*, 1(1):157–161, 2017.
25. A. Buscarino, M. Frasca, M. Branciforte, L. Fortuna, and J.C. Sprott. Synchronization of two Rössler systems with switching coupling. *Nonlinear Dynamics*, 88(1):673–683, 2016.

Chapter 9
A Systematic Approach for Multi-switching Compound Synchronization of Nonidentical Chaotic Systems Using Optimal Control

Jessica Zaqueros-Martinez, Gustavo Rodriguez-Gomez, and Felipe Orihuela-Espina

9.1 Introduction

Currently, there is no easy and systematic solution to the problem of synchronizing three or more chaotic systems. The existing solutions face issues such as lack of scalability when the number of chaotic systems increases or when the chaotic systems are nonidentical.

The synchronization of chaotic systems can be approached from two perspectives. One involves a master-slave problem, where the slave system follows the master system, and the other is based on bidirectional coupling, where all systems involved adapt to each other. Synchronizing two chaotic systems has become somewhat routine in both scenarios. However, extending this synchronization to three or more chaotic systems is feasible in specific cases but generally remains as an unsolved challenge.

The multi-system synchronization proposed to employ three chaotic systems, two as master systems and one as slave, and was called combination synchronization in the seminal work [1].

Conversely, the multi-switching synchronization scheme used distinct state variables from both master and slave systems to generate different synchronization

J. Zaqueros-Martinez (✉)
Computer Sciences, Instituto Nacional de Astrofísica, Óptica y Electrónica (INAOE), Puebla, Mexico

G. Rodriguez-Gomez
INAOE, Puebla, Mexico
e-mail: grodrig@inaoep.mx

F. Orihuela-Espina
University of Birmingham (UoB), Birmingham, UK
e-mail: f.orihuela-espina@bham.ac.uk

E. Campos-Cantón et al. (eds.), *Complex Systems and Their Applications*,
https://doi.org/10.1007/978-3-031-51224-7_9

errors. For the first time, this approach was proposed to synchronize two identical chaotic systems [2].

From these seminal works, other studies proposed mixing these approaches and increasing the number of involved chaotic systems [3–5]. Other synchronization schemes were developed like compound synchronization, which considers scaling and base drive systems [6, 7]. Some studies have even proposed hybrid methods to synchronize three or more chaotic master systems and one or more chaotic slave systems using a combination of multi-switching and compound-compound synchronization [8–11].

It is worth noting that the synchronization techniques mentioned above are primarily designed to work with either identical or different chaotic systems. However, when presenting numerical results, researchers typically opt for synchronizing identical systems because achieving synchronization with different systems often requires complex solutions. Additionally, the control methods utilized in these synchronization efforts are primarily based on active control or backstepping control, which rely on the Lyapunov or Hurwitz theorems to ensure convergence of synchronization errors, but their generalization can be challenging.

On the other hand, optimal control is a general framework that can be used to solve a variety of problems effectively [12], including scenarios involving tracking problems where chaotic synchronization problems are found. We suggest that this versatility of optimal control can be utilized to facilitate the synchronization of more than two chaotic systems.

In our study, we investigate multi-switching compound synchronization among nonidentical chaotic systems, employing optimal control in a systematic manner. A critical prerequisite for ensuring the convergence of synchronization errors through optimal control is that the slave chaotic systems must exhibit complete controllability. To confirm this condition, we propose linearizing the slave chaotic systems within a neighborhood of their equilibrium points and then obtaining the controllability matrix and calculating its rank.

Under this chapter, we committed to the following goals:

- The linearization of chaotic systems taken as slaves
- The design of controls to perform multi-switching compound synchronization by means of optimal control theory
- Evidencing the convergence to zero of the synchronization error

In this chapter, as we will show, our findings indicate that there are no theoretical limitations to employing optimal control in this class of synchronization. As anticipated, the numerical outcomes align with the theoretical framework.

This chapter is based on the report of Ramsay project with code 60389 at the University of Birmingham.

9.2 Multi-switching Compound Synchronization

The following notation is used throughout this chapter. Letters (a) denote scalar values, bold letters (**a**) denote vectors, and upper letters (A) denote a constant matrix.

The multi-switching compound synchronization involves a single scaling drive system, multiple base drive systems, and one response system [1, 8, 13]. Let the scaling drive systems be given by

$$\frac{\mathrm{d}\mathbf{x}}{\mathrm{d}t} = \mathbf{f}_1(\mathbf{x}). \tag{9.1}$$

The two base drive systems are

$$\frac{\mathrm{d}\mathbf{y}}{\mathrm{d}t} = \mathbf{f}_2(\mathbf{y}), \tag{9.2}$$

$$\frac{\mathrm{d}\mathbf{z}}{\mathrm{d}t} = \mathbf{f}_3(\mathbf{z}), \tag{9.3}$$

and the response system is given by

$$\frac{\mathrm{d}\mathbf{w}}{\mathrm{d}t} = \mathbf{g}(\mathbf{w}) + \mathbf{u}, \tag{9.4}$$

where $\mathbf{x} = (x_1, x_2, x_3, \ldots, x_n)^T, \mathbf{y} = (y_1, y_2, y_3, \ldots, y_n)^T, \mathbf{z} = (z_1, z_2, z_3, \ldots, z_n)^T$, and $\mathbf{w} = (w_1, w_2, w_3, \ldots, w_n)^T$ are state space vectors of the systems, $\mathbf{f}_1$, $\mathbf{f}_2$, $\mathbf{f}_3$, $\mathbf{g} : \mathbb{R}^n \to \mathbb{R}^n$ are continuous vector functions, and $\mathbf{u} = (u_1, u_2, u_3, \ldots, u_q)^T$ with $1 \le q \le n$ is a nonlinear control function. All the systems have the same number of state variables.

Definition 9.1 ([1, 8, 13]) The drive systems (9.1)–(9.3) are said to be in compound synchronization with the response system (9.4) if there exist four constant diagonal matrices A, B, C, and $D \in \mathbb{R}^n \to \mathbb{R}^n$ and $D \neq 0$ such that

$$\lim_{t\to\infty} \|\mathbf{e}\| = \lim_{t\to\infty} \|A\mathbf{x}\,(B\mathbf{y} + C\mathbf{z}) - D\mathbf{w}\| = 0,$$

where $\|\cdot\|$ is there matrix norm, $\mathbf{e} = AX(BY + CZ) - DW$ is the synchronization error vector and $X = \mathrm{diag}(x_1, x_2, x_3, \ldots, x_n)$, $Y = \mathrm{diag}(y_1, y_2, y_3, \ldots, y_n)$, $Z = \mathrm{diag}(z_1, z_2, z_3, \ldots, z_n)$, and $W = \mathrm{diag}(w_1, w_2, w_3, \ldots, w_n)$. The matrices A, B, C, and D are called scaling matrices.

Following the definition (9.1), synchronization errors are obtained as a diagonal matrix:

$$\mathbf{e} = \mathrm{diag}(e_1, e_2, e_3, \ldots, e_n) = AX(BY + CZ) - DW. \tag{9.5}$$

If $A = \text{diag}(\alpha_1, \alpha_2, \alpha_3, \ldots, \alpha_n)$, $B = \text{diag}(\beta_1, \beta_2, \beta_3, \ldots, \beta_n)$, $C = \text{diag}(\gamma_1, \beta_2, \gamma_3, \ldots, \gamma_n)$, and $D = \text{diag}(\delta_1, \delta_2, \delta_3, \ldots, \delta_n)$ and if errors (9.5) are operated by left multiplying both sides by a vector $v = (1, 1, \ldots, 1)_{1\times n}$, then the components of the error vector can be rewritten as

$$e_{ijkl} = \alpha_i x_i(\beta_j y_j + \gamma_k z_k) - \delta_l w_l. \tag{9.6}$$

Remark 1 ([8, 13]) In compound synchronization, the indices of the error states (9.6) are strictly chosen to satisfy $i = j = k = l$, with $i, j, k, l = 1, 2, \ldots, n$.

Definition 9.2 ([8, 13]) If error states (9.6) are redefined such that $i = j = k \neq l$ or $i = j = l \neq k$ or $i = k = l \neq j$ or $j = k = l \neq i$ or $i = j \neq k = l$ or $i = k \neq j = l$ or $i = l \neq j = k$ or $i = j \neq k \neq l$ or $i = k \neq j \neq l$ or $i = l \neq k \neq j$ or $i \neq j = k \neq l$ or $i \neq j \neq k = l$ or $i \neq k \neq j = l$ or $i \neq j \neq k \neq l$ and

$$\lim_{t\to\infty} e_{ijkl} = \lim_{t\to\infty} [\alpha_i x_i \left(\beta_j y_j + \gamma_k z_k\right) - \delta_l w_l] = 0.$$

where $i, j, k, l = 1, 2, \ldots, n$, then systems (9.1)–(9.4) are said to be in multi-switching compound synchronization.

Remark 2 ([8]) If $\alpha_i \neq 0$, $\beta_j = 0$, or $\gamma = 0$ and $\delta_l \neq 0$ for all $i, j, k, l = 1, 2, \ldots, n$, then it is possible to synchronize the compound of two drive systems and a response system using multi-switching.

9.3 Optimal Control

In optimal control theory, chaotic synchronization is viewed as a tracking problem where the master system is the reference to follow and the slave system is the process to be controlled.

The objective of optimal control theory is to determine signal control that satisfies physical restrictions while minimizing (or maximizing) a performance measure.

The formulation of a problem in terms of optimal control requires

- A mathematical model of the process to be controlled
- The proposal of physical constraints of the problem
- The specification of a performance measure

If a control $\mathbf{u} = [u_1, u_2, \ldots, u_q]$ with $1 \leq q \leq n$ satisfies physical restrictions on time interval $[t_0, t_f]$, then it is called the admissible control [12].

The problem of optimal control is to find an admissible control $\mathbf{u}^*$ that forces the system given by the nonlinear function $\mathbf{a}(\mathbf{x}(t), \mathbf{u}(t), t)$ with n-dimensional state vector $\mathbf{x}$

$$\frac{d\mathbf{x}}{dt}(t) = \mathbf{a}(\mathbf{x}(t), \mathbf{u}(t), t),$$

to follow an admissible trajectory $\mathbf{x}^*$ which minimizes the performance measure.

In this work, we utilize the next performance measure

$$J(\mathbf{u}(t)) = \int_{t_0}^{t_f} g(\mathbf{x}(t), \mathbf{u}(t), t)dt, \tag{9.7}$$

where g is defined by

$$\begin{aligned} g(\mathbf{x}(t), \mathbf{u}(t), t) =& \frac{1}{2}\Big((x_1 - r_1)^2 + (x_2 - r_2)^2 + (x_3 - r_3)^2 + \\ & R(u_1^2 + u_2^2 + u_3^2)\Big), \end{aligned} \tag{9.8}$$

where R is a positive constant; r_1, r_2, and r_3 are state variables of the master system; and x_1, x_2, and x_3 are state variables of the slave system.

To solve this problem, we need the Hamiltonian. In optimal control theory, the Hamiltonian $\mathcal{H}$ is defined by

$$\mathcal{H}(\mathbf{x}(t), \mathbf{u}(t), \mathbf{p}(t), t) = g(\mathbf{x}(t), \mathbf{u}(t), t) + \mathbf{p}^T(t)[\mathbf{a}(\mathbf{x}(t), \mathbf{u}(t), t)], \tag{9.9}$$

where $\mathbf{p} = [p_1, p_2, \ldots, p_n]$ are Lagrange multipliers also called co-states.

The necessary conditions to solve the optimal control problem are determined from the variational calculus and are those in the following equations:

$$\begin{aligned} \frac{d\mathbf{x}^*}{dt}(t) &= \frac{\partial \mathcal{H}}{\partial \mathbf{p}}(\mathbf{x}^*(t), \mathbf{u}^*(t), \mathbf{p}^*(t), t), \\ \frac{d\mathbf{p}^*}{dt}(t) &= \frac{\partial \mathcal{H}}{\partial \mathbf{x}}(\mathbf{x}^*(t), \mathbf{u}^*(t), \mathbf{p}^*(t), t), \\ 0 &= \frac{\partial \mathcal{H}}{\partial \mathbf{u}}(\mathbf{x}^*(t), \mathbf{u}^*(t), \mathbf{p}^*(t), t) \text{ with } t \in [t_0, t_f]. \end{aligned} \tag{9.10}$$

for all $t \in [t_0, t_f]$. In this research, the initial condition $\mathbf{x}(t_0)$ and the final time t_f are specified; the final state is free; the boundary conditions of this problem are given by

$$\begin{aligned} \mathbf{x}^*(t_0) &= \mathbf{x}_0, \\ \mathbf{p}^*(t_f) &= \mathbf{0}. \end{aligned} \tag{9.11}$$

So, the optimal control $\mathbf{u}$ is found by solving equations (9.10) and (9.11). In consequence, this problem becomes a two-point boundary value problem (TBVP) and needs to be solved with specific numerical methods for TBVP.

The sufficient condition that ensures that the Hamiltonian in (9.9) has a minimum in **u** and, in consequence, the performance measure (9.7) is also minimized into the optimal control problem is the Legendre-Clebsch condition

$$\frac{\partial^2 \mathcal{H}}{\partial \mathbf{u}^2}(\mathbf{x}^*(t), \mathbf{u}^*(t), \mathbf{p}^*(t), t) = \mathcal{H}_{\mathbf{uu}} > 0, \tag{9.12}$$

for all $t \in [t_0, t_f]$.

9.3.1 *Controllability*

Controllability is a requirement placed on a chaotic system when it is considered a slave system, with the aim of achieving chaotic synchronization through optimal control. Below, we provide the definition of this property.

Definition 9.3 ([12]) If there is a finite time $t_1 \geq t_0$ and a control $\mathbf{u}(t)$, $t \in [t_0, t_1]$, which transfers the estate $\mathbf{x}_0$ to the origin at time t_1, the state $\mathbf{x}_0$ is said to be *controllable at time* t_0. If all the values of $\mathbf{x}_0$ are controllable for all t_0, the system is *completely controllable* or simply *controllable*.

Also, let the continuous time system be

$$\frac{\mathrm{d}\mathbf{x}}{\mathrm{d}t} = A\mathbf{x} + B\mathbf{u},$$

where $\mathbf{x}$ is the vector of states (vector of dimension n), $\mathbf{u}$ is the control (dimension q), A is a matrix $n \times n$, and B is a matrix $n \times q$. A time-invariant linear system is controllable if and only if the matrix $n \times nq$

$$E = \left[B|AB|A^2B|\cdots|A^{n-1}B \right],$$

has rank n [14].

9.4 Global Error

When conducting numerical simulations, it is essential to take into account the errors arising from the use of numerical methods. A well-established result in the existing literature pertains to how the global error, denoted as $e_n(t)$, in the numerical approximation of a solution for ordinary differential equations with initial values is bounded by

$$|e_n(t)| < \{A\delta + Bh^p + Ch^q\}.$$

Here, the constants A, B, and C are contingent on the coefficients of the methods and the Lipschitz constant of ordinary differential equations, while δ represents the bound on initial conditions, as described in references such as [15, 16].

The bounds of the global error are then given in terms of the error in the initial value δ, the discretization or truncation error, and the roundoff error. The choice of the integration step ought to maintain a compromise between the later two. Besides, the error in the initial values in the floating point arithmetic ought to be of the same order as that of the integration method.

The limitations on the global error can be expressed in relation to the error in initial values (denoted as δ), the discretization or truncation error, and the roundoff error. When selecting the integration step, it is crucial to maintain a balance between the latter two factors. Furthermore, in the context of floating point arithmetic, the error in initial values should ideally be of the same order as that of the integration method.

For dynamic chaotic systems, the integration step (h) and the error in the representation become exceptionally significant due to their sensitivity to initial conditions. If one chooses an integration step that effectively controls the discretization error but fails to bound the growth of roundoff error, the latter can disrupt the chaotic system. This disruption may lead to outcomes such as the loss of chaos (referred to as *superstability*) or the system transitioning to a different attractor, thus altering the system's behavior.

Even variable step numerical integration methods with error control encounter challenges in determining an appropriate integration step to limit the growth of the error, like a phenomenon so-called *computational chaos*. This causes that systems that were originally non-chaotic may exhibit chaotic behavior in computational simulations. For instance, one study reported the emergence of computational chaos when employing the Runge-Kutta-Fehlberg method with error control and a variable step (`ode45` in MATLAB) to simulate a system modeling the coupling of two oscillators when this system is non-chaotic [17].

9.4.1 *Special Numerical Method*

In the literature, there are a number of special numerical methods that exploit the known characteristics or properties of the dynamical systems. Such is the case of the trigonometric polynomial methods (TPM) proposed by Gautschi [18].

TPM has been described and analyzed in detail in [16, 18], but briefly, it is an explicit method of trigonometric order $q = 1$ and algebraic order $p = 2$, given by

$$y_{n+2} - y_{n+1} = h(\beta_1 f_{n+1} + \beta_0 f_n),$$

where

$$\beta_0(\nu) = -\frac{1}{2}(1 + \frac{1}{12}\nu^2 + \frac{1}{120}\nu^4 + \ldots),$$
$$\beta_1(\nu) = \frac{3}{2}(1 - \frac{1}{4}\nu^2 + \frac{1}{120}\nu^4 + \ldots),$$

with $\nu = \omega h$, the constant integration step h, and the truncation error $\frac{5}{12}h^3(y^{(3)} + \omega^2 y^{(1)})$. If the frequency ω is unknown, it is safer to underestimate it than to overestimate it [16].

The TPM method has been used to integrate chaotic systems, giving competent results comparable to Runge-Kutta 4 (RK-4) but with a computational cost similar to forward Euler (FE) [19–22].

9.5 Chaotic Systems

For the purposes of exemplification of the proposed approach, we chose an illustrative scenario employing the Chen, Lorenz, and Lu as driving systems. These systems were chosen due to their extensive study and classification into three different families [23]. Notably, Chen and the Lu systems inspired the development of new methods for chaotic system analysis [24]. We designated the Rossler system as the slave system. Note that we claim that our methodology should work for other configurations.

The Chen system [25] is

$$\begin{aligned}
\frac{dx_1}{dt} &= a_1(x_2 - x_1), \\
\frac{dx_2}{dt} &= (c_1 - a_1)x_1 - x_1x_3 + c_1x_2, \\
\frac{dx_3}{dt} &= x_1x_2 - b_1x_3,
\end{aligned} \tag{9.13}$$

where $a_1, b_1,$ and c_1 are constant parameters.

Lorenz system [26] is

$$\begin{aligned}
\frac{dy_1}{dt} &= a_2(y_2 - y_1), \\
\frac{dy_2}{dt} &= b_2y_1 - y_1y_3 - y_2, \\
\frac{dy_3}{dt} &= y_1y_2 - c_2y_3,
\end{aligned} \tag{9.14}$$

where $a_2, b_2,$ and c_2 are constant parameters.

Lu system [27] is

$$\begin{aligned}\frac{dz_1}{dt} &= a_3(z_2 - z_1),\\ \frac{dz_2}{dt} &= b_3 z_2 - z_1 z_3,\\ \frac{dz_3}{dt} &= z_1 z_2 - c_3 z_3,\end{aligned} \tag{9.15}$$

where a_3, b_3, and c_3 are constant parameters.

Rossler system [28] with added controls is

$$\begin{aligned}\frac{dw_1}{dt} &= -(w_2 + w_3) + u_1,\\ \frac{dw_2}{dt} &= w_1 + a_4 w_2 + u_2,\\ \frac{dw_3}{dt} &= w_3(w_1 - c_4) + b_4 + u_3,\end{aligned} \tag{9.16}$$

where u_1, u_2, and u_3 are controls to be designed and a_4, b_4, and c_4 are constant parameters.

The sensitivity of chaotic systems to initial conditions can be measured using Lyapunov exponents.

Theorem 9.1 ([29]) *Assume that x_0 is an equilibrium point of the differential equation $dx/dt = f(x)$. Then the Lyapunov exponents at the equilibrium point are the real parts of the eigenvalues of the equilibrium point.*

The next definition is also useful to calculate the Lyapunov exponents.

Definition 9.4 ([30]) An equilibrium point p^* in phase (or state) space of an autonomous ordinary differential equation (ODE) is a point at which all derivatives of the state variables are zero, also known as a stationary point.

9.6 Synchronization Using Optimal Control

In this section, we present the proposed methodology to achieve synchronization in nonidentical chaotic systems through the use of optimal control:

1. *Verifying the controllability of the slave system* To verify the controllability of the slave system first is necessary to linearize it. For this purpose, we find the equilibrium points of the slave system. Then, we obtain the Jacobian matrix of the system and evaluate it at each equilibrium point. Next, we calculate the controllability matrix of each Jacobian matrix and obtain its rank. Finally, the system is controllable if the rank is complete according to Sect. 9.3.1.

2. *Obtaining numerical vectors of master systems using the TPM* To obtain numerical vectors of the new master system, we start by obtaining numerical vectors for each drive chaotic system. To achieve this, we propose to integrate each chaotic system individually using the TPM. In order to do this, we must first underestimate or determine the frequency of each chaotic system. Once we have the frequency, we can apply the TPM and obtain and save vectors of state variables for each system. Finally, we operate the resulting state vectors of all the drive systems involved, based on the positive term of switching state error equation (9.6).
3. *Designing the controller using optimal control theory and slave system equations* To design the control to synchronize multi-switching compound nonidentical chaotic systems employing optimal control theory and slave system equations, we propose to follow the next steps. First, we obtain the Hamiltonian (9.9) by considering the integrand of performance measure (9.8) and the system of equations of the slave chaotic system with added controls $d\mathbf{x}/dt = \mathbf{a}(\mathbf{x}, \mathbf{u}(t), t)$, as explained in Sect. 9.3. Second, we build the system equations (9.10) with boundary conditions (9.11). Finally, we solve this system equations to find the equations of co-states $\mathbf{p}$ and the seeking control $\mathbf{u}$.
4. *Performing synchronizations using a numerical method to solve TBVP* To perform synchronizations, we propose employing a specific numerical method to solve TVBP. First, we carefully select one validated numerical strategy. Then, using this numerical method, the obtained vector of the new master system, the found co-states, and the designed control, we realize the numerical simulations. Finally, we verify that synchronization errors converge to zero.

9.7 Results

In this section, we present the numerical outcomes of synchronization. To begin, we will outline the selected switching state errors to realize synchronizations in this research. Subsequently, we show the synchronization results achieved through optimal control and after with active control. Lastly, we conduct a comparative analysis of the outcomes obtained with these control strategies.

9.7.1 Multi-switching Compound Synchronization of Four Nonidentical Chaotic Systems

We realize the multi-switching compound synchronization with four chaotic systems using optimal control. The chaotic systems used are in Sect. 9.5. According to definition (9.2), different switching combinations exist for defining synchronization error states. For no particular reason, we select the next switching state errors

$$\begin{cases} e_{2121} = \alpha_2 x_2(\beta_1 y_1 + \gamma_2 z_2) - \delta_1 w_1, \\ e_{1232} = \alpha_1 x_1(\beta_2 y_2 + \gamma_3 z_3) - \delta_2 w_2, \text{ Switch 1} \\ e_{3313} = \alpha_3 x_3(\beta_3 y_3 + \gamma_1 z_1) - \delta_3 w_3, \end{cases} \tag{9.17}$$

and

$$\begin{cases} e_{1321} = \alpha_1 x_1(\beta_3 y_3 + \gamma_2 z_2) - \delta_1 w_1, \\ e_{3232} = \alpha_3 x_3(\beta_2 y_2 + \gamma_3 z_3) - \delta_2 w_2, \text{ Switch 2} \\ e_{2113} = \alpha_2 x_2(\beta_1 y_1 + \gamma_1 z_1) - \delta_3 w_3, \end{cases} \tag{9.18}$$

The value of the scaling parameters in switched error states are $\alpha_1 = \alpha_2 = \alpha_3 = 1$, $\beta_1 = \beta_2 = \beta_3 = 1$, $\gamma_1 = \gamma_2 = \gamma_3 = 1$, and $\delta_1 = \delta_2 = \delta_3 = 1$. To carry out the numerical simulations, we use MATLAB R2021b (The MathWorks, Inc. Natick, MA, USA) on a personal computer with machine precision 2.2204×10^{-16}. A free implementation of the synchronizations can be downloaded at https://github.com/Jessica-ZM/MultiSwitchingSynch#multiswitchingsynch.

9.7.2 *Initial Conditions and Equilibrium Points for the Chaotic Systems*

Next, we present the parameters and initial conditions of the chaotic systems utilized in this research, which are listed in Table 9.1. The selection of parameters and initial conditions for the driving chaotic systems is based on sources provided in Table 9.1 to ensure the generation of chaotic attractors. We use TPM with an underestimated frequency $\omega = 10$ and step $h = 1/1000$ to integrate each chaotic system and $t_f = 10$. Phase spaces of each drive chaotic system are in Fig. 9.1.

Secondly, the equilibrium points of each drive system are detailed in Table 9.2. As per Theorem 9.1, the Lyapunov exponents for the Chen, Lorenz, and Lu systems at each equilibrium point can also be found in Table 9.2.

Table 9.1 Parameters and initial conditions used in each chaotic system. Those values were taken from [8], or we specified the source in each value

Chaotic systems	Parameters	Initial conditions
Chen	$a_1 = 35, b_1 = 3, c_1 = 28$	$(x_1(0), x_2(0), x_3(0)) = (0.5, 1, 5)$ [31]
Lorenz	$a_2 = 10, b_2 = 28, c_2 = 8/3$	$(y_1(0), y_2(0), y_3(0)) = (10, 10, 10)$ [32]
Lu	$a_3 = 36, b_3 = 20, c_3 = 3$	$(z_1(0), z_2(0), z_3(0)) = (12, 26, 36)$ [33]
Rossler	$a_4 = 0.2, b_4 = 0.2, c_4 = 5.7$	$(w_1(0), w_2(0), w_3(0)) = (40, -20, 30)$

Fig. 9.1 Phase space of selected drive chaotic systems. (**a**) Chen. (**b**) Lorenz. (**c**) Lu

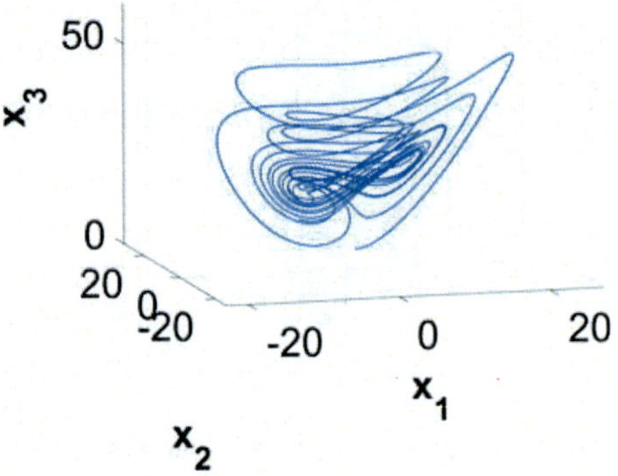

(a)

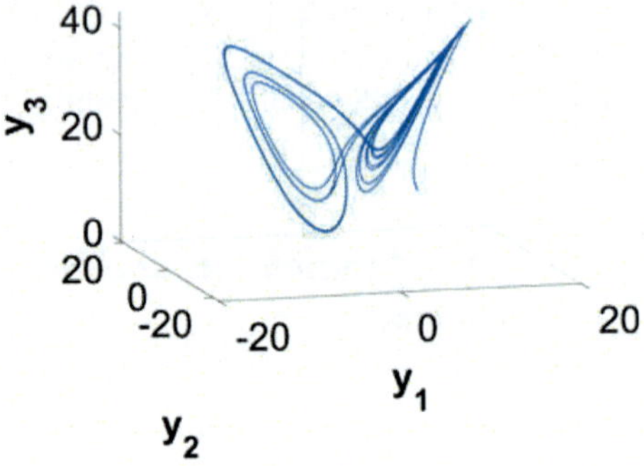

(b)

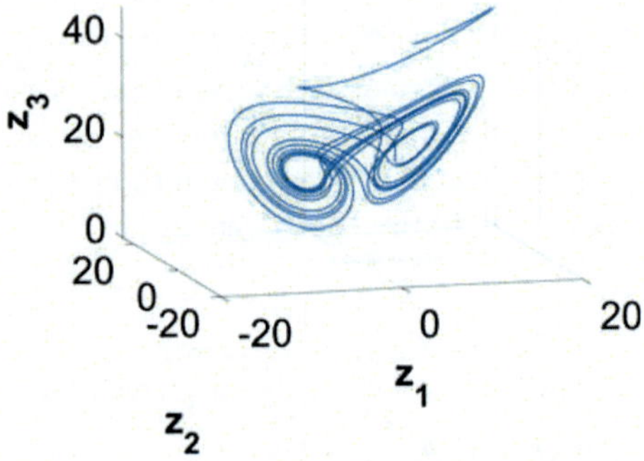

(c)

Table 9.2 Equilibrium points, eigenvalues, and Lyapunov exponents of drive systems

Chaotic system	Equilibrium points	Eigenvalues	Lyapunov exponents L_1	L_2	L_3
Chen	$(0, 0, 0)$	$\{-30.836, 23.836, -3.000\}$	-30.836	23.856	-3.000
	$(-3\sqrt{7}, -3\sqrt{7}, 21)$	$\{-18.428, 4.214 + 14.885i, 4.214 - 14.885i\}$	-18.428	4.214	4.214
	$(3\sqrt{7}, 3\sqrt{7}, 21)$	$\{-18.428, 4.214 + 14.885i, 4.214 - 14.885i\}$	-18.428	4.214	4.214
Lorenz	$(0, 0, 0)$	$\{-22.828, 11.828, -2.667\}$	-22.828	11.828	-2.667
	$(-6\sqrt{2}, -6\sqrt{2}, 27)$	$\{-13.855, 0.094 + 10.194i, 0.094 - 10.194i\}$	-13.855	0.094	0.094
	$(6\sqrt{2}, 6\sqrt{2}, 27)$	$\{-13.855, 0.094 + 10.194i, 0.094 - 10.194i\}$	-13.855	0.094	0.094
Lu	$(0, 0, 0)$	$\{-36, 20, -3\}$	-36.000	20.000	-3.000
	$(-2\sqrt{15}, -2\sqrt{15}, 20)$	$\{-22.652, 1.826 + 13.689i, 1.826 - 13.689i\}$	-22.652	1.826	1.826
	$(2\sqrt{15}, 2\sqrt{15}, 20)$	$\{-22.652, 1.826 + 13.689i, 1.82 - 13.689i\}$	-22.652	1.826	1.826

9.7.3 *Optimal Control Synchronizations*

Following the proposed methodology in Sect. 9.6, we present numerical simulations.

9.7.3.1 Controllability of the Slave System

Since the Rossler system is the slave system, we will verify if it is controllable. Firstly, we compute the equilibrium points of the Rossler system (9.16) without added controls; those are

$$\mathbf{eq}_1 = \left(\frac{4}{5(57 + \sqrt{3233})}, \frac{4}{-57 - \sqrt{3233}}, \frac{4}{57 + \sqrt{3233}}\right),$$

and

$$\mathbf{eq}_2 = \left(\frac{57 + \sqrt{3233}}{20}, \frac{-57 - \sqrt{3233}}{4}, \frac{57 + \sqrt{3233}}{4}\right).$$

Next, we linearized the system through the Jacobian matrix, which is

$$\mathcal{J}(w_1, w_2, w_3) = \begin{pmatrix} 0 & -1 & -1 \\ 1 & \frac{1}{5} & 0 \\ z & 0 & -\frac{57}{10} + w_1 \end{pmatrix}. \tag{9.19}$$

Following, we evaluate each equilibrium point in the Jacobian matrix (9.19), we use MATLAB for this step, and we have the following matrices:

$$\mathcal{J}(\mathbf{eq}_1) = \begin{pmatrix} 0 & -1 & -1 \\ 1 & \frac{1}{5} & 0 \\ \frac{4}{57+\sqrt{3233}} & 0 & -\frac{57}{10} + \frac{4}{5(57+\sqrt{3233})} \end{pmatrix},$$

and

$$\mathcal{J}(\mathbf{eq}_2) = \begin{pmatrix} 0 & -1 & -1 \\ 1 & \frac{1}{5} & 0 \\ \frac{57+\sqrt{3233}}{4} & 0 & -\frac{57}{10} + \frac{57+\sqrt{3233}}{20} \end{pmatrix}.$$

The linearized slave system at each equilibrium point has the following form $\frac{d\mathbf{w}}{dt} = A\mathbf{w} + B\mathbf{u}$ where $A = \mathcal{J}(\mathbf{eq}_1)$ or $A = \mathcal{J}(\mathbf{eq}_2)$ and B is the identity matrix.

The next step is to calculate the controllability matrices of the linearized system in each equilibrium point. Those are

$$\begin{pmatrix} 1 & 0 & 0 & 0 & -1 & -1 & -1.0351 & -0.2 & 5.6930 \\ 0 & 1 & 0 & 1 & 0.2 & 0 & 0.2 & -0.9600 & -1 \\ 0 & 0 & 1 & 0.0351 & 0 & -5.6930 & -0.2 & -0.0351 & 32.3748 \end{pmatrix},$$

and

$$\begin{pmatrix} 1 & 0 & 0 & 0 & -1 & -1 & -29.4649 & -0.2 & 0.0070 \\ 0 & 1 & 0 & 1 & 0.2 & 0 & 0.2 & -0.96 & -1 \\ 0 & 0 & 1 & 28.4649 & 0 & -0.007 & -0.2 & -28.4649 & -28.4649, \end{pmatrix},$$

respectively. The rank of both matrices is 3. According to Definition 9.3, the Rossler system is controllable at each equilibrium point.

9.7.3.2 Numerical Vectors of Master Systems Using the TPM

Following the second part of the methodology and taking account the equations of switching state errors (9.17) and (9.18), we obtain new master systems. The phase spaces are plotted in Fig. 9.2. In this case, we have two new master systems corresponding with each switching state error.

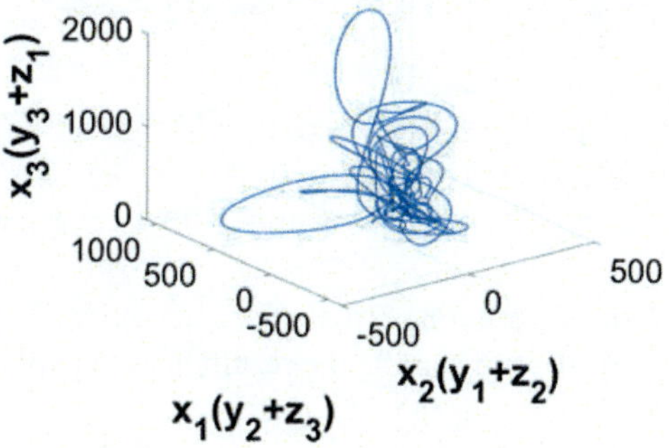

(a)

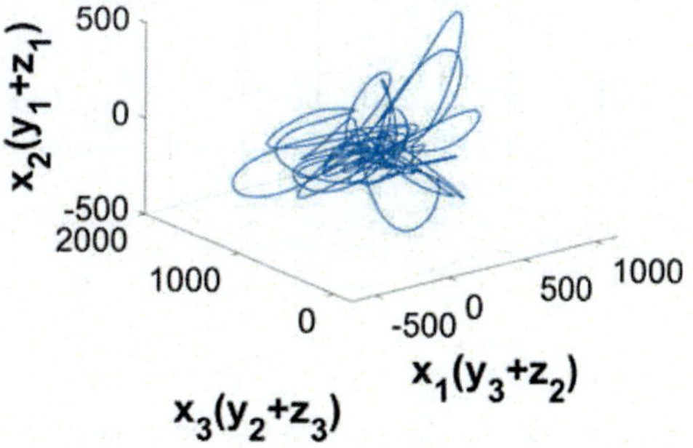

(b)

Fig. 9.2 Phase space of new master systems. (**a**) Master system associated with switch 1. (**b**) Master system associated with switch 2

9.7.3.3 Optimal Controller Design

Continuing with the methodology, our next step involves the design of optimal control. To achieve this, we calculate the Hamiltonian as outlined in Sect. 9.3, which is

$$\begin{aligned}\mathcal{H}(\mathbf{x}(t), \mathbf{u}(t), \mathbf{p}(t), t) =& \frac{1}{2}\Big((x_1 - r_1)^2 + (x_2 - r_2)^2 + (x_3 - r_3)^2 + \\ & R(u_1^2 + u_2^2 + u_3^2)\Big) + p_1(-(w_2 + w_3) + u_1) \\ & + p_2(w_1 + a_4 w_2 + u_2) \\ & + p_3(w_3(w_1 - c_4) + b_4 + u_3).\end{aligned} \tag{9.20}$$

Next, solving TVBP system equations (9.10) and (9.11), we obtain the following co-states:

$$\frac{dp_1}{dt} = -p_2 + r_1 - w_1 - p_3 w_3,$$
$$\frac{dp_2}{dt} = p_1 - a_4 p_2 + r_2 - w_2, \tag{9.21}$$
$$\frac{dp_3}{dt} = p_1 + r_3 - p_3(-c_4 + w_1) - w_3.$$

Now, substituting Rossler system (9.16), Hamiltonian (9.20), and co-states (9.21) in (9.10) and solving the resulting system of equations, the control is

$$u_1 = -p_1/R,$$
$$u_2 = -p_2/R, \tag{9.22}$$
$$u_3 = -p_3/R.$$

Moreover, substituting the found control (9.22) in Hamiltonian (9.20) and calculating the hessian of the Hamiltonian, we obtain

$$\mathcal{H}_{\mathbf{uu}} = \begin{pmatrix} 1/R & 0 & 0 \\ 0 & 1/R & 0 \\ 0 & 0 & 1/R \end{pmatrix},$$

where $\mathcal{H}_{\mathbf{uu}} > 0$ because the parameter $R > 0$. Hence, the Hamiltonian (9.20) satisfies the Legendre-Clebsch condition (9.12).

9.7.3.4 Numerical Results

The next step of the methodology is to perform the numerical simulations. For this purpose, we use the numerical method `bvp4c` included in MATLAB because it is validated and is specific to solving boundary value problems.

Considering the switching state errors (9.17), we implement and execute the synchronizations. The resulting phase space and state variable plots are depicted in Fig. 9.3, where the slave system follows the drive system. Further, this is confirmed with relative errors of the synchronization, as presented in Fig. 9.4. From the first time unit of simulation, the synchronization errors decrease to less than 10^{-5}. Analogously, we present the phase space and state variable plots of the synchronizations involving the switching state errors (9.18). These plots are presented in Fig. 9.5 showing that the slave system follows the new drive system. The relative synchronization errors are displayed in Fig. 9.6. As in the case above, the synchronization errors become less than 10^{-5} from the first time unit of the simulation.

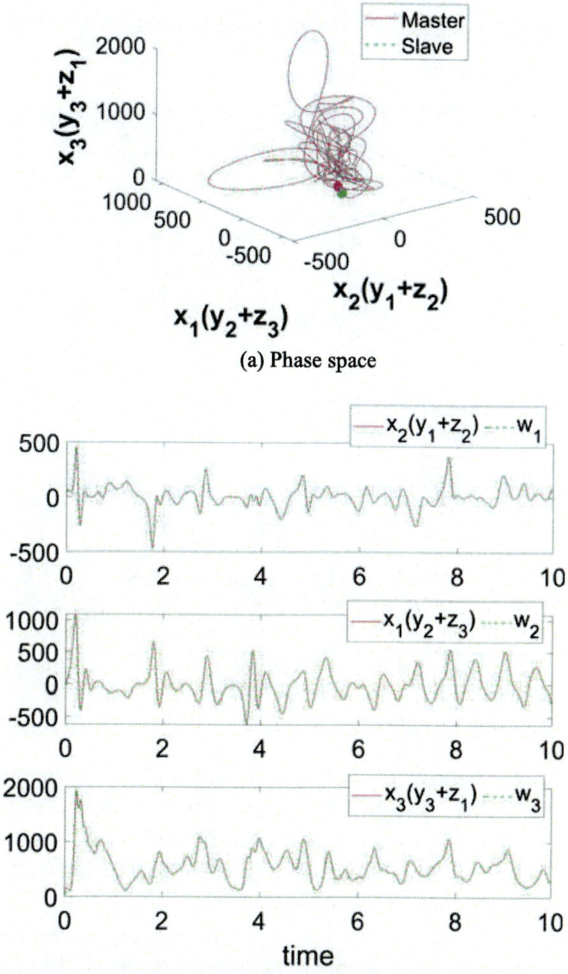

(a) Phase space

(b) State variables

Fig. 9.3 Phase space and state variables for synchronization of switch 1 using optimal control. (**a**) In phase space, the slave system (green dotted line) follows the master system (magenta solid line). (**b**) Each state variable of the slave system (green dotted line) follows the master system (magenta solid line)

9.7.4 Active Control Synchronizations

Because synchronization using active control is widely studied [1, 8, 13], in this section, we briefly explain the steps to carry it out. Following the active control designed in [8], we synchronize the selected four chaotic systems Sect. 9.5 with the same parameters and initial conditions reported in Table 9.1. Steps to synchronize with active control can be summarized as

1. *Selecting one switched state error.* As mentioned in Sect. 9.7.1, we selected the switching state errors (9.17) and (9.18) to realize the synchronizations.

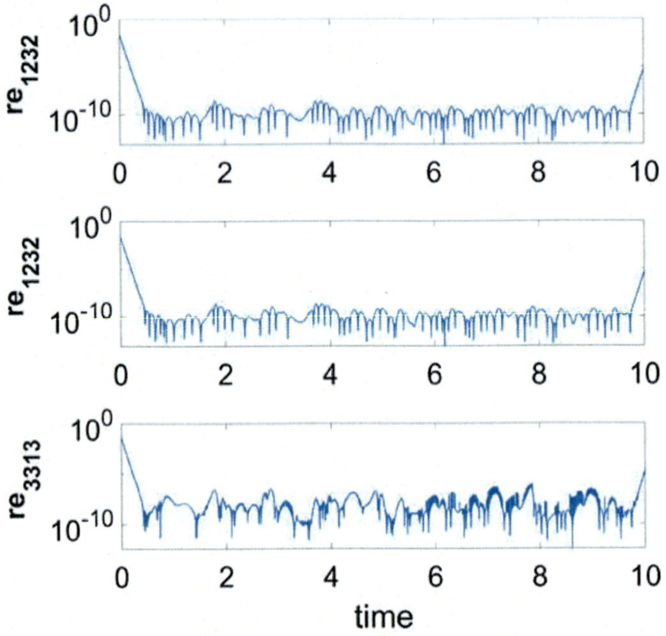

Fig. 9.4 Relative errors of synchronizations of switch 1 using optimal control. The synchronization errors decrease to less than 10^{-5} from the first time of simulation

2. *Designing the controller for each selected switching state error.* To achieve this, we take into account the equations of drive and slave chaotic systems given in Sect. 9.5. Hence, using the derivative of each switching state error and a specific Lyapunov function, it is possible to find the control to achieve multi-switching compound synchronization.
3. *Proving that this controller allowed achieving multi-switching compound synchronization using a Lyapunov function.* By construction, this condition is satisfied.
4. *Performing numerical simulations using TPM.* To realize the numerical simulations, we propose to use TPM using an underestimated frequency $\omega = 10$ and integration step $h = 1/1000$.

Now, applying the methodology described above involving switching state errors (9.17), we obtain the next results. The phase space and state variable plots are depicted in Fig. 9.7 where the slave system follows the drive system. The relative synchronization errors are plotted in Fig. 9.8. In this case, the synchronization errors become less than 10^{-5} from the eighth time unit of simulation.

Similarly, we realize the synchronization using switching state errors (9.18). The phase space and variable state plots are in Fig. 9.9; here, the slave system follows the master system. The relative synchronization errors are depicted in Fig. 9.10. The synchronization errors need all time of simulation to become less than 10^{-5}.

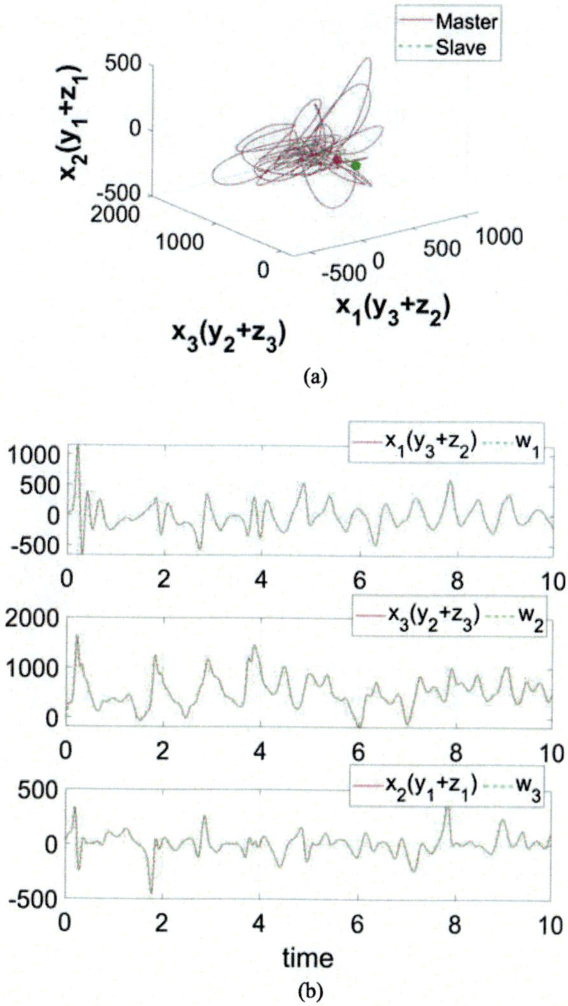

Fig. 9.5 Phase space and state variables for synchronization of switch 2 using optimal control. In both plots, the slave system (green dotted line) follows the master system (magenta solid line). (**a**) Phase space. (**b**) State variables

9.7.5 Comparison

The relative synchronization error plots presented oscillations in both controls, optimal and active. However, the oscillations using the active control are quicker than the optimal control. This implies that active control requires greater effort compared to optimal control. The above may affect actual applications. For example, these oscillations can cause premature wear of the gears of a motor.

Besides, we calculate the time from which all the relative switching state errors of each synchronization are less than 2%. For Switch 1 using optimal control, the time was 0.02 time units and using active control was 1.6760 time units, whereas for Switch 2 using optimal control the time was 0.05 time units and using active control

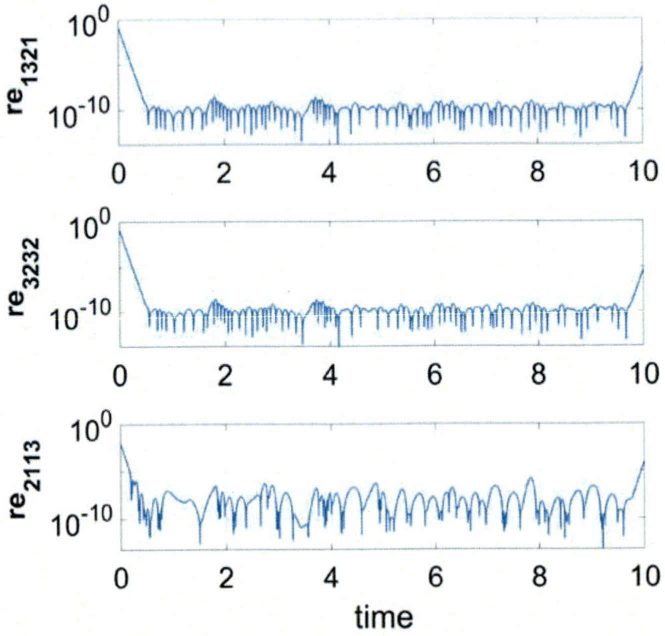

Fig. 9.6 Relative errors of synchronizations of switch 2 using optimal control. Analogous to switch 1, the synchronization errors need one unit of time to become less than 10^{-5}

was 2.7390 time units. We also compare the convergence time of active control and optimal control, taking as 100% the time obtained from active control. We found that optimal control only needs 1.82% of active control time or less to converge. These percentages are reported in Table 9.3.

9.8 Discussion

In this study, we have established how to use the optimal control in achieving multi-switching compound synchronization. One of the key insights of our work is that a single control design can be adapted for use across various switching state error configurations.

We have also elucidated a theoretical approach to derive the optimal control design, ensuring the convergence of synchronization errors to zero. However, when transitioning from theory to practical implementation, we encounter certain challenges, particularly in the realm of numerical simulations. These challenges often stem from the sensitivity of the numerical methods employed for synchronization. To address this, one practical solution is to explore alternative numerical methods for solving boundary value problems.

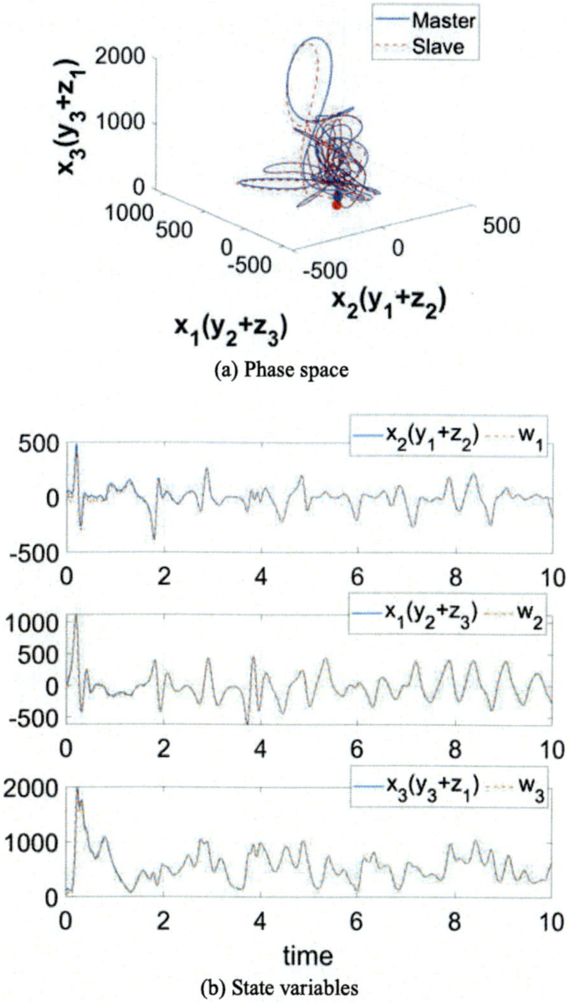

Fig. 9.7 Phase space and state variables for synchronization of switch 1 using active control. (**a**) The slave system (orange dotted line) follows the master system (blue solid line), but there is a separation between the two systems at the beginning of the simulation. This is confirmed in the plots of the first and third state variables in plots of state variables in (**b**)

On the other hand, we have synchronized with active control with the aim of comparing their results with optimal control results. With active control, numerical simulations are straightforward to implement, and we have proposed utilizing specialized numerical methods like TPM. However, the primary drawback of employing this control method is that whenever the switching state error configuration is altered, a new controller must be designed, and its convergence must be demonstrated. This necessitates a repetitive and resource-intensive process that can be a significant limitation.

In this study, we have intentionally omitted the consideration of simulation runtime, as it falls outside the scope of our project objectives. However, it is important to note that in certain practical applications, simulation runtime can

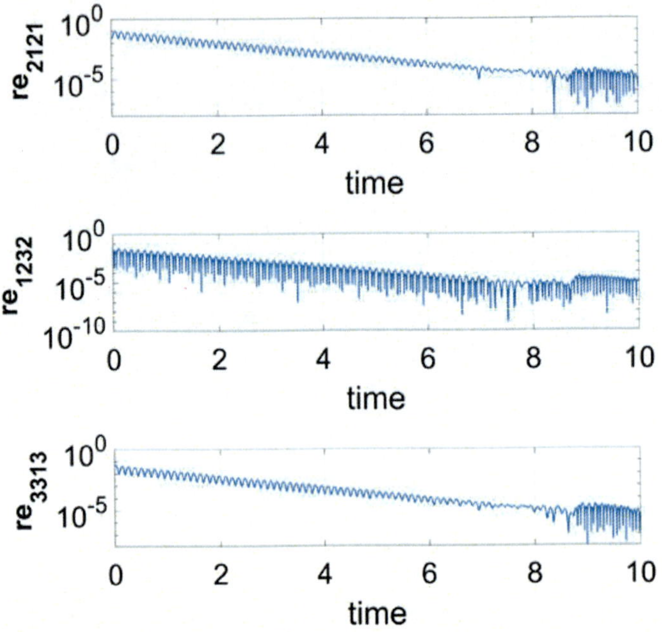

Fig. 9.8 Relative errors of synchronizations of switch 1 using active control. The synchronization errors become less than 10^{-5} after 8 units of time. These plots present quick oscillations compared with error plots using optimal control

be a critical factor. Specifically, for the configurations employed in this project, simulations conducted using active control take about 1 minute, whereas those utilizing optimal control require around 4 minutes to run.

9.9 Conclusions

In this research, we have explored the application of optimal control to achieve the synchronization of nonidentical chaotic systems through a multi-switching compound approach. Our investigation involved both theoretical analysis and numerical simulations.

The application of optimal control theory has demonstrated its efficacy in ensuring the convergence of synchronization errors, a fact substantiated by our numerical simulations. Additionally, we compared these results with those obtained using active control synchronization, a widely employed method in this field of research. Our findings indicate that the use of optimal control leads to a remarkable

Fig. 9.9 Phase space and state variables for synchronization of switch 2 using active control. (**a**) Phase space of the slave system (orange dotted line) and the master system (blue solid line). (**b**) Plots of state variables of both systems. For the first two units of time of the simulation, the slave system exhibits different behavior from the master system until the control forces it to follow. So in the phase space, it appears as a separation between the two systems

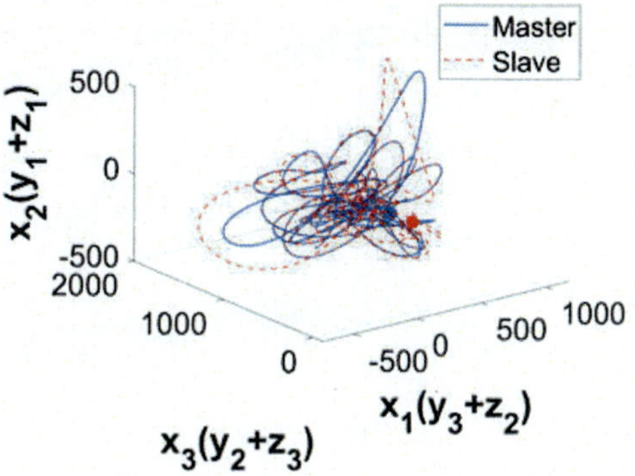

(a) Phase space

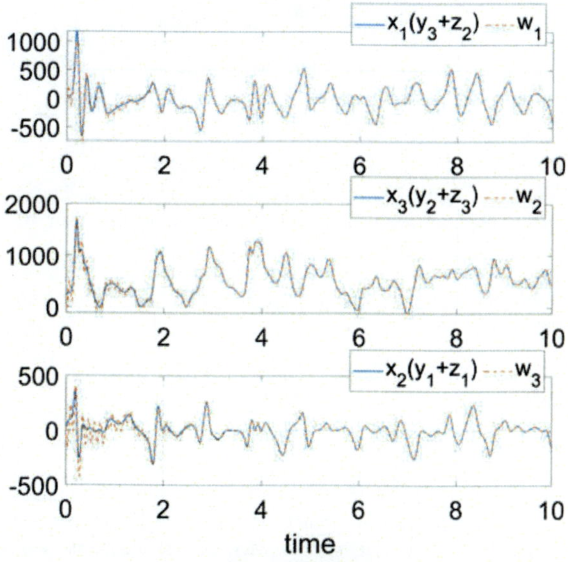

(b) State variables

improvement, with synchronization errors converging up to 98.81% faster than when employing active control, as observed in the studies reviewed within this chapter.

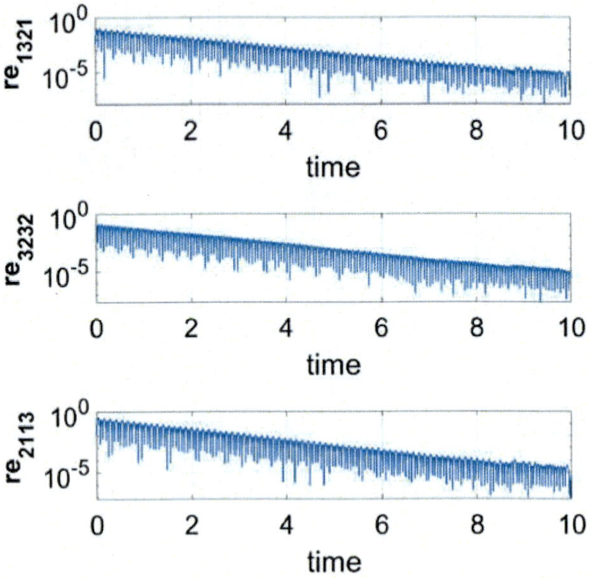

Fig. 9.10 Relative errors of synchronizations of switch 2 using active control. The synchronization errors need all simulation time to decrease to less than 10^{-5}. These plots also present quick oscillations

Table 9.3 Comparison of convergence of synchronization errors using optimal control and active control

	Optimal control	Active control
Switch 1	1.19%	100%
Switch 2	1.82%	100%

Acknowledgments We want to acknowledge the generosity of Paul and Yuanbi Ramsay Research Fund. Thanks to that, student Mrs. Jessica Zaqueros was able to spend 10 -week research secondment at the University of Birmingham to realize the work presented in this chapter.

References

1. Luo Runzi, Wang Yinglan, and Deng Shucheng. Combination synchronization of three classic chaotic systems using active backstepping design. *Chaos: An Interdisciplinary Journal of Nonlinear Science*, 21(4):043114, 10 2011.
2. Ahmet Uçar, Karl E. Lonngren, and Er-Wei Bai. Multi-switching synchronization of chaotic systems with active controllers. *Chaos, Solitons & Fractals*, 38(1):254–262, 2008.
3. Zhaoyan Wu and Xinchu Fu. Combination synchronization of three different order nonlinear systems using active backstepping design. *Nonlinear Dynamics*, 73(3):1863–1872, August 2013.

4. Song Zheng. Multi-switching combination synchronization of three different chaotic systems via nonlinear control. *Optik*, 127(21):10247–10258, 2016.
5. Ayub Khan, Mridula Budhraja, and Aysha Ibraheem. Multi-switching Synchronization of Four Non-identical Hyperchaotic Systems. *International Journal of Applied and Computational Mathematics*, 4(2):71, March 2018.
6. Junwei Sun, Yi Shen, Quan Yin, and Chengjie Xu. Compound synchronization of four memristor chaotic oscillator systems and secure communication. *Chaos: An Interdisciplinary Journal of Nonlinear Science*, 23(1):013140, 03 2013.
7. Ailong Wu and Jine Zhang. Compound synchronization of fourth-order memristor oscillator. *Advances in Difference Equations*, 2014(1):1–16, April 2014.
8. Nitish Prajapati, Ayub Khan, and Dinesh Khattar. On multi switching compound synchronization of non identical chaotic systems. *Chinese Journal of Physics*, 56(4):1656–1666, 2018.
9. Ayub Khan, Mridula Budhraja, and Aysha Ibraheem. Multi-switching compound synchronization of four different chaotic systems via active backstepping method. *International Journal of Dynamics and Control*, 6(3):1126–1135, September 2018.
10. A. Khan, D. Khattar, and N. Agrawal. Anti Difference Multiswitching Compound–Compound Combination Synchronization of Seven Chaotic Systems. *Differential Equations and Dynamical Systems*, September 2021. Sin número de pp.
11. Ayub Khan, Mridula Budhraja, and Aysha Ibraheem. Multiswitching compound–compound synchronisation of six chaotic systems. *Pramana*, 91(6):73, September 2018.
12. Donald E. Kirk. *Optimal Control Theory, An Introduction*. Dover Publications, Inc., 1970.
13. U. E. Vincent, A. O. Saseyi, and P. V. E. McClintock. Multi-switching combination synchronization of chaotic systems. *Nonlinear Dynamics*, 80(1):845–854, April 2015.
14. Katsuhiko Ogata. *Ingeniería de control moderna*. Pearson Educación, Madrid, Espaa, 5 edition, 2010. ISBN: 978-84-8322-660-5.
15. Peter Henrici. *Discrete Variable Methods in Ordinary Differential Equations*. John Wiley & Sons, Inc., Hoboken, NJ, USA,, 1962.
16. J. D. Lambert. *Computational Methods in Ordinary Differential Equations*. John Wiley & Sons, Inc., Hoboken, Nueva Jersey, EUA, 1973.
17. Joseph D. Skufca. Analysis still matters: A surprising instance of failure of Runge–Kutta–Fehlberg ode solvers. *SIAM Review*, 46(4):729–737, 2004.
18. W Gautschi. Numerical integration of ordinary differential equations based on trigonometric polynomials. *Numer.\ Math.*, 3:381–397, 1961.
19. A D Pano-Azucena, E Tlelo-Cuautle, G Rodriguez-Gomez, and L G de la Fraga. FPGA-based implementation of chaotic oscillators by applying the numerical method based on trigonometric polynomials. *AIP Advances*, 8(7):75217, 2018.
20. Jessica Zaqueros-Martinez, Gustavo Rodriguez-Gomez, Esteban Tlelo-Cuautle, and Felipe Orihuela-Espina. Trigonometric polynomials methods to simulate oscillating chaotic systems. *AIP Conference Proceedings*, 2425(1):420035, 2022.
21. Jessica Zaqueros-Martinez, Gustavo Rodriguez-Gomez, Esteban Tlelo-Cuautle, and Felipe Orihuela-Espina. *Synchronization of Chaotic Electroencephalography (EEG) Signals*, pages 83–108. Springer International Publishing, Cham, 2022.
22. Jessica Zaqueros-Martinez, Gustavo Rodriguez-Gomez, Esteban Tlelo-Cuautle, and Felipe Orihuela-Espina. Fuzzy synchronization of chaotic systems with hidden attractors. *Entropy*, 25(3), 2023.
23. A. Vaněček and S. Čelikovský. *Control Systems: From Linear Analysis to Synthesis of Chaos*. Prentice-Hall international series in systems and control engineering. Prentice Hall, 1996.
24. G.A. Leonov and N.V. Kuznetsov. On differences and similarities in the analysis of Lorenz, Chen, and Lu systems. *Applied Mathematics and Computation*, 256:334–343, 2015.
25. Guanrong Chen and Tetsushi Ueta. Yet another chaotic attractor. *International Journal of Bifurcation and Chaos*, 9(7):1465–1466, 1999.
26. Edward N. Lorenz. Deterministic nonperiodic flow. *Journal of Atmospheric Sciences*, 20(2):130–141, 1963.

27. Jinhu Lü and Guanrong Chen. A new chaotic attractor coined. *International Journal of Bifurcation and chaos*, 12(03):659–661, 2002.
28. O.E. Rössler. An equation for continuous chaos. *Physics Letters A*, 57(5):397–398, 1976.
29. Huaguang Zhang, Derong Liu, and Zhiliang Wang. *Controlling Chaos: Suppression, Synchronization and Chaotification.* Springer London, 2009.
30. William E. Boyce and Richard C. DiPrima. *Elementary differential equations and boundary value problems.* John Wiley and Sons, 7 edition, 2001. ISBN: 0-471-31999-6.
31. Yanwu Wang, Zhi-Hong Guan, and Xiaojiang Wen. Adaptive synchronization for Chen chaotic system with fully unknown parameters. *Chaos, Solitons & Fractals*, 19(4):899–903, 2004.
32. Teh-Lu Liao and Sheng-Hung Lin. Adaptive control and synchronization of lorenz systems. *Journal of the Franklin Institute*, 336(6):925–937, 1999.
33. V Sundarapandian and R Karthikeyan. Anti-synchronization of lü and pan chaotic systems by adaptive nonlinear control. *European Journal of Scientific Research*, 64(1):94–106, 2011.

Chapter 10
Limit Cycle Generation by Inducing the Controllable Hopf Bifurcation

Jesus R. Pulido-Luna, Nohe R. Cazarez-Castro, Selene L. Cardenas-Maciel, and Jorge A. López-Rentería

10.1 Introduction

In the study of dynamical systems, many scientists address their research into the analysis of the topological structures of the solutions, as well as how they change their structure as a parameter varies through a critical value. This phenomenon is called *bifurcation*, and the parameter is said to be the *bifurcation parameter* [1]. In literature, there exist several types of bifurcations, for example, the pitchfork bifurcation, the saddle-node bifurcation, the transcritical bifurcation, and the Poincaré-Andronov-Hopf bifurcation (commonly known as the Hopf bifurcation) [2], to mention the most studied. Of the aforementioned bifurcations, the most utilized and studied is the Hopf bifurcation, which refers to the appearance or vanishing of periodic orbits.

A periodic orbit is a behavior in the solution trajectories of a system, with the property of being closed curves. If every trajectory in a neighborhood around the

J. R. Pulido-Luna · N. R. Cazarez-Castro
Departamento de Ingeniería Eléctrica y Electrónica, Tecnológico Nacional de México – Instituto Tecnológico de Tijuana, Calzada Tecnológico S/N, Fraccionamiento Tomas Aquino, Tijuana, México
e-mail: rogelioxpulido@ieee.org; nohe@ieee.org

S. L. Cardenas-Maciel
Departamento de Ciencias Básicas, Tecnológico Nacional de México – Instituto Tecnológico de Tijuana, Calzada Tecnológico S/N, Fraccionamiento Tomas Aquino, Tijuana, México
e-mail: lilettecardenas@ieee.org

J. A. López-Rentería (✉)
Departamento de Ingeniería Eléctrica y Electrónica, CONAHCYT/Tecnológico Nacional de México – Instituto Tecnológico de Tijuana, Calzada Tecnológico S/N, Fraccionamiento Tomas Aquino, Tijuana, México
e-mail: jorge.lopez@tectijuana.edu.mx

E. Campos-Cantón et al. (eds.), *Complex Systems and Their Applications*,
https://doi.org/10.1007/978-3-031-51224-7_10

periodic orbit approaches it as the time tends to infinity, it is called a stable or attractive periodic orbit or a stable limit cycle. If the phenomenon occurs as the time approaches negative infinity, then it is said to be an unstable limit cycle. Similarly, if there is a set of trajectories in the neighborhood that approaches the periodic orbit as the time approaches infinity and a set of trajectories that approaches the periodic orbit as the time approaches minus infinity, then it is called a semi-stable periodic orbit or a semi-stable limit cycle [3].

On the other hand, in a differential equation system, if a pair of complex conjugate eigenvalues travel through the complex plane, as the bifurcation parameter varies, crossing the imaginary axis and changing the stability of the equilibrium point, then a unique limit cycle emerges from the equilibrium in an interval around the critical parameter value, and then we say that the system undergoes the Hopf bifurcation in the beforementioned interval. This implies that the Hopf bifurcation can only take place in systems of dimension two or higher [4]. The Hopf bifurcation can be found naturally in systems coming from biology phenomena [5], chemistry [6], pattern formation [7], and chaotic systems [8], among many others [9, 10]. The Hopf bifurcation is said to be controllable [11, 12] if it is possible to control the orientation and the stability of the emerging periodic orbits by modifying one or multiple parameters. Moreover, the normal form of the Hopf bifurcation represents a specific configuration where parameters can explicitly influence the system's amplitude, frequency, direction, and stability. This unique form simplifies the identification and control of these system characteristics.

Given that the study of periodic orbits (also known as self-sustained oscillations) is a vast field with many possible applications, from mechanical systems [13–15] and electronic systems [16–18] among many others [19–21], recently, many works have shown the interest in the generation of robust oscillations of underactuated dynamical systems [22], specifically with the Hopf bifurcation analysis of a controlled inertia wheel inverted pendulum. In [23], the authors present a robust controller for the generation of stable limit cycles in multi-input mechanical systems subjected to model uncertainties, based on port-controlled Hamiltonian (PCH) model and energy-based control by considering the Hamiltonian function as the Lyapunov function. In [24], the authors studied inducing robust stable oscillations in nonlinear systems of any order through creating stable limit cycles in the closed-loop system using the Lyapunov stability theory. Also in [25], the authors deal with the tracking approach of a self-generated stable limit cycle for an underactuated mechanical system using passivity-based control.

The main contribution of this work is the design of a control law that induces the normal form of the Hopf bifurcation, which is completely controllable, on a class of second-order differential equations. To achieve it, it is necessary to establish an equivalence between the second-order differential equation and a system of differential equations; once established the equivalence, the controller is designed in vectorial form, to be implemented in a scalar form in the differential equation.

The rest of this work is organized as follows: In Sect. 10.2, the Hopf bifurcation preliminaries are presented. In Sect. 10.3, the system description and main problem are formulated. Section 10.4 contains the methodology to design the controller which is going to induce the controllable Hopf bifurcation in a second-order differential equation, and in Sect. 10.5, numerical results are presented to exhibit the methodology to induce the controllable Hopf bifurcation. Finally, in Sect. 10.6, conclusions and comments are provided.

10.2 Problem Statement

Consider the dynamical system governed by

$$\ddot{\theta} + h(\theta, \dot{\theta}) = \epsilon(\theta, \dot{\theta}), \tag{10.1}$$

with $h(\theta, \dot{\theta})$ being a smooth function. The main objective of this work is to design the scalar controller $\epsilon(\theta, \dot{\theta})$ to induce limit cycles with predefined amplitude A and frequency ω. The controller design will be done due to the simplicity of the design in the state-space system associated with the differential equation (10.1).

Specifically, it is known that the homogeneous differential equation (10.1) is equivalent to the state-space representation

$$\begin{pmatrix} \dot{x_1} \\ \dot{x_2} \end{pmatrix} = \begin{pmatrix} x_2 \\ -h(x) \end{pmatrix}, \tag{10.2}$$

when the controller $\epsilon(\theta, \dot{\theta}) = 0$, thanks to the smooth change of variable $(x_1, x_2)^T = (\theta, \dot{\theta})^T$. It is in this state-space representation that the controller $u = (u_1, u_2)^T$ will be added to (10.2) in order to transform the system into

$$\begin{pmatrix} \dot{x_1} \\ \dot{x_2} \end{pmatrix} = \begin{pmatrix} x_2 \\ -h(x) \end{pmatrix} + \begin{pmatrix} u_1 \\ u_2 \end{pmatrix}. \tag{10.3}$$

In addition, the design of the controller u will be inspired in the leader-follower (previously known as *master-slave*) synchronization scheme presented in [26, 27]. However, it is important to note that u is a vector function in the plane, while $\epsilon(\theta, \dot{\theta})$ is a scalar function. In consequence, the first step is to design the controller u and then establish an equivalence with the scalar function $\epsilon(\theta, \dot{\theta})$ for the use in the differential equation (10.1) in terms of the vector function u from (10.3). This method of design will force the system (10.3) to follow the dynamics of the normal form of the Hopf bifurcation. That is, the differential equation (10.1) will present the Hopf bifurcation, and in consequence, it will be possible to ensure the existence of limit cycles with predefined amplitude A and frequency ω.

The following section will present the necessary preliminaries concerning the Hopf bifurcation to base the reasoning that will be used to induce the behavior in the beforementioned differential equation.

10.3 The Hopf Bifurcation on the Plane

Consider the two-dimensional nonlinear system given by

$$\dot{y} = g(y, \eta), \tag{10.4}$$

where $y = (y_1, y_2)^T$ is the state vector, $\eta \in \mathbb{R}$, and g is a smooth function. Suppose that y_0 is an equilibrium point such that $g(y_0, \eta)$ has one pair of complex eigenvalues $\lambda_{1,2}(\eta) = \alpha(\eta) \pm i\beta(\eta)$. That becomes purely imaginary when $\eta = 0$. Then generically, as η passes through $\eta = 0$, the equilibrium changes stability, and a unique limit cycle emerges from it.

Theorem 10.1 (Two–dimensional Hopf bifurcation [4]) *Consider the nonlinear system (10.4), and suppose that for some parameter value η_0, it satisfies the following conditions:*

(i) $d = \left.\dfrac{d\alpha(\eta)}{d\eta}\right|_{\eta=\eta_0} \neq 0$ *(known as the* ***nondegeneracy condition****)*

(ii) $l_1 \neq 0$ *(known as the* ***transversality condition****)*

where l_1 is the first Lyapunov coefficient of the system. Thus, a unique limit cycle emerges from the equilibrium y_0 for $\eta > 0$ if $l_1 d < 0$ or $\eta < 0$ if $l_1 d > 0$.

Theorem 10.1 considers the nondegeneracy and the transversality conditions, but in order to apply them to a specific system, it is necessary to compute d (which is pretty straightforward) and l_1. To calculate the first Lyapunov coefficient, the following theorem is considered:

Theorem 10.2 (First Lyapunov coefficient [11]) *Consider the equivalent system*

$$\dot{y} = Jy + G(y), \tag{10.5}$$

with $J = \begin{pmatrix} 0 & -\omega \\ \omega & 0 \end{pmatrix}$, $G(y) = \begin{pmatrix} G_1(y) \\ G_2(y) \end{pmatrix}$, $G(y_0) = 0$, *and* $D\left(G(y_0)\right) = 0$. *Then the linear function l_1 is called the first Lyapunov coefficient and can be computed as*

$$l_1 = \frac{1}{16\omega}\left(R_1 + \omega R_2\right), \tag{10.6}$$

where

$$\begin{aligned} R_1 &= G_{1y_1y_2}\left(G_{1y_1y_1} + G_{1y_2y_2}\right) - G_{2y_1y_2}\left(G_{2y_1y_1} + G_{2y_2y_2}\right) - G_{1y_1y_1}G_{2y_1y_2} \\ &\quad + G_{1y_2y_2}G_{2y_2y_2}, \\ R_2 &= G_{1y_1y_1y_1} + G_{1y_1y_2y_2} + G_{2y_1y_1y_2} + G_{2y_2y_2y_2}. \end{aligned}$$

If the conditions of Theorem 10.1 are fulfilled, then the system (10.4) is locally topologically equivalent near the origin to the normal form of the Hopf bifurcation, defined as

$$\dot{\gamma} = \begin{pmatrix} \mu & -\omega \\ \omega & \mu \end{pmatrix} \gamma + \sigma \left(\gamma_1^2 + \gamma_2^2 \right) \gamma, \tag{10.7}$$

where $\gamma = (\gamma_1, \gamma_2)^T$ is the state vector, $\mu \in \mathbb{R}$ is the bifurcation parameter, and $\sigma = \text{sign}(l_1) = \pm 1$ represents the sign of the first Lyapunov coefficient. Four possibilities can be derived from the normal form of the Hopf bifurcation [28]; however, it is important to note that in all cases the origin is a fixed point which is stable at $\mu = 0$ for $\sigma = -1$ and unstable at $\mu = 0$ for $\sigma = 1$:

1. When $d > 0$ and $\sigma = 1$, the origin is an unstable fixed point for $\mu > 0$ and an asymptotically stable fixed point for $\mu < 0$, with an unstable periodic orbit for $\mu < 0$.
2. For $d > 0$ and $\sigma = -1$, the origin is an asymptotically stable fixed point for $\mu < 0$ and an unstable fixed point for $\mu > 0$, with an asymptotically stable periodic orbit for $\mu > 0$.
3. If $d < 0$ and $\sigma = 1$, the origin is an unstable fixed point for $\mu < 0$ and asymptotically stable fixed point for $\mu > 0$, with an unstable periodic orbit for $\mu > 0$.
4. Finally, in the case that $d < 0$ and $\sigma = -1$, the origin is an asymptotically stable fixed point for $\mu < 0$ and an unstable fixed point for $\mu > 0$, with an asymptotically stable periodic orbit for $\mu < 0$.

Remark 10.1 For $\sigma = -1$, it is possible for the periodic orbit to exist for either $\mu > 0$ (Case 2) or $\mu < 0$ (Case 4); however, in each case, the periodic orbit is asymptotically stable. Similarly, for $\sigma = 1$, it is possible for the periodic orbit to exist for either $\mu > 0$ (Case 3) or $\mu < 0$ (Case 1); however, in each case, the periodic orbit is unstable. Thus, the case $\sigma = -1$ is referred to as *supercritical* bifurcation, and the case $\sigma = 1$ is referred to as *subcritical* bifurcation.

10.4 Controller Design

This section will present the design of the controller $u(\xi, \mu)$, and the equivalence with the scalar controller $\epsilon(\theta, \dot{\theta})$ will be presented. Namely, consider the nonlinear system given by

$$\dot{\gamma} = g(\gamma, \mu), \tag{10.8}$$

where the vector field $g(\gamma, \mu)$ is the normal form of the Hopf bifurcation on the plane (10.7) presented in Sect. 10.3, with periodic solution $\gamma = \gamma(t)$ and $\mu \in \mathbb{R}$ as the bifurcation parameter.

Hence, consider the vectorial controller $u(\xi, \mu)$ as

$$u(\xi, \mu) = -Q(\xi)\xi - f(x) + g(\gamma, \mu), \tag{10.9}$$

recalling that $\xi = x - \gamma$, with $Q(\xi)$ being governed by

$$Q(\xi) = \begin{pmatrix} q_1(\xi_1) & 0 \\ 0 & q_2(\xi_2) \end{pmatrix}, \tag{10.10}$$

where $q_1(\xi_1)$ and $q_2(\xi_2)$ are smooth functions on $\xi = (\xi_1, \xi_2)^T$.

Considering that, the differential equation (10.1) is equivalent to the vectorial representation given by

$$\dot{x} = \begin{pmatrix} x_2 \\ -h(x) \end{pmatrix}, \tag{10.11}$$

when the input is turned off (i.e., when $u = (u_1, u_2)^T = (0, 0)^T$). All of the aforementioned is considering the previously presented smooth change of variables

$$\begin{pmatrix} x_1 \\ x_2 \end{pmatrix} = \begin{pmatrix} \theta \\ \dot{\theta} \end{pmatrix}. \tag{10.12}$$

Taking into consideration all of the above, it is clear that is necessary to design the scalar control law $\epsilon(\theta, \dot{\theta})$ for (10.1) in terms of the vector control $u = (u_1(\xi_1, \mu), u_2(\xi_2, \mu))^T$ of (10.2) with $\xi = (\xi_1, \xi_2)^T$ being functions on t and μ, all of this assuming the equivalence between both of these expressions.

With all the preceding discussion, it is possible to establish the following result:

Lemma 10.1 *The nonlinear differential equation*

$$\ddot{\theta} + h(\theta, \dot{\theta}) = \epsilon(\theta, \dot{\theta}),$$

is equivalent to the state-space control system

$$\dot{x} = f(x) + u(\xi, \mu), \tag{10.13}$$

where $f(x) = \begin{pmatrix} x_2 \\ -h(x) \end{pmatrix}$ *and* $u(\xi, \mu) = \begin{pmatrix} u_1(\xi, \mu) \\ u_2(\xi, \mu) \end{pmatrix}$ *if and only if*

$$\epsilon(\theta, \dot{\theta}) = -q_2(\dot{\theta} - \gamma_2)(\dot{\theta} - \gamma_2) + h(\theta, \dot{\theta}) + \ddot{\theta} - \frac{\partial g_1}{\partial \gamma_1} g_1(\gamma, \mu) + \left(1 - \frac{\partial g_1}{\partial \gamma_2}\right) g_2(\gamma, \mu)$$
$$+ \left(q_1(\theta - \gamma_1) + \dot{q}_1(\theta - \gamma_1)(\theta - \gamma_1)\right)\left(\dot{\theta} - g_1(\gamma_1, \mu)\right),$$

where $q_1(\xi_1)$ and $q_2(\xi_2)$ are smooth functions, $\gamma_1 = \gamma_1(t)$ and $\gamma_2 = \gamma_2(t)$ are the solutions of the normal form of the Hopf bifurcation, and $g_1(\gamma, \mu)$ and $g_2(\gamma, \mu)$ are the vector fields of the normal form of the Hopf bifurcation.

Proof Consider both of the components from the state-space control system (10.13) as

$$\dot{x}_1 = f_1(x) + u_1(\xi_1, \mu), \tag{10.14}$$

$$\dot{x}_2 = f_2(x) + u_2(\xi_2, \mu). \tag{10.15}$$

The derivative of the first component (10.14) gives us

$$\ddot{x}_1 = D_t\left(f_1(x)\right) + D_t\left(u_1(\xi_1, \mu)\right), \tag{10.16}$$

where $D_t(\cdot)$ represents the time derivative. This allows to match (10.15) and (10.16) since $\ddot{x}_1 = \dot{x}_2$, and according to the smooth change of variables (10.12), it is possible to obtain

$$D_t\left(f_1(x)\right) + D_t\left(u_1(\xi_1, \mu)\right) = f_2(x) + u_2(\xi_2, \mu), \tag{10.17}$$

and with the rearrangement of (10.17), it is possible to reach

$$D_t\left(f_1(x)\right) - f_2(x) = -D_t\left(u_1(\xi, \mu)\right) + u_2(\xi, \mu). \tag{10.18}$$

Note that $D_t(f_1(x)) = D_t(x_2) = \dot{x}_2$ and $f_2(x) = -h(x)$, in consequence (10.18), can be rewritten as

$$\dot{x}_2 + h(x) = -D_t\left(u_1(\xi_1, \mu)\right) + u_2(\xi_2, \mu). \tag{10.19}$$

Thence, a proposal for the vectorial controller is given by

$$u_1(\xi_1, \mu) = -q_1(\xi_1)\xi_1 - f_1(x) + g_1(\gamma, \mu), \tag{10.20}$$

$$u_2(\xi_2, \mu) = -q_2(\xi_2)\xi_2 - f_2(x) + g_2(\gamma, \mu), \tag{10.21}$$

which allows to compute $D_t(u_1(\xi_1, \mu))$ as

$$D_t(u_1(\xi_1, \mu)) = -\left(q_1(\xi_1) + \dot{q}_1(\xi_1)\xi_1\right)\dot{\xi}_1 - \dot{x}_2 + \frac{\partial g_1}{\partial \gamma_1}\dot{\gamma}_1 + \frac{\partial g_1}{\partial \gamma_2}\dot{\gamma}_2. \tag{10.22}$$

By substituting (10.21) and (10.22) into (10.19) and reverting the change of variables defined in (10.12), it is possible to obtain

$$\ddot{\theta} + h(\theta, \dot{\theta}) = \epsilon(\theta, \dot{\theta}), \tag{10.23}$$

where the scalar controller is finally defined as

$$\epsilon(\theta,\dot{\theta}) = -q_2(\dot{\theta}-\gamma_2)(\dot{\theta}-\gamma_2)+h(\theta,\dot{\theta})+\ddot{\theta}-\frac{\partial g_1}{\partial \gamma_1}g_1(\gamma,\mu)+\left(1-\frac{\partial g_1}{\partial \gamma_2}\right)g_2(\gamma,\mu)$$
$$+\left(q_1(\theta-\gamma_1)+\dot{q}_1(\theta-\gamma_1)(\theta-\gamma_1)\right)\left(\dot{\theta}-g_1(\gamma_1,\mu)\right),$$

as previously claimed, where $q_1(\xi_1)$ and $q_2(\xi_2)$ are smooth functions, $\gamma_1 = \gamma_1(t)$ and $\gamma_2 = \gamma_2(t)$ are the solutions of the normal form of the Hopf bifurcation, and $g_1(\gamma,\mu)$ and $g_2(\gamma,\mu)$ are the vector fields of the normal form of the Hopf bifurcation. ■

Lemma 10.1 permits to analyze the solutions in the controlled state-space system, whose dynamics come to be equivalent to the dynamics of the differential equation (10.1). Therefore, the following result derives the convergence of their solutions toward the predefined solutions of (10.8).

Lemma 10.2 *The solution $x = x(t)$ of the state-space system* (10.3) *tends to the solution of the normal form of the Hopf bifurcation $\gamma = \gamma(t)$ from* (10.8) *as $t \to \infty$ under the control law* (10.9).

Proof To demonstrate the convergence and considering that the difference between γ and x is defined as $\xi = x - \gamma$, define the quadratic Lyapunov candidate function given by

$$V(\xi) = \frac{1}{2}\left(\xi_1^2 + \xi_2^2\right), \tag{10.24}$$

whose time derivative can be computed as

$$\dot{V}(\xi) = \xi_1\dot{\xi}_1 + \xi_2\dot{\xi}_2. \tag{10.25}$$

Note that given the definition $\xi = x - \gamma$, it is clear that $\dot{\xi} = \dot{x} - \dot{\gamma}$, which makes it possible to substitute (10.8) and (10.11) into (10.25) and, in consequence, it is possible to obtain

$$\dot{V}(\xi) = \xi_1\left(x_2 - g_1(\gamma,\mu) + u_1(\xi,\mu)\right) + \xi_2\left(-h(x) - g_2(\gamma,\mu) + u_2(\xi,\mu)\right), \tag{10.26}$$

and by substituting the control law (10.9) in the Lyapunov function derivative (10.26), it is possible to achieve

$$\begin{aligned}\dot{V}(\xi) &= \xi_1\left(-q_1(\xi_1)\xi_1\right) + \xi_2\left(-q_2(\xi_2)\xi_2\right),\\ &= -q_1(\xi_1)\xi_1^2 - q_2(\xi_2)\xi_2^2.\end{aligned} \tag{10.27}$$

Since $q_i(\xi_i) \geq 0$ and $\forall \xi_i \in \mathbb{R}$, with $i = 1, 2$, and according to the LaSalle's invariance principle [29], it is possible to ensure the asymptotic stability of the system, and in consequence it is assured that the solution x tends to the solution γ, since $\xi \to 0$ as $t \to \infty$. ∎

Remark 10.2 Note that the aforementioned Lemma 10.2 shows that the dynamics of

$$\dot{x} = f(x) + u(\xi, \mu) \tag{10.28}$$

tends to the dynamics of

$$\dot{\gamma} = g(\gamma, \mu), \tag{10.29}$$

as $t \to \infty$, which is the normal form of the Hopf bifurcation, due to the controller $u(\xi, \mu)$ substituting the original dynamics of the state-space system (10.11) with the dynamics of the normal form of the Hopf bifurcation (10.8), which assures that $Q(\xi)\xi$ vanishes as $t \to \infty$.

Finally, with all the previously mentioned, it is possible to establish that the differential equation (10.1) undergoes the controllable Hopf bifurcation since the state-space system (10.3) presents the controllable form of the Hopf bifurcation. This fact is formally established in the following result:

Theorem 10.3 *The nonlinear differential equation*

$$\ddot{\theta} + h(\theta, \dot{\theta}) = \epsilon(\theta, \dot{\theta}),$$

undergoes the controllable form of the Hopf bifurcation under the control law

$$\epsilon(\theta, \dot{\theta}) = -q_2(\dot{\theta}-\gamma_2)(\dot{\theta}-\gamma_2)+h(\theta, \dot{\theta})+\ddot{\theta}-\frac{\partial g_1}{\partial \gamma_1} g_1(\gamma, \mu)+\left(1-\frac{\partial g_1}{\partial \gamma_2}\right) g_2(\gamma, \mu)$$
$$+ \left(q_1(\theta - \gamma_1) + \dot{q}_1(\theta - \gamma_1)(\theta - \gamma_1)\right)\left(\dot{\theta} - g_1(\gamma_1, \mu)\right),$$

recalling that $g(\gamma, \mu) = (g_1(\gamma, \mu), g_2(\gamma, \mu))^T$ is the vector field from (10.7).

Proof According to Lemma 10.1, the state-space system

$$\dot{x} = f(x) + u(\xi, \mu),$$

with $u(\xi, \mu) = -Q(\xi)\xi - f(x) + g(\gamma, \mu)$ is equivalent to the nonlinear differential equation

$$\ddot{\theta} + h(\theta, \dot{\theta}) = \epsilon(\theta, \dot{\theta}),$$

with

$$\epsilon(\theta,\dot{\theta}) = -q_2(\dot{\theta}-\gamma_2)(\dot{\theta}-\gamma_2)+h(\theta,\dot{\theta})+\ddot{\theta}-\frac{\partial g_1}{\partial \gamma_1}g_1(\gamma,\mu)+\left(1-\frac{\partial g_1}{\partial \gamma_2}\right)g_2(\gamma,\mu)$$

$$+\left(q_1(\theta-\gamma_1)+\dot{q}_1(\theta-\gamma_1)(\theta-\gamma_1)\right)\left(\dot{\theta}-g_1(\gamma_1,\mu)\right).$$

Since Lemma 10.1 establishes that the dynamics of (10.3) are topologically equivalent to the one of the controlled nonlinear differential equations (10.1) via the scalar control function (10.9). In addition, Lemma 10.2 and Remark 10.2 demonstrate that $\dot{\xi} \to 0$ due to $\dot{x} \to \dot{\gamma}$ as $t \to \infty$. Therefore, the second-order differential equation (10.1) undergoes the controllable Hopf bifurcation as $t \to \infty$. ■

The fact that the nonlinear differential equation (10.1) undergoes the controllable Hopf bifurcation, which is achieved from its normal form (10.7), allows, due to the structure of the normal form, to manipulate adequately the parameters to generate limit cycles with suitable amplitude and frequency. Besides, it is important to mention that it is also possible to choose the stability of the raising periodic orbits, even when in most of the applications stable limit cycles are needed, as shown in the following section.

10.5 Inducing the Hopf Bifurcation to the Duffing Equation Circuit

This section will utilize the control law designed in the Duffing nonlinear differential equation. The Duffing nonlinear differential equation is given by

$$\ddot{\theta}+\delta\dot{\theta}+\alpha\theta+\beta\theta^3=\epsilon(\theta,\dot{\theta}), \tag{10.30}$$

with $\delta, \alpha, \beta \in \mathbb{R}$, where δ controls the amount of damping, α the linear stiffness, and β the amount of nonlinearity in the restoring force.

Now, recalling the smooth change of variables $x_1 = \theta$ and $x_2 = \dot{\theta}$, it allows to represent the Duffing equation in its state-state representation given by

$$\dot{x} = \begin{pmatrix} 0 & 1 \\ -\alpha & -\delta \end{pmatrix} x + \begin{pmatrix} 0 \\ -\beta x_1^3 \end{pmatrix}, \tag{10.31}$$

and the dynamics of the Duffing equation for $\alpha = -1$, $\delta = 0.3$, and $\beta = 1$ are shown in Fig. 10.1.

Moreover, the circuit implementation of the Duffing equation is given in Figs. 10.2, 10.3, and 10.4, Fig. 10.2 gives the solution $x_1 = x_1(t)$, Fig. 10.3 computes the solution $x_2 = x_2(t)$, and Fig. 10.4 computes the cubic term needed in the circuit. It is important to consider that Figs. 10.2, 10.3, and 10.4 represent a single circuit, divided only to let the reader appreciate better the design. Taking

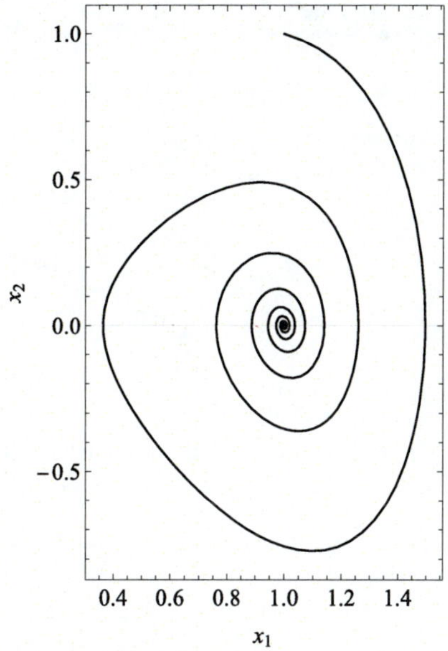

Fig. 10.1 Numerical simulation of the Duffing equation

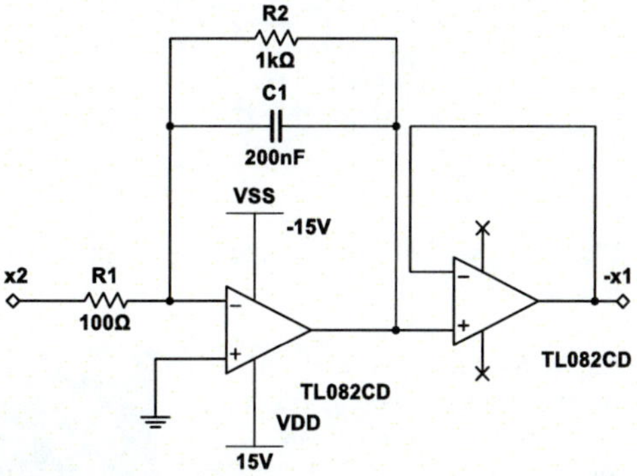

Fig. 10.2 Part 1 of 3 of the circuit implementation of the Duffing equation

into consideration the appropriate selection of *off-the-shelf* components to attain the same coefficients $\delta, \alpha, and\beta$ as in the numerical simulations, the response of the implemented system can be observed in Fig. 10.5 taking as initial conditions the point $x_0 = (1, 1)^T$.

Now, the use of Theorem 10.3 allows to obtain an expression for the controller in its vectorial form, which is expressed as

$$u(\xi, \mu) = -\begin{pmatrix} q_1(\xi_1) & 0 \\ 0 & q_2(\xi_2) \end{pmatrix} \xi - \begin{pmatrix} 0 & 1 \\ -\alpha & -\delta \end{pmatrix} x - \begin{pmatrix} 0 \\ -\beta x_1^3 \end{pmatrix}$$
$$+ \begin{pmatrix} \mu & -\omega \\ \omega & \mu \end{pmatrix} \gamma + \sigma \left(\gamma_1^2 + \gamma_2^2\right) \gamma, \qquad (10.32)$$

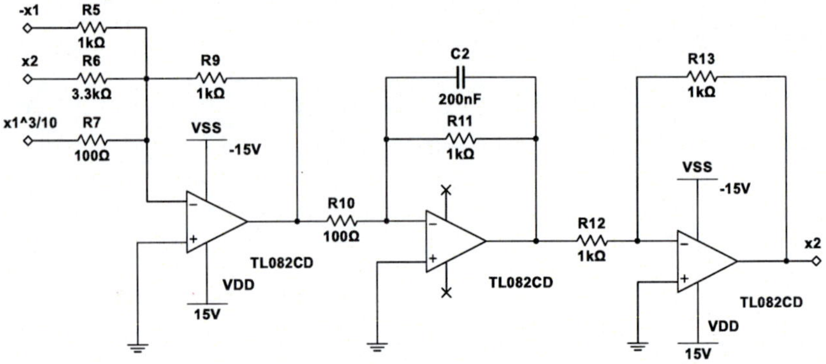

Fig. 10.3 Part 2 of 3 of the circuit implementation of the Duffing equation

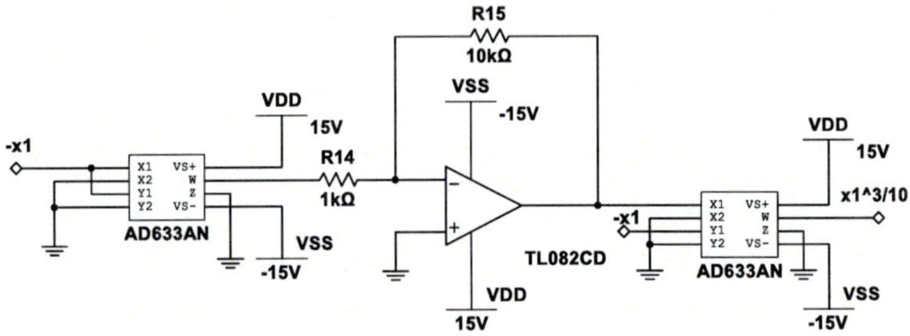

Fig. 10.4 Part 3 of 3 of the circuit implementation of the Duffing equation

Fig. 10.5 Output signal of the implemented Duffing circuit with initial conditions $x_0 = (1, 1)^T$

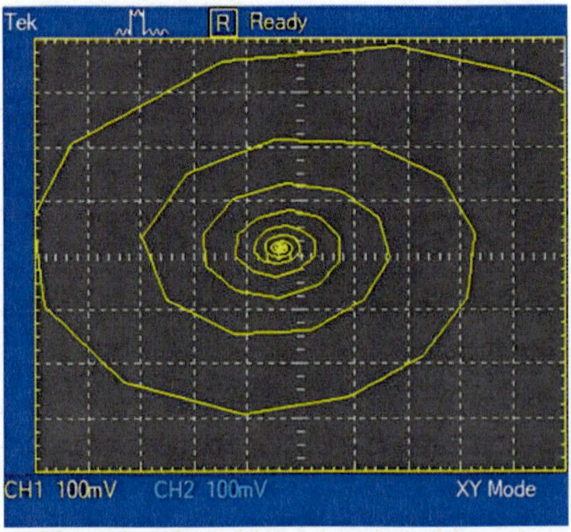

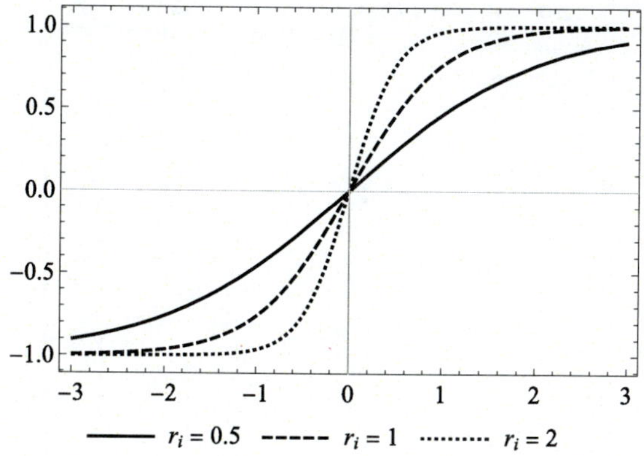

Fig. 10.6 Hyperbolic tangent function ($h(\xi_i) = p_i \tanh(r_i \xi_i)$), evaluated at multiple values of r_i

where $\xi = x - \gamma$ is the difference between the dynamics of the Duffing equation and the normal form of the Hopf bifurcation. The function $q_i(\xi_i)$ with $i = 1, 2$ is the adaptation function designed in the same spirit than in [27] but using a smooth function in order to avoid abrupt changes in the solutions. The aforementioned adaptation function is defined as

$$q_i(\xi_i) = (p_i \tanh(r_i \xi_i))^2, \tag{10.33}$$

where p_i represents the maximum value of $q_i(\xi_i)$ and r_i represents the steepness of the curve. When the magnitude of r_i increases, the steepness expands as well, and when the magnitude of p_i is incremented the amplitude that is induced to the controller grows accordingly, which ensures a faster adaptation (see Fig. 10.6). Substituting the controller in the state-space representation of the Duffing equation, it is possible to obtain

$$\dot{x} = -\begin{pmatrix} q_1(\xi_1) & 0 \\ 0 & q_2(\xi_2) \end{pmatrix} \xi + \begin{pmatrix} \mu & -\omega \\ \omega & \mu \end{pmatrix} \gamma + \sigma \left(\gamma_1^2 + \gamma_2^2\right) \gamma, \tag{10.34}$$

where $\mu = 1$, $\omega = 1$, and $\sigma = -1$. According to Lemma 10.1, the controlled system (10.34) is equivalent to the Duffing differential equation, and in consequence, the normal Hopf bifurcation is being induced into the differential equation.

The circuit implementation of the controlled system can be appreciated from Figs. 10.7, 10.8 10.9, 10.10, 10.11, 10.12, and 10.13, and, once again, they are part of the same circuit (but given the size of the circuit is presented in multiple figures). Figure 10.7 presents the computing process of the now controlled solution $x_1 = x_1(t)$, Fig. 10.8 presents the controlled solution $x_2 = x_2(t)$, Fig. 10.9 presents the implementation of the adaptation function $q_1(\xi_1)$, Fig. 10.10 presents the implementation of the adaptation function $q_2(\xi_2)$, Figs. 10.11 and 10.12 present the calculation of the solution from the normal form of the Hopf bifurcation, and, finally, Fig. 10.13 presents the implementation of the interaction between the adaptation functions and the system states.

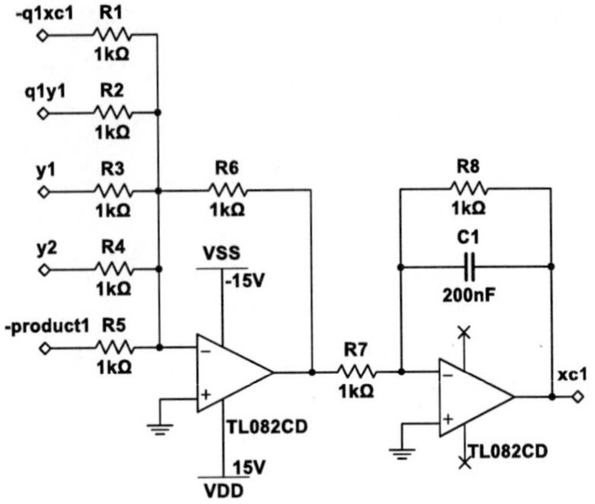

Fig. 10.7 Part 1 of 7 of the circuit implementation of the Duffing equation

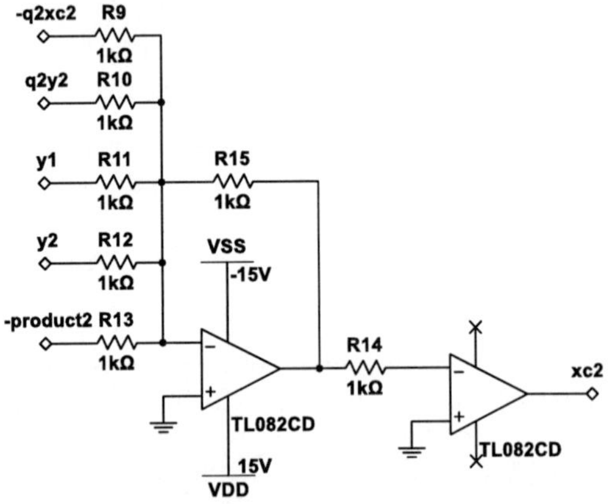

Fig. 10.8 Part 2 of 7 of the circuit implementation of the Duffing equation

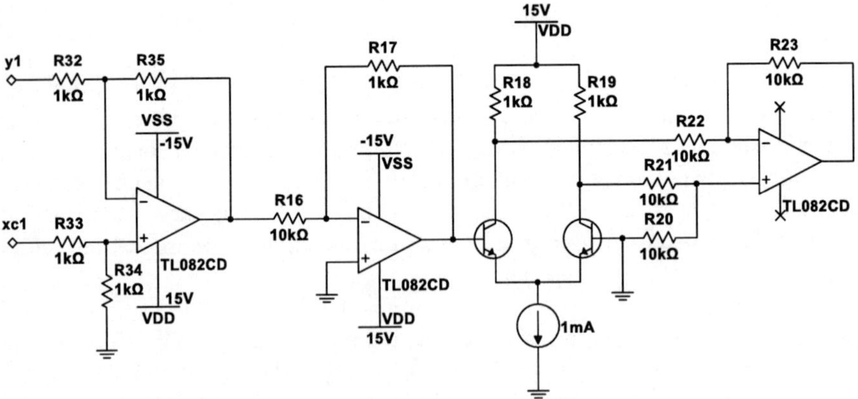

Fig. 10.9 Part 3 of 7 of the circuit implementation of the Duffing equation

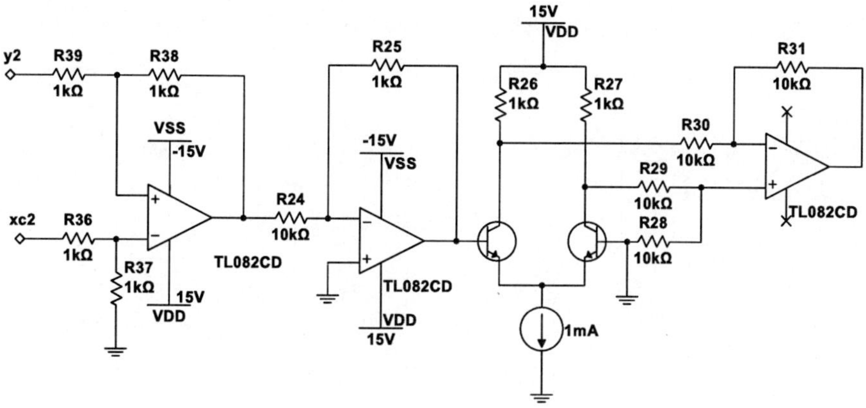

Fig. 10.10 Part 4 of 7 of the circuit implementation of the Duffing equation

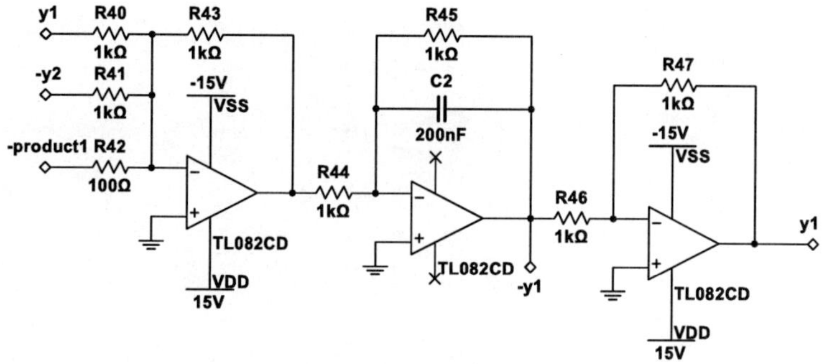

Fig. 10.11 Part 5 of 7 of the circuit implementation of the Duffing equation

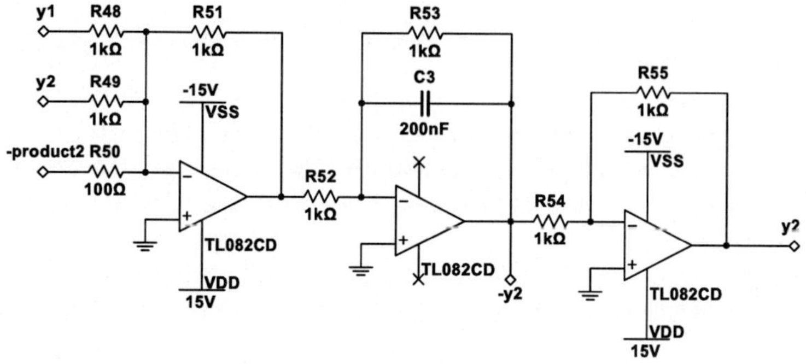

Fig. 10.12 Part 6 of 7 of the circuit implementation of the Duffing equation

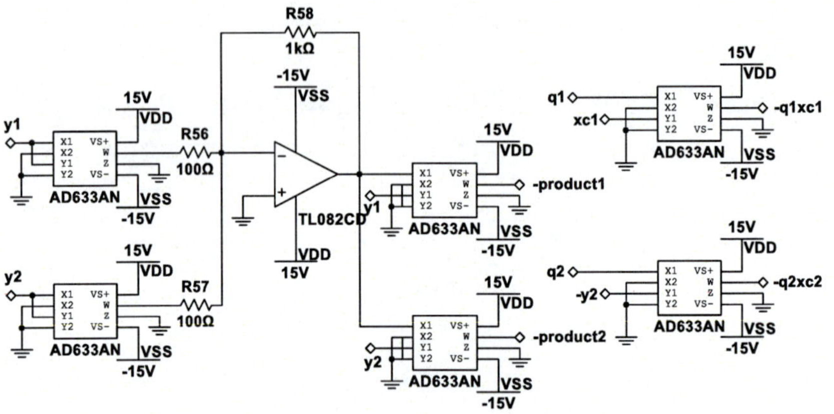

Fig. 10.13 Part 7 of 7 of the circuit implementation of the Duffing equation

Fig. 10.14 Solutions of the controlled system with initial conditions outside the limit cycle $x_0 = (2, 1.5)^T$

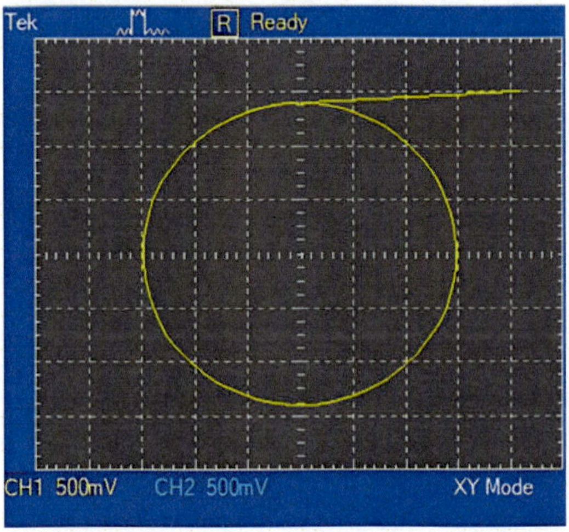

The controlled behavior of the Duffing circuit is shown in Figs. 10.14 and 10.15. It is easy to observe that the periodic motion of the solution presents a stable limit cycle with amplitude $A = 1$ and frequency $\omega = 1$. Furthermore, it is important to note that if the initial condition x_0 is outside the limit cycle, the solution tends to the limit cycle from the outside in a smooth form (thanks to the adaptation function $q_i(\xi_i)$), without crossing the limit cycle as presented in Fig. 10.14. The same applies when the initial conditions of the system are inside the limit cycle which can be also appreciated in Fig. 10.15.

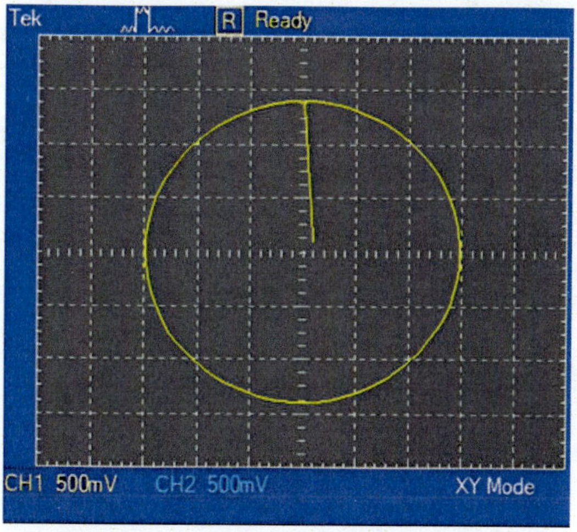

Fig. 10.15 Solutions of the controlled system with initial conditions inside the limit cycle $x_0 = (0.1, 0.1)^T$

10.5.1 The Equivalent Controller in the Duffing Differential Equation

Given that the controller was originally designed for application to the differential equation, it is now appropriate to establish the equivalence between the vectorial controller and the scalar controller. To accomplish this, essential to utilize the results provided in Lemma 10.1, which arises

$$\begin{aligned}\epsilon(\theta,\dot{\theta}) = {}& \gamma_1\left[-3\sigma\left(\gamma_1^2+\mu\gamma_2^2\right)+\omega\left(\omega+1\right)\right]+\gamma_2\left[\sigma\left(\omega\gamma_1^2+\gamma_2^2\right)+2\omega\mu\right]\\ &+\left(\gamma_1^2+\gamma_2^2\right)\left[-3\sigma\gamma_1\left(\gamma_1^2+\gamma_2\right)+\sigma\gamma_2\left(\omega-\mu+1\right)\right]+\Delta_1\left(\dot{\theta}-\gamma_2\right)\\ &+\left(\Delta_2+\Delta_3\right)\left[\dot{\theta}-\mu\gamma_1+\omega\gamma_2-\sigma\gamma_1\left(\gamma_1^2+\gamma_2^2\right)\right]\left(\theta-\gamma_1\right)+\ddot{\theta}+\delta\dot{\theta}\\ &+\alpha\theta+\beta\theta^3, \end{aligned} \tag{10.35}$$

where

$$\Delta_1 = -p_2^2\tanh^2\left(r_2(\dot{\theta}-\gamma_2)\right), \tag{10.36}$$

$$\Delta_2 = p_1^2\tanh^2\left(r_1(\theta-\gamma_1)\right), \tag{10.37}$$

$$\Delta_3 = 2p_1^2\,\mathrm{sech}^2\left(r_1(\theta-\gamma_1)\right)\tanh\left(r_1(\theta-\gamma_1)\right), \tag{10.38}$$

which allows to write the Duffing differential equation as

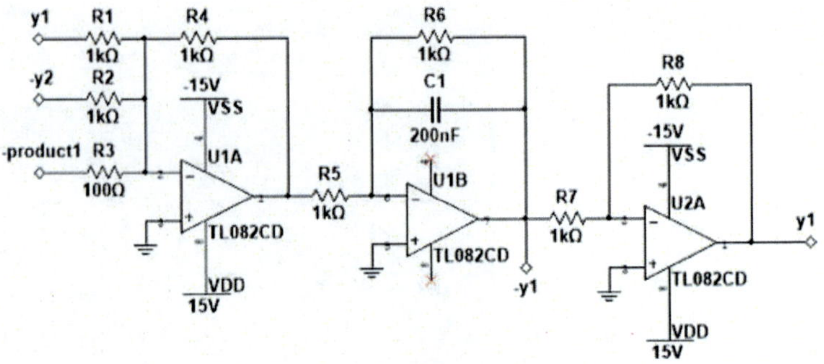

Fig. 10.16 Part 1 of 5 of the circuit implementation of the Duffing equation

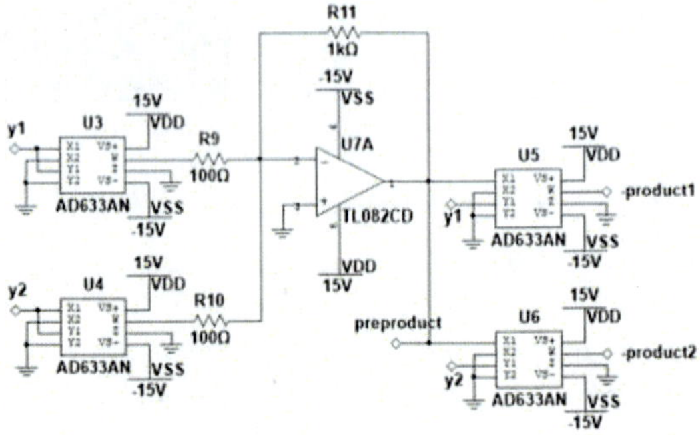

Fig. 10.17 Part 2 of 5 of the circuit implementation of the Duffing equation

$$
\begin{aligned}
0 = &\gamma_1 \left[-3\sigma\left(\gamma_1^2 + \mu\gamma_2^2\right) + \omega(\omega+1)\right] + \gamma_2\left[\sigma\left(\omega\gamma_1^2 + \gamma_2^2\right) + 2\omega\mu\right] \\
&+ \left(\gamma_1^2 + \gamma_2^2\right)\left[-3\sigma\gamma_1\left(\gamma_1^2 + \gamma_2\right) + \sigma\gamma_2(\omega - \mu + 1)\right] + \Delta_1\left(\dot{\theta} - \gamma_2\right) \\
&+ (\Delta_2 + \Delta_3)\left[\dot{\theta} - \mu\gamma_1 + \omega\gamma_2 - \sigma\gamma_1\left(\gamma_1^2 + \gamma_2^2\right)\right](\theta - \gamma_1),
\end{aligned} \tag{10.39}
$$

and since $\Delta_1, \Delta_2, \Delta_3 \to 0$ when $t \to \infty$, it is possible to obtain

$$
\begin{aligned}
&\gamma_1\left[-3\sigma\left(\gamma_1^2 + \mu\gamma_2^2\right) + \omega(\omega+1)\right] + \gamma_2\left[\sigma\left(\omega\gamma_1^2 + \gamma_2^2\right) + 2\omega\mu\right] + \\
&\qquad \left(\gamma_1^2 + \gamma_2^2\right)\left[-3\sigma\gamma_1\left(\gamma_1^2 + \gamma_2\right) + \sigma\gamma_2(\omega - \mu + 1)\right] = 0,
\end{aligned} \tag{10.40}
$$

which is the controlled form of the Duffing differential equation.

Taking into account that a limit cycle of amplitude $A = \sqrt{\mu}$ and frequency ω is desired, the circuit implementation of the controlled differential equation is given in Figs. 10.16, 10.17, 10.18, 10.19, and 10.20, and this is considering the same parameters $A = 1$, $\omega = 1$, and $\sigma = -1$. Finally the solution of the Duffing

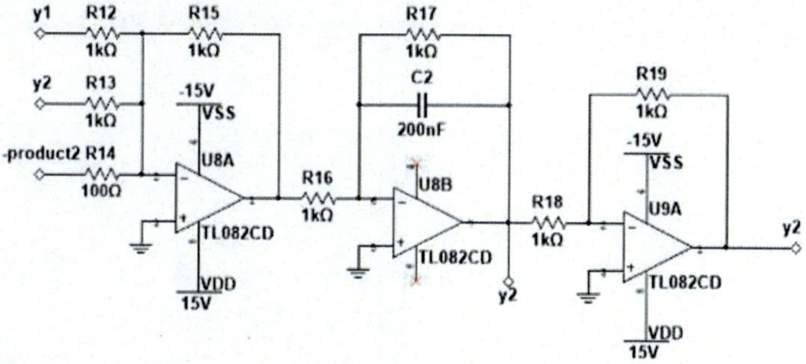

Fig. 10.18 Part 3 of 5 of the circuit implementation of the Duffing equation

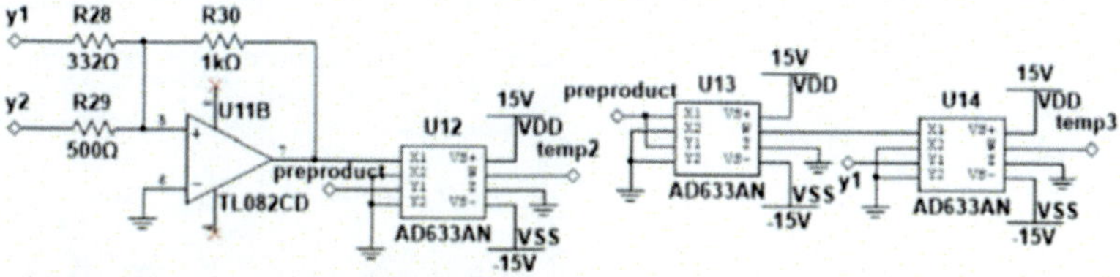

Fig. 10.19 Part 4 of 5 of the circuit implementation of the Duffing equation

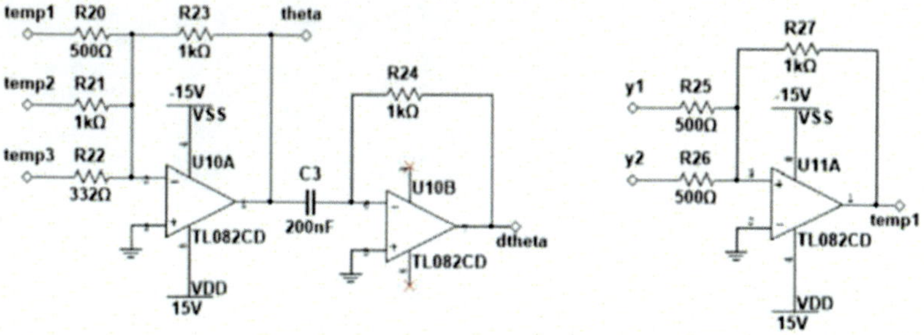

Fig. 10.20 Part 5 of 5 of the circuit implementation of the Duffing equation

differential equation under the scalar controller can be appreciated in Figs. 10.21 and 10.22. Figure 10.21 presents a case when the initial condition is outside the limit cycle, and Fig. 10.22 presents the case when the initial conditions are on the inside of the limit cycle; in both cases, the solutions tend in a smooth way to the predefined limit cycle.

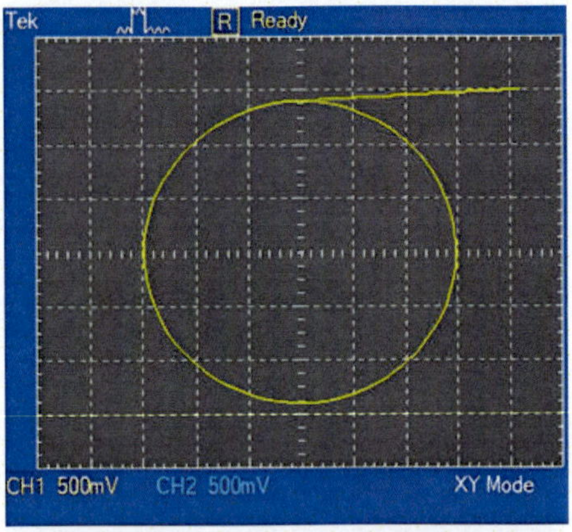

Fig. 10.21 Controlled Duffing differential equation with initial conditions outside the limit cycle $x_0 = (1, 1.5)^T$

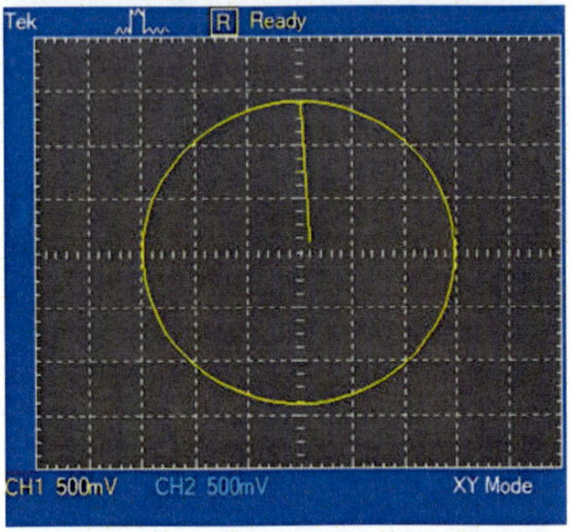

Fig. 10.22 Controlled Duffing differential equation with initial conditions inside the limit cycle $x_0 = (0.1, 0.1)^T$

10.6 Conclusion

This work presented a methodology to induce a controllable form of the Hopf bifurcation into a class of second-order differential equations, which allows to define a priori the amplitude and frequency of the desired limit cycles. The controller design is inspired by a leader-follower synchronization scheme and takes advantage of its benefits: the solution of the second-order differential equation can enter a limit cycle of amplitude and frequency desired despite the initial conditions without crossing the limit cycle and entering in a smooth way.

For higher-dimensional systems, the center manifold can be computed for the controlled system in order to reduce the problem to the plane.

Acknowledgments This work received funding from CONAHCyT (Project No. A1–S–32341) and from TecNM (Project Nos. 11122.21-P, 14492.22-P, 15171.22-P). Jesus R. Pulido-Luna is supported by CONAHCyT with a D.Sc. Scholarship.

References

1. J. Guckenheimer and P. Holmes. *Nonlinear Oscillations, Dynamical Systems, and Bifurcations of Vector Fields*, volume 42 of *Applied Mathematical Sciences*. Springer New York, New York, USA, 1 edition, 2002.
2. L. Perko. *Differential Equations and Dynamical Systems*, volume 7 of *Texts in Applied Mathematics*. Springer New York, New York, USA, 3 edition, 2001.
3. A. A. Andronov, A. A. Vitt, and S. E. Khaikin. *Theory of oscillators*, volume 4 of *International Series of Monographs in Physics*. Pergamon Press, 2 edition, 1966.
4. J. E. Marsden and M. McCracken. *The Hopf Bifurcation and Its Applications*, volume 19 of *Applied Mathematical Sciences*. Springer–Verlag New York, New York, USA, 1976.
5. W. Liu, C. Fu, and B. Chen. Hopf bifurcation and center stability for a predator–prey biological economic model with prey harvesting. *Communications in Nonlinear Science and Numerical Simulation*, 17(10):3989–3998, 2012.
6. H. Errami, M. Eiswirth, D. Grigoriev, W. M. Seiler, T. Sturm, and A. Weber. Detection of Hopf bifurcations in chemical reaction networks using convex coordinates. *Journal of Computational Physics*, 291:279–302, 2015.
7. S. Djilali, B. Ghanbari, S. Bentout, and A. Mezouaghi. Turing–Hopf bifurcation in a diffusive mussel–algae model with time–fractional–order derivative. *Chaos, Solitons & Fractals*, 138:109954, 2020.
8. J. Castro, J. Alvarez, F. Verduzco, and J. E. Palomares-Ruiz. Chaotic behaviour of driven, second–order, piecewise linear systems. *Chaos, Solitons & Fractals*, 105:8–13, 2017.
9. J. Castillo, J. Llibre, and F. Verduzco. The pseudo–Hopf bifurcation for planar discontinuous piecewise linear differential systems. *Nonlinear Dynamics*, 90:1829–1840, 2017.
10. J. M. Islas, J. Castillo, B. Aguirre-Hernandez, and F. Verduzco. Pseudo-Hopf bifurcation for a class of 3d Filippov linear systems. *International Journal of Bifurcation and Chaos*, 31(02):2150025, 2021.
11. Y. A. Kuznetsov. *Elements of Applied Bifurcation Theory*, volume 112 of *Applied Mathematical Sciences*. Springer–Verlag New York, New York, USA, 3 edition, 2004.
12. J. A. López-Rentería, F. Verduzco, and B. Aguirre-Hernández. Control of the Hopf bifurcation by a linear feedback control. *International Journal of Bifurcation and Chaos*, 25(1):1550006, 2015.
13. M. Azhdari and T. Binazadeh. Generation of stable limit cycles in nonlinear sandwich systems with dead–zone nonlinearity and input saturation. *Journal of the Franklin Institute*, 358(18):10029–10051, 2021.
14. J. A. López-Rentería, L. Herrera-Garcia, S. L. Cardenas-Maciel, L. T. Aguilar, and N. R. Cazarez-Castro. Self–sustaining oscillations with an internal two–fuzzy inference system based on the Poincaré–Bendixon method. *IEEE Transactions on Fuzzy Systems*, 30:2563–2573, 2022.
15. L. Herrera-Garcia, N. R. Cazarez-Castro, S. L. Cardenas-Maciel, J. A. López-Rentería, and L. T. Aguilar. Self-excited periodic motion in underactuated mechanical systems using two-fuzzy inference system. *Fuzzy Sets and Systems*, 438:25–45, 2022.
16. X. Liao and N. Mu. Self-sustained oscillation in a memristor circuit. *Nonlinear Dynamics*, 96:1267–1281, 2019.
17. M. F. Rahman, T. Pang, S. Sapper, E. Shoubaki, S. Jimenez, A. Lemmon, and M. Manjrekar. Analysis and mitigation of self-sustained turn-off oscillations in sic jfet supercascode circuits. In *2021 IEEE Applied Power Electronics Conference and Exposition (APEC)*, pages 1531–1536, 2021.
18. S. A. Kurkin, D. D. Kulminskiy, V. I. Ponomarenko, M. D. Prokhorov, S. V. Astakhov, and A. E. Hramov. Central pattern generator based on self-sustained oscillator coupled to a chain of oscillatory circuits. *Chaos*, 32(3):033117, 2022.
19. P. Kongkhambut, J. Skulte, L. Mathey, J. G. Cosme, A. Hemmerich, and H. Keßler. Observation of a continuous time crystal. *Science*, 377:670–673, 2022.

20. J. Bisquert and A. Guerrero. Chemical inductor. *Journal of the American Chemical Society*, 144(13):5996–6009, 2022.
21. X. Liang, Z. Chen, L. Zhu, and K. Li. Light-powered self-excited oscillation of a liquid crystal elastomer pendulum. *Mechanical Systems and Signal Processing*, 163:108140, 2022.
22. N. K. Haddad, S. Belghith, H. Gritli, and A. Chemori. From Hopf bifurcation to limit cycles control in underactuated mechanical systems. *International Journal of Bifurcation and Chaos*, 27(7):1750104, 2017.
23. T. Binazadeh and M. Karimi. Robust stable limit cycle generation in multi–input mechanical systems. *Robotica*, 39(7):1316–1327, 2021.
24. A. R. Hakimi and T. Binazadeh. Robust generation of limit cycles in nonlinear systems: Application on two mechanical systems. *Journal of Computational and Nonlinear Dynamics*, 12(4):041013, 2017.
25. H. Gritli, N. Khraief, A. Chemori, and S. Belghith. Self–generated limit cycle tracking of the underactuated inertial wheel inverted pendulum under IDA–PBC. *Nonlinear Dynamics*, 89(3):2195–2226, 2017.
26. J. R. Pulido-Luna, J. A. López-Rentería, and N. R. Cazarez-Castro. Design of a non-homogeneous nonlinear synchronizer and its implementation in reconfigurable hardware. *Mathematical and Computational Applications*, 25(3):51, 2020.
27. J. R. Pulido-Luna, J. A. López-Rentería, and N. R. Cazarez-Castro. Mamdani–type fuzzy–based adaptive nonhomogeneous synchronization. *Complexity*, 2021(9913114):1–11, 2021.
28. S. Wiggins. *Introduction to Applied Nonlinear Dynamical Systems and Chaos*, volume 2 of *Texts in Applied Mathematics*. Springer–Verlag New York, New York, USA, 2 edition, 2003.
29. H. K. Khalil. *Nonlinear Systems*. Prentice Hall, New Jersey, USA, 3 edition, 2002.

Chapter 11
Cascading Timers

The New Method to Solve the Sequential Problems Including Timers in Automation Systems

Saturnino Soria-Tello

11.1 Introduction

The chronogram of functions is one of the analytical tools used to solve sequential systems based on the time function. It is a totally graphical method in which it is graphed at the time of ignition all output functions, input functions, and memory functions. Figure 11.1 shows an example of the application of the chronogram of functions [1]. It is an analytical method used by some manufacturers of equipment and industrial machinery; its contribution is in the design and identification of failures; with the interpretation of this diagram, it is easy to understand a system; and this diagram represents a cycle of the system.

Only the switching on of the functions and the turning on of the preset of the timers are represented in the schedule. From the time diagram, you can get the logical equation of each output function called (F) and each of the timers called (T); the discrete input variable is (A).

This chapter proposes a new method developed by **Soria-Tello Saturnino**, to solve in an analytical, didactic, systematic, and effective way; sequential systems of the industrial type that include in their logical structure the time function, using the method called "cascading timers," can be solved from simple problems where a single timer is applied to complex systems that use more timers, from single-cycle problems to cyclic systems, including systems with multiple main lines and with external input signals.

The simplicity of this methodology makes it very versatile in the solution for any type of sequential problem that includes the time function. When solving sequential systems based on the time function using analytical methods, there is the advantage

S. Soria-Tello (✉)
FIME of the UANL, Pedro de Alba S/N CD Universitaria, San Nicolás de los Garza, CP, Mexico
e-mail: saturnino.soriatl@uanl.edu.mx

E. Campos-Cantón et al. (eds.), *Complex Systems and Their Applications*,
https://doi.org/10.1007/978-3-031-51224-7_11

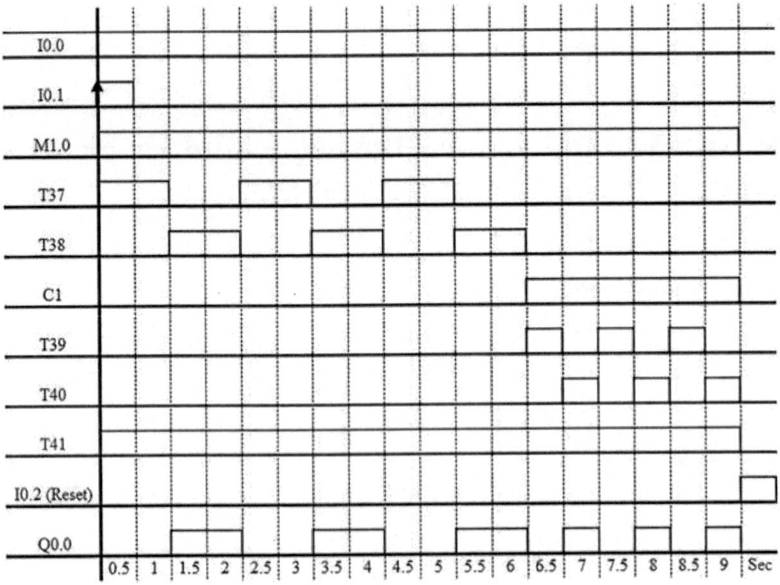

Fig. 11.1 Chronogram of functions in a sequential system based on timers

that a reliable and standard solution can be arrived at without having accumulated years of practical experience.

11.2 Limits of Application of the Method

As any method is convenient to recognize the limits of application, the method is based on the timer on delay (TON). The reason is how simple it is to operate this timer; its application embraces from wired logic systems, called electromechanical control systems with relays; and it is possible to use it in programmed logic systems, called systems based on a PLC. It does not apply to systems based on electronic components, such as time integrated circuits or generic array logic (GAL).

11.3 Structure of the Cascading Timer Method

The method is based on that every sequential system requires to be activated through a starting device, which is how it happens in all process and industrial machinery. The timers must be turned on one by one, and when the starting device is turned on, the coil and the preset value of timer 1 (T1) are activated immediately; when the preset time value is finished, the contacts change state by turning on the coil and the preset value of timer 2 and so on until the last timer (Tn) is reached. A graph

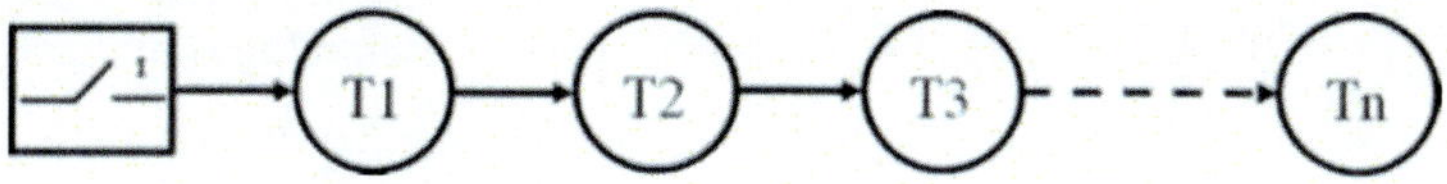

Fig. 11.2 Schematic of a cascading timer system *n* timers

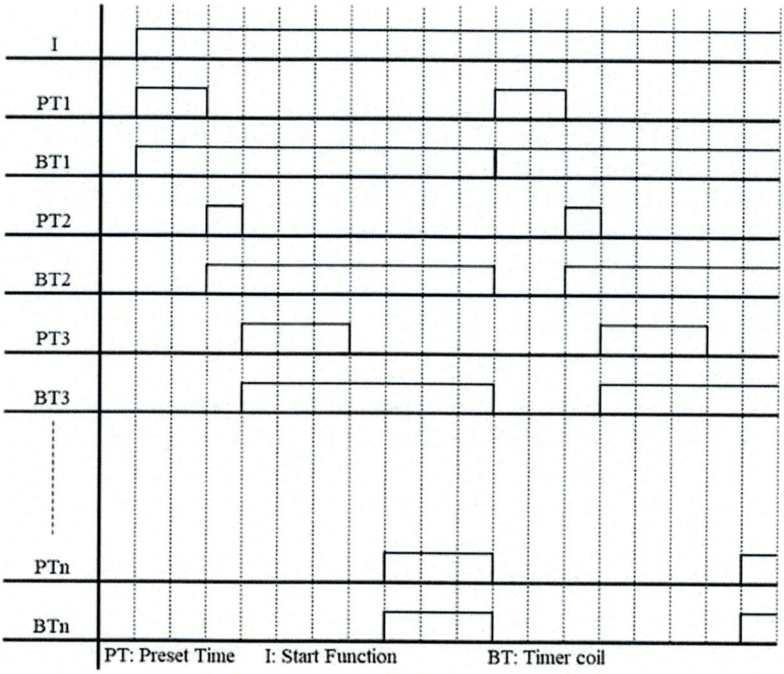

Fig. 11.3 Chronogram of time functions

of a system with timers on the main line is shown in the diagram in Fig. 11.2. The timer coils, once activated, should be kept on until the sequential system cycle is completed.

The arrangement of the timers is monotonous ascending [1] in such a way that it must be fulfilled that the timer T1 turns on the timer T2, and this, in turn, turns on the timer T3 and so on until reaching the last timer of the waterfall. From the schematic of Fig. 11.2, it is possible to obtain a schedule of times, and as shown in Fig. 11.3, the diagram shows the preset values of the timers (PT1) and the coils of the same timers (BT1), and the coils are kept on until the operation cycle is finished.

The timer coils are ignited at same time that the preset value, and they remain on until the cycle ends, in such a way the T1 timer coil is on all the cycle time only turns off for a very short instant of time. The basis of "cascading timers" is to turn on the system timers one by one; from the T2 timer, each timer is turned on by the immediately previous one giving a cascade type effect. Once the timer coil is turned

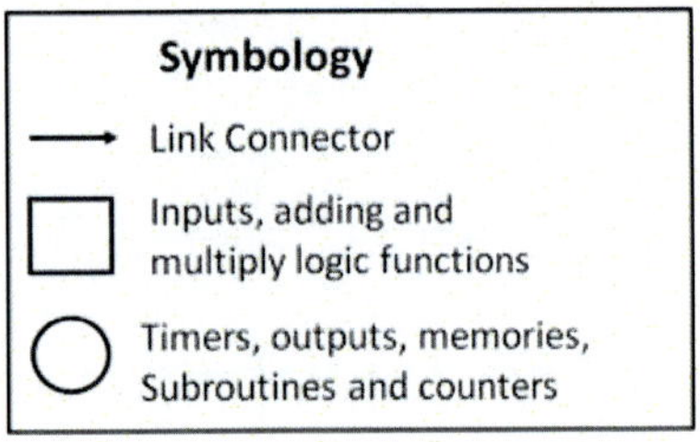

Fig. 11.4 Schematic of a cascading timer system *n* timers

on, it will remain so until the cycle is completed, ensuring that all timers operate at the time destined for each of them.

Figure 11.4 shows the symbology adopted by "cascading timers". The output and time functions are represented by a circle, and the input, addition, and multiplication functions are represented by a square. The interconnection between the functions is made by a link connector represented by an arrow with eight possible directions defined by the contact quadrant shown in Fig. 11.5.

Link Connector
The link connector is responsible for defining the path of the flow of information between timers and output functions. It is also used as a result of the functions of addition and multiplication of functions. It is represented in a logical equation as denied or affirmed of a function. This is defined by the direction of the link connector itself.

Input, Addition, and Multiplication Function
As shown in Fig. 11.4, the interconnection of the functions located in a table is done by a link connector, and the address of this link connector is not affected by the contact quadrant, meaning that the logical level of the input function is the same that will be represented in the logical equation.

Time, Output, and Memory Functions
These functions are represented by a circle. They arrive and leave link connectors as interconnection devices of the function with the rest of the system. When they enter, they become ignition conditions affecting the logical equation of the function to which it is entering, and when it leaves, they become condition of the function from where it leaves conditioning the function to which it enters.

11.4 Quadrant of Contacts

One of the fundamental tools of the method in its application is to interpret as open and closed contacts, through the address, the link connectors, to apply to a logical equation. The contact quadrant is shown in Fig. 11.5, which identifies the type of contacts that correspond to each of the quadrants.

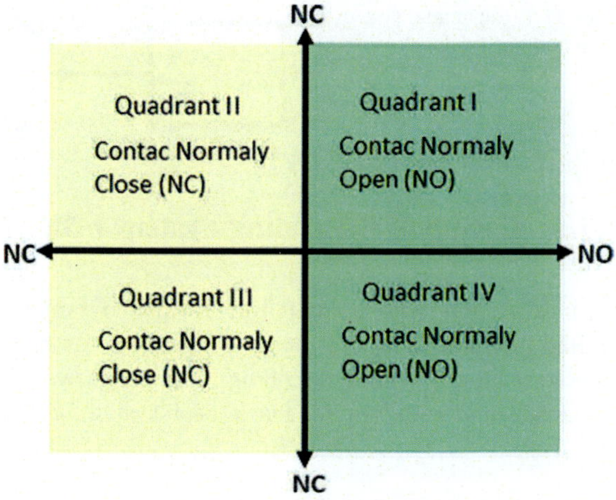

Fig. 11.5 Quadrant of contacts as a special tool of the cascading timer method

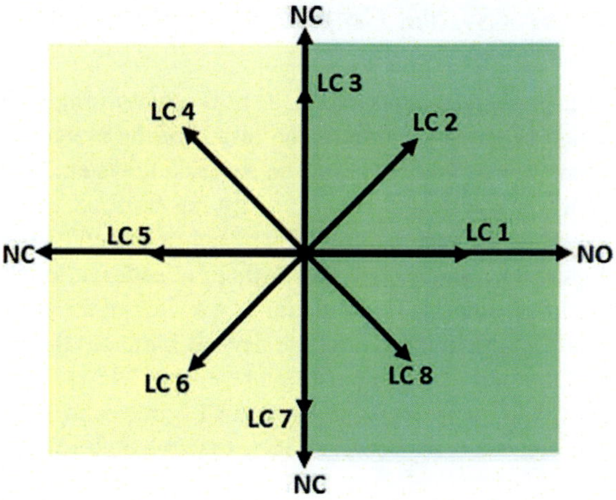

Fig. 11.6 The eight possible directions of link connector

The link connectors are transferred to a logical variable identifying which function it comes from, this being the variable that will identify the link connector. It is located in one of the contact quadrants to identify if it is a normally open contact (NA) or a normally closed contact (NC), defining with this if the function is denied or affirmed.

The link connectors are located in the contact quadrant according to the connector's own address. It can have only eight addresses, and these are shown in Fig. 11.6 where LC1, LC2, and LC8 are link connectors that result in normally open contacts. In an equation, they are expressed with affirmed values, the LC3 link connectors. LC4, LC5, LC6, and LC7 are link connectors that result in a normally closed contact in a ladder logic diagram and in an equation. These are represented by the negation of the function from which the link connector comes.

Important: Link connectors that do not apply the contact quadrant are those that come from an input function, an addition, or a multiplication of functions.

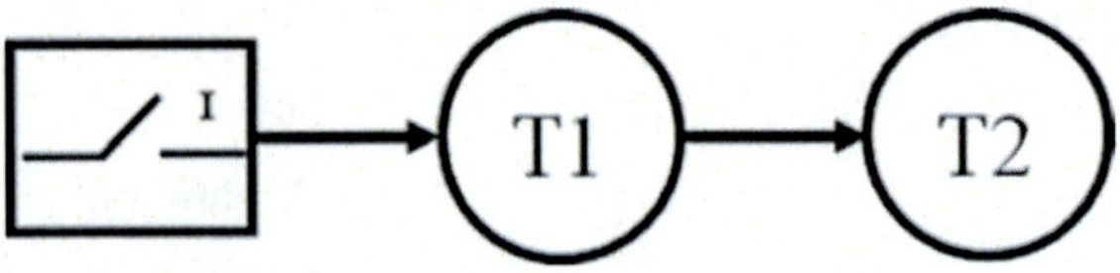

Fig. 11.7 Sequential system without loop feedback

11.5 Graph of Cascading System with a Main Line

It is a system with a main line formed by time functions. This main line is called the main line, and the time functions can be connected to output functions where required. With a cascading timer graph, you can represent systems without feedback, and with loop feedback, each system has its field of application.

11.5.1 System Graphic Without Loop Feedback

This graph represents a cascade of times as a single network, and without backward union between the timers, the restart of the system is done with the same external input device that turns on the sequential system. This is because the coils of the timers are activated even after having finished counting their preset time. In this example, the system is reset by means of switch I, resulting in a single-cycle system. Figure 11.7 shows a system without a feedback loop. The system is composed of a start function (I). This function is a switch of the sustained type and two timers, T1 and T2. The link between the devices indicates the path of the waterfall, indicating that the start function must first be enabled. Immediately after the T1 coil is enabled as well as its preset value, once the T1 time ends, the contacts must change the state and activate the T2 coil and its respective time value.

11.6 Graph of System with *n* Timers and *q* Functions

A system of "cascading timers" is structured by timers and input functions, memories, and output functions. The latter are the devices that change their on or off. These functions are in charge of link connectors from the timers. These enter the output functions generating a logical equation for each function. This education is represented in a contact logic diagram. The diagram in Fig. 11.8 shows a system with n timers and q output functions. The system can be as extensive as required by the sequential system to be solved.

The output functions are dependent on the components of the main line. In the case of the function F1, it is dependent on the input function (I) and the time function T1, until reaching the q-esima function which is dependent on a timer Tn-1 and the

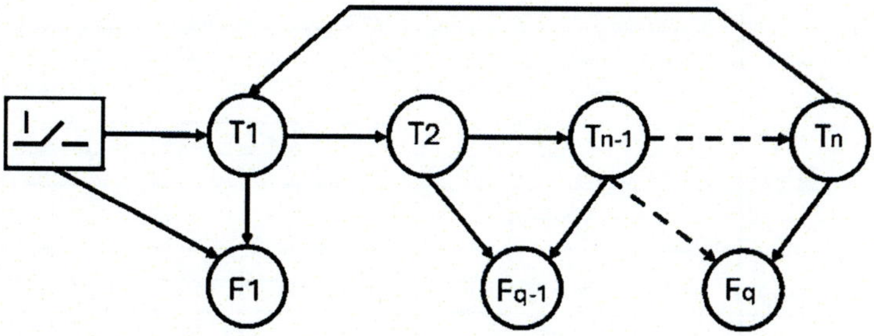

Fig. 11.8 System with ***n*** timers and ***q*** output functions

Fig. 11.9 Sequential system with an output

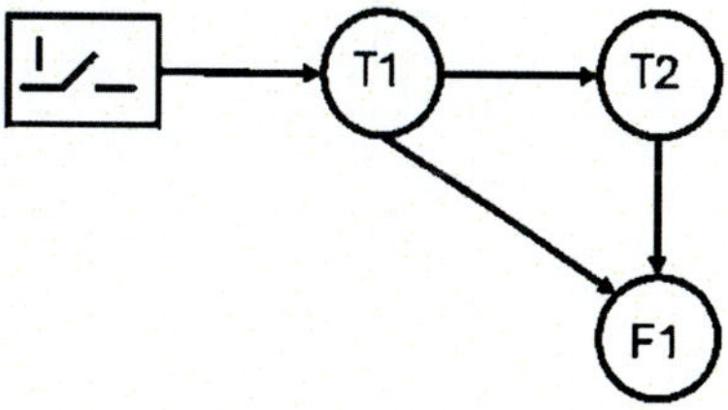

timer Tn. Figure 11.8 shows that an output function can be obtained from each timer located on the main line; however, this depends on the automatic system.

Graph with an Output Function (F1)

So far, the cascade of timers does not generate an output function that makes any changes in a process or a machine. Functions dependent on these times can only be added to the cascading timer diagram. The location depends on what the sequence requires and at what time the function should be turned on and off, for example, being analyzed. An F1 function is added, as shown in Fig. 11.9, which is dependent on both T1 and T2 timers. One timer turns it on and the other timer turns it off.

Assigning preset values to the times of T1 and T2, called PT1 and PT2:

$$PT1 = 5Sec \tag{11.1}$$

$$PT2 = 4Sec \tag{11.2}$$

From the diagram in Fig. 11.9, you can define the sequence of system operation by simply following the direction of the link connectors:

- I1 startup function is enabled and remains enabled.
- The coil of T1 and the preset (PT1) are lit.
- When you finish counting, the T1 timer changes its state and turns on the T2 coil, the preset (PT2), and the F1 function.
- When the preset value of T2 is finished, the contacts change state, turning off the F1 function and waiting for the startup device I to turn off to reset to the preset values of the timers and restart the system.

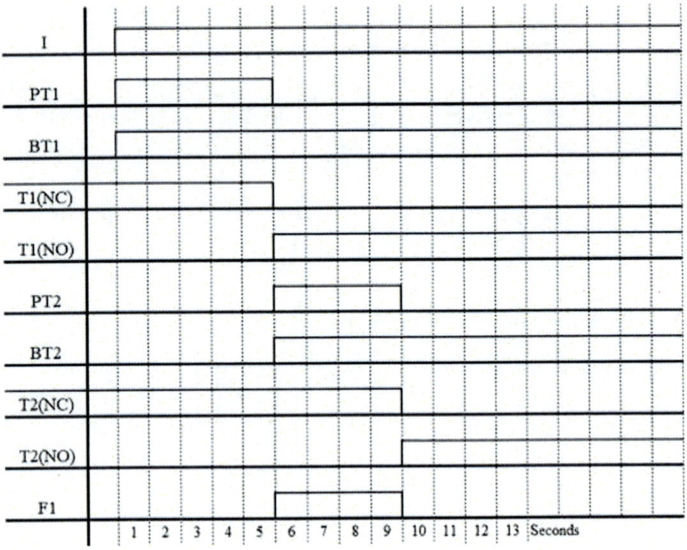

Fig. 11.10 Chronogram of functions of one cycle of sequence that includes timers

Fig. 11.11 Sequential system with loop feedback

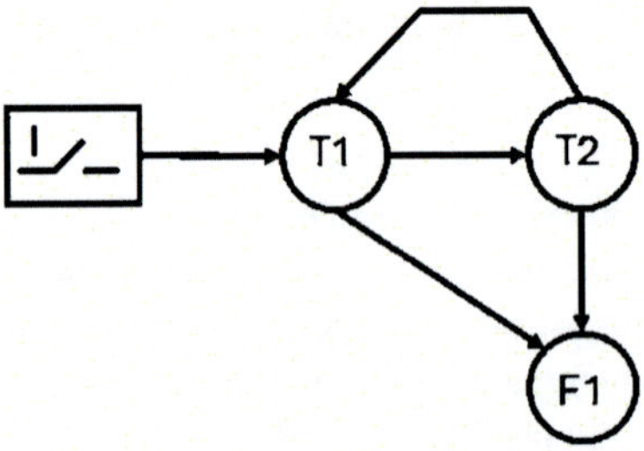

The sequence explained can be transferred to a time schedule, shown in Fig. 11.10, checking that the cascading timer diagram can be interpreted using this diagram.

System Graphic with Loop Feedback

A system of cascading timers with feedback is the representation of a cyclic system, shown in Fig. 11.11. In this type of systems, the device responsible for resetting it is the last timer of the cascade of timers, and that is where the feedback or regressive union comes from. The start of the system is usually an action generated by an input.

Loop feedback restart the sequence starting with the first timer of cascade, it becomes cyclical for an indefinite number of cycles and when it is fed back to the start it becomes a single cycle system. The sequence of the system is depicted in Fig. 11.12, and now, the system shows more than one cycle.

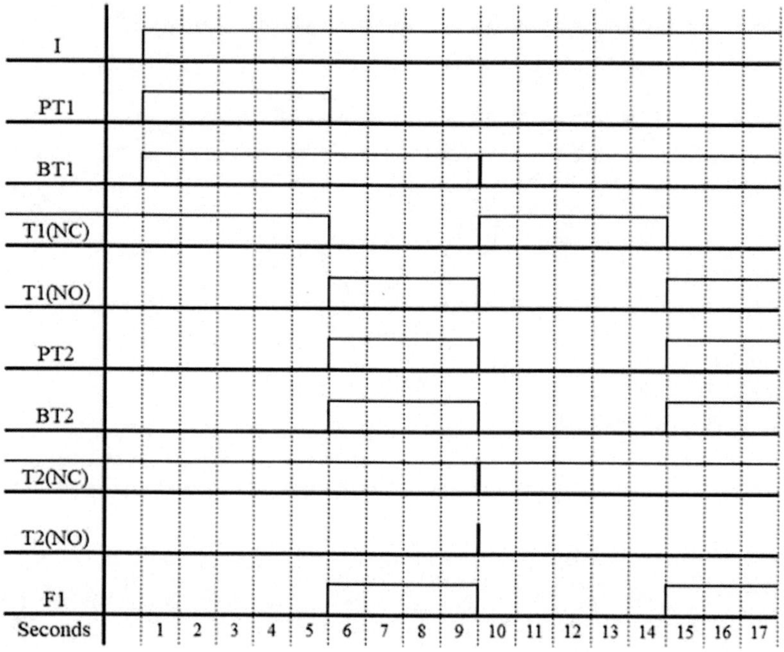

Fig. 11.12 Time schedule of the sequence of operation of a cyclic system

11.7 System Graphic with Feedback Loop

A main line in a cascading timer system is the main time path of the system. This main line depends on the activation and deactivation of the functions of the system, and this main line depends on the system working. If one of the links is interrupted, the system is disabled, and a system with a main line has a trajectory. The system shown in Fig. 11.13 shows a system with a main line composed of the start functions I and the time functions T1 and T2. So far, only the cascading time graph has been studied. It has been emphasized that the foundation of the methodology is to keep the timer coils on until a cycle is completed, and the next step is to obtain the equations of the system one for each timer and one for each output function. The supporting tool for transferring a cascading timer diagram to logical equations is the "contact quadrant." The location in the contact quadrant of the system link connectors is done using the simple steps explained below. To explain each of the link connectors listed individually, the diagram shown in Fig. 11.13 identifying each link connector is analyzed.

The system has five link connectors, of which two have the same addresses, CE1 and CE2, a horizontal CE4 connector, the CE5 connector located in quadrant III, and the CE3 connector located in quadrant IV.

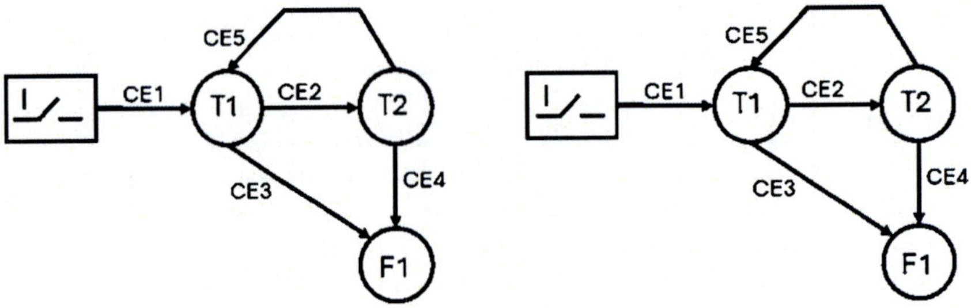

Fig. 11.13 Identification of link connectors according to the quadrant of contacts

Fig. 11.14 Link connectors with horizontal right direction

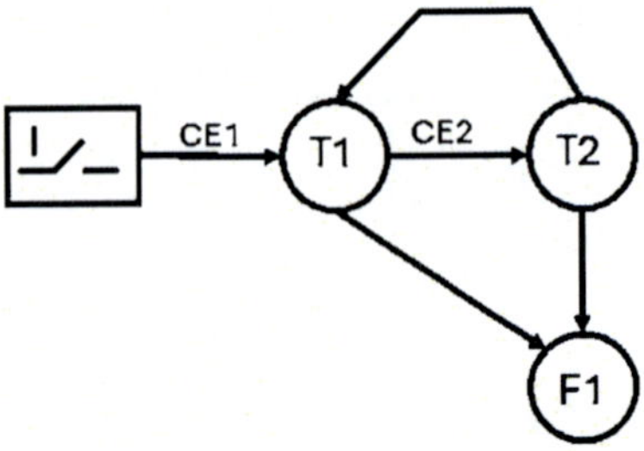

Location of a Horizontal Link Connector

The horizontal link connector can have two left or right directions: the left-hand connector according to the contact quadrant is equal to a normally closed contact, and the right-hand link connector is equal to a normally open contact. The CE1 link connector is not passed through the contact quadrant because it comes from an input function, while the CE2 connector must be passed through the contact quadrant. These are shown in Fig. 11.14.

The CE2 link connector is located in the quadrant and starts at the origin, and the direction is to the right. This location is shown in Fig. 11.15, resulting in a normally open contact when represented in a ladder logic diagram, and is represented as an affirmed variable in a logical equation.

Vertical Link Connector

This link connector can also have two up or down directions with no tilt angle, and in both directions, it becomes a normally closed contact when represented in a contact logic diagram and is represented as a negated function in an equation. In the example, the connector CE4, shown in Fig. 11.16, is a contact normally closed when transferred to a ladder logic diagram from T2 that conditions the function F1, and when obtaining the equation of F1, it is considered as the negation of T2.

The location of this link connector starts at the origin of the quadrant, and the direction is downward (Fig. 11.17). It can be seen that if you change the direction up, the result is the same. When two link connectors arrive at F1, both conditions

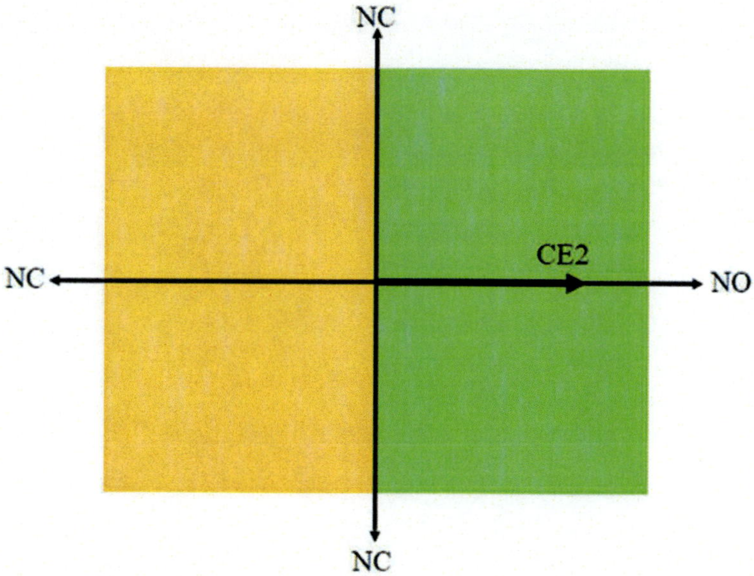

Fig. 11.15 Location of the CE2 link connector in the contact quadrant

Fig. 11.16 Link connector with vertical down direction

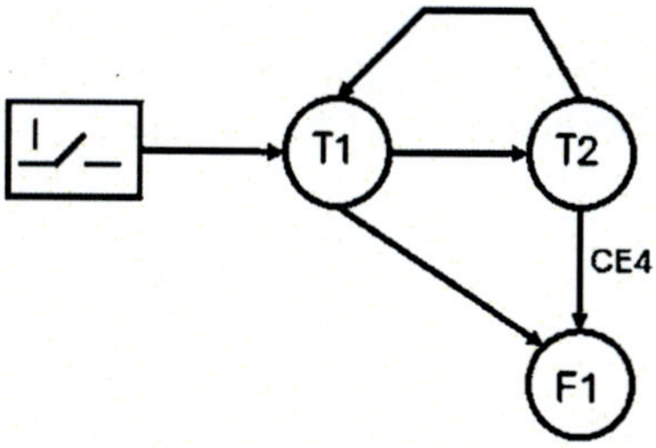

are multiplied to give the result of F1, because they are conditions that must be met at the same time for F1 to turn on.

Link Connector with Multiple Directions on Its Path

A link connector can have several directions in its path, from departure to arrival to the function to be conditioned, such as the CE5 connector shown in Fig. 11.18. The direction with which it enters the function is the one that defines the type of contact and the logical level of the link connector.

In the example, it is observed that CE5 has three changes of direction in its trajectory; however, the one that defines the link connector is with which direction it enters the time function T1. It is precisely to the function that will condition. As an important point, a link connector should only be located in one of the four quadrants. If we transport the last direction of the connector path to the contact quadrant, the origin of the quadrant is located at the origin of the connector of the last section,

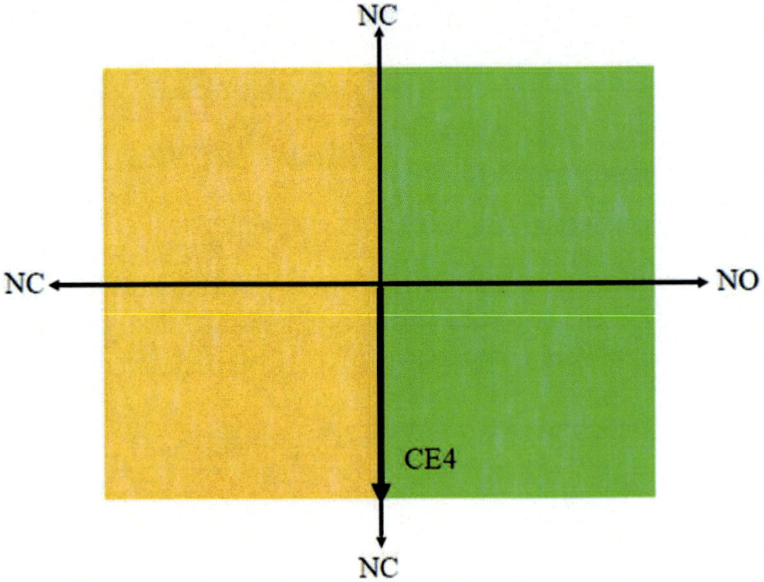

Fig. 11.17 Link connector with vertical down direction

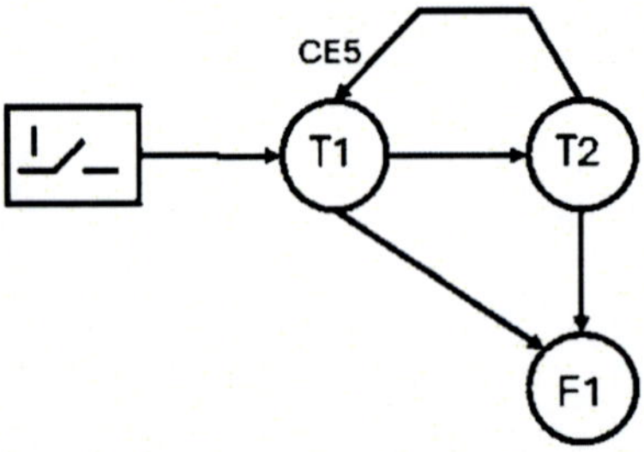

Fig. 11.18 Link connector with several directions on its path

shown in Fig. 11.19, resulting in the location of the connector in the third quadrant. If the decision is made to connect to T1 from T2 but with an opposite direction to the current one, the connector would be in the II quadrant not changing the logical levels of representation both in a ladder logic diagram and in the logical equation that would represent T1.

Fourth Quadrant Link Connector

The CE3 connector is the last one missing to complete the example that is being analyzed. According to the direction shown, it is in the IV quadrant shown in Fig. 11.20, resulting in a normally open contact when transferring this connector to a ladder logic diagram and becoming an affirmed function when transferred to a logical equation.

If the function F1 were located at the top of the function T2 and not at the bottom as it is located in Fig. 11.20, then the connector would be located in the first quadrant

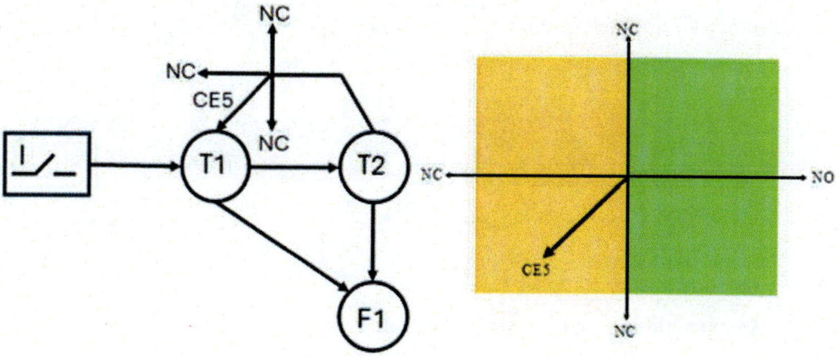

Fig. 11.19 Location of the CE5 link connector in the contact quadrant

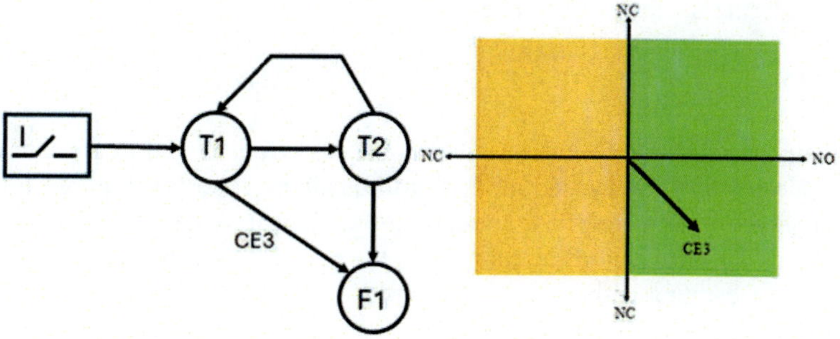

Fig. 11.20 Link connector located in the IV quadrant

(I). The result when converting it into a contact and transferring it to an equation is the same in both quadrants.

11.8 Equations of the Logical Variables of the System

Once the link connectors are located in the contact quadrant and after identifying each of them, the next step is to transfer each of the link connectors to a logical equation that represents each output and time function. To transfer to an equation, the link connectors must be identified as affirmed or negated, the negation is to be represented with a higher line applied to the function to be represented by the link connector, and the affirmed is only represented as the function without any additional symbol. The number of conditions that forms the equations of the functions will depend on the number of link connectors that enter the function from whom the equation is being obtained.

The cascading timer diagram shown in Fig. 11.21 has four output functions; five functions of time, forming a main line; and an input function called I which is

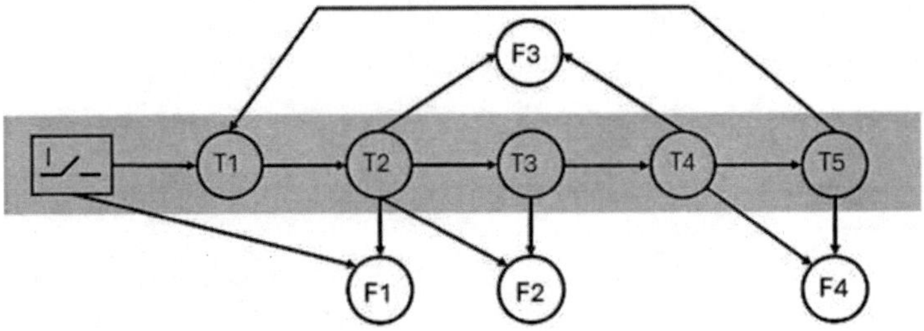

Fig. 11.21 An example of cascading timer diagram

responsible for turning on the system, finding the logical equations of the output and time functions, and transferring the equations to a ladder logic diagram.

Considering the next preset time:

$$PT1 = 20S,\ PT2 = 10S,\ PT3 = 5S,\ PT4 = 10S,\ PT5 = 15S \qquad (11.3)$$

To obtain the equation of F1, the two link connectors from T1 and T2 are considered; the connector of T1 is located in the fourth quadrant, resulting in a true function; and the link connector of T2 is located on the vertical axis resulting in a negated connector. The equation of the function is the result of multiplying these two link connectors, resulting in the following equation:

$$F1 = I\bar{T}2 \qquad (11.4)$$

The operating conditions of F2 come from the T2 and T3 timers, the T2 link connector is located in the fourth quadrant, and the T3 link connector is located on the vertical axis, resulting in the following equation:

$$F2 = T2\bar{T}3 \qquad (11.5)$$

The F3 function has as ignition conditions, a T2 connector and a T4 connector, located in the first quadrant and on the vertical axis, respectively. Considering these conditions results in the following equation:

$$F2 = T2\bar{T}4 \qquad (11.6)$$

For function F4, the operating conditions come from T4 and T5, located in the fourth quadrant and the vertical axis, resulting in the following logical equation:

$$F2 = T4\bar{T}5 \qquad (11.7)$$

Equations of Time Functions

Five equations must be obtained. It starts with the timer T1. This timer enters two link connectors: one of the input function I and another connector of the time function T5. The link connector of the input function I does not enter the contact quadrant. This function is considered affirmed by being disconnected switch I. The feedback is located in the third quadrant resulting in the following equation:

$$T1 = I\bar{T}5 \tag{11.8}$$

The time functions T2, T3, T4, and T5 are conditioned by a link connector, the link connectors come from the immediately preceding timer, and the equations are as follows:

$$T2 = T1 \tag{11.9}$$

$$T3 = T2 \tag{11.10}$$

$$T4 = T3 \tag{11.11}$$

$$T5 = T4 \tag{11.12}$$

These are the last logical equations of the system, with the support of the FluidSim[2]. The equations are transferred to a ladder logic diagram shown in Fig. 11.22.

With the variables transferred to the Siemens PLC [3, 4], the ladder logic diagram shown in Fig. 11.23 is obtained.

Some systems may have more than one main line, but only one must be active. This type of system will be explained later. From each system, you can obtain information on cycle time and the time that the functions are on through calculations made by simple formulas of the method.

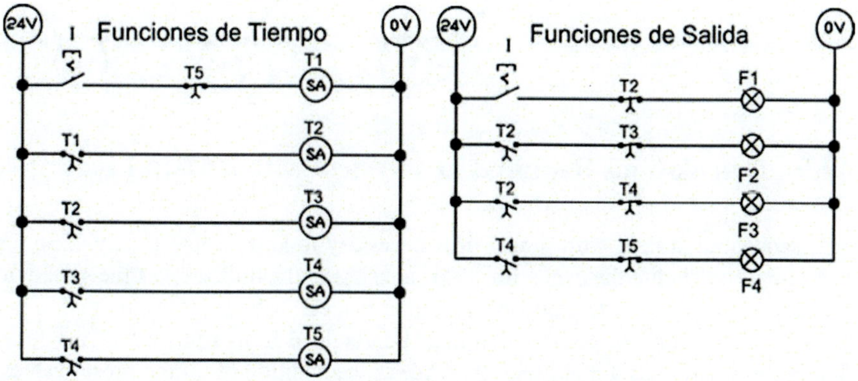

Fig. 11.22 Diagram of ladder logic resulting from the transfer of logic equations with FluidSim®

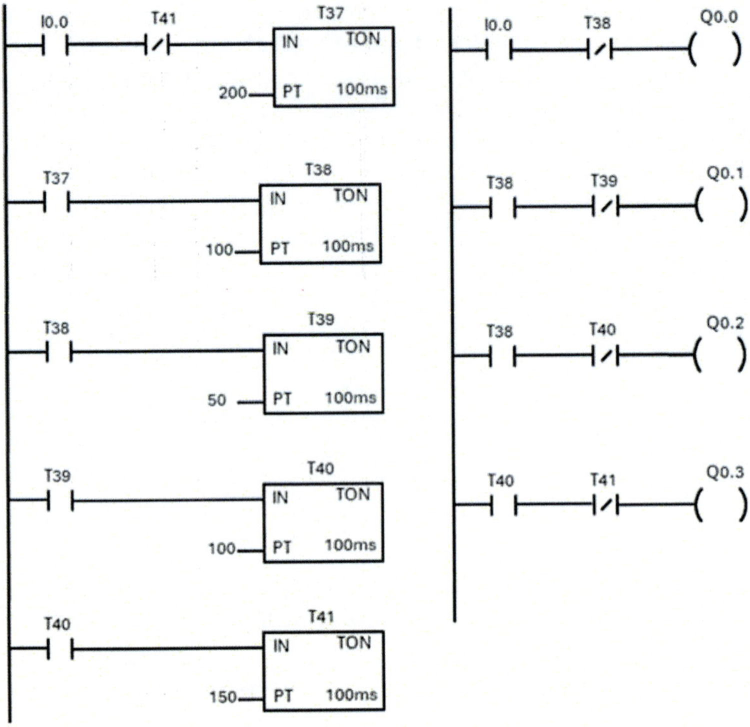

Fig. 11.23 Ladder logic diagram with the software MIcroWin®

Fig. 11.24 System with a main line with ***n*** timers and ***q*** output functions

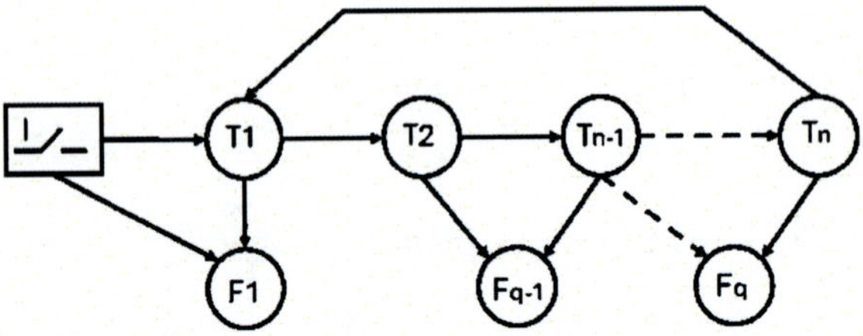

11.9 Calculations Required in a System with a Main Line

In a sequential system with timers, it is necessary to know some time values, the most important being the cycle time. This consists of the time it takes the system to develop a complete cycle, knowing this time you can calculate the production per hour, per work shift, or per day that must be obtained from a process or machine. When the system has a main line, this time is the addition of all the preset values. Figure 11.24 shows a system with a main line.

The cycle time of the system is defined as the addition of the preset values of all timers located on this line, from the timer T1 to the umpteenth timer (Tn).

Equation (Eq. 1)
HERECYCLE TIME OF A MAINLINE SYSTEM

$$T_{Ciclo} = \sum_{t=i}^{n} P_{Tt}$$

Where
Pr = Timer preset
t = *t-esimo* timer
n = Maximum number of timers on the main line
i = Time cascade start timer
T_{Ciclo} = System cycle time

Fig. 11.25 Variables for Eq. 11.1

This addition can be expressed by the following equation. This addition can be represented by a general equation where all the timers of a main line are considered, where the result of the formula is in units of time, and where the timers have the same time scale. The general formula that solves the calculation of the cycle time of a sequential system with a main line is then expressed (Fig. 11.25).

Applying the equation to the system in Fig. 11.21, this system has already been analyzed. Only the cycle time needs to be obtained. Applying the formula directly, the following calculations are obtained:

$$Tciclo = \sum_{t=1}^{5} PTt = PT1 + PT2 + PT3 + PT4 + PT5 \tag{11.13}$$

where $i = 1$ and $n = 5$, resulting in the next cycle time

$$Tciclo = PT1 + PT2 + PT3 + PT4 + PT5 \tag{11.14}$$

and substituting values

$$Tciclo = 20S + 10S + 5S + 10S + 15S = 60S. \tag{11.15}$$

11.10 Calculations of the Ignition Time of the Output Functions

In addition to the calculation of the cycle time, other data can be obtained that give more information about the sequential system, such as the time that an output function is kept on or off. This is achieved by taking advantage of the fact that

Equation (Eq. 2)
CALCULATING THE POWER-ON TIME OF
A MAINLINE-DEPENDENT FUNCTION

$$T_{Fq} = \sum_{t=c+1}^{a} P_{T\,t}$$

Where:
a= Timer with link connector in the II or III quadrant, left horizontal line or on the vertical line in both directions.
c= Timer with link connector in quadrant I, IV or right horizontal line.
TFq= Ignition time of the *q-esima* function (*Fm*).
Cf= q-esima function.
t= t-esimo timer.

Fig. 11.26 Variables of Eq. 11.2

two link connectors arrive at the output function, one in charge of turning it on and another in charge of turning it off. To obtain the time that an output function is on, the link connectors responsible for turning the function on and off are taken into account. These link connectors come from the time functions. The link connector that turns on the output function is in the first or fourth quadrant or right horizontal line, while the connector that turns off the function is called, located in the second or third quadrant, with vertical up or down direction and horizontal direction with left direction. In such a way that an addition of preset time values is made from to the timer that turns on the output function, the time value of this timer does not interfere with the time that the function is on since it just starts its on, so it takes the time of the immediate upper timer added 1. Equation 11.2 shows the application of the described explanation to obtain the ignition time of an output function (Fig. 11.26).

The diagram shown in the example in Fig. 11.21 has four output functions, and applying Eq. 11.2 gives the ignition time of each of the functions during an operation cycle:

$$Tciclo = \sum_{t=c+1}^{a} PTt \tag{11.16}$$

F1 has two link connectors one is coming from I, it is the input, and the second one is coming from T2. Where, and $\mathbf{q} = \mathbf{1}$, the value of zero is because the link connector located in the fourth quadrant comes from the input function, to have a non-zero value requires that the link connector comes from a timer, these values are substituted in Eq. 11.2 to obtain the time that the function F1 is on in each cycle.

$$TF1 = \sum_{t=0+1}^{2} PTt = \sum_{t=1}^{2} PTt = PT1 + PT2 \tag{11.17}$$

Substituting the preset time values of timers 1 and 2,

$$TF1 = PT1 + PT2 = 20S + 10S = 30S \tag{11.18}$$

For the function F2, the values of the link connectors are $c = 2$, $a = 3$, and $q = 2$. These values correspond to the timers where the link connectors come from. Substituting these values in Eq. 11.2 gives the time that the function F2 is on:

$$TF2 = \sum_{t=2+1}^{3} PTt = \sum_{t=3}^{3} PTt = PT1 + PT2 = 5S \tag{11.19}$$

The ignition time of the F3 function is in charge of the link connectors from the T2 and T4 timers, giving the values of $c = 2$, $a = 4$, and $q = 3$. Substituting these values, you have the following result:

$$TF3 = \sum_{t=2+1}^{4} PTt = \sum_{t=3}^{4} PTt = PT3 + PT4 = 5S + 10S = 15S \tag{11.20}$$

For the function F4, we have the values of $c = 4$, $a = 5$, and $q = 4$, obtained from the link connectors from the timers T4 and T5. Substituting these values, you have the following result:

$$TF4 = \sum_{t=4+1}^{5} PTt = \sum_{t=5}^{5} PTt = PT5 = 15S \tag{11.21}$$

11.11 Future Investigations

For future investigations, there are problems with more than one main line, including inputs to permit turning on the timers, including memory links affecting the turning on of the outputs. Some industrial problems need more than one timer link so it means it is necessary to use subroutines of timers.

Figure 11.27 shows a system with more than one main line. It is necessary to select which main line needs to be working, and it depends on which process needs to be worked. The process can have n main lines, and every main line means a different process or recipe.

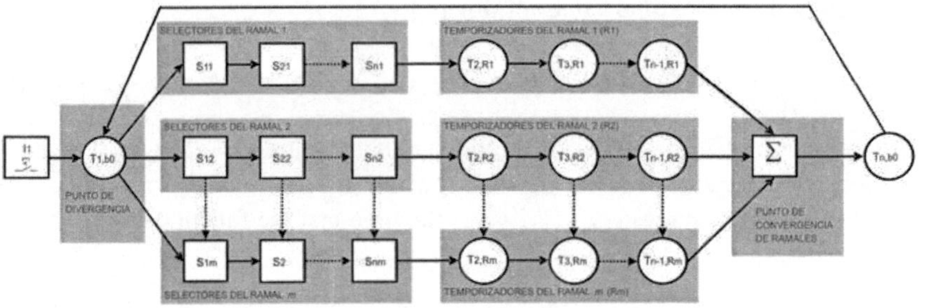

Fig. 11.27 Cascading timer method with ***n*** main lines

References

1. Ernesto Torres Quiroga. *Impacto de las corrientes geomagnéticas inducidas en el sistema eléctrico de potencia.* PhD thesis, Universidad Autónoma de Nuevo León, 2021.
2. Jairo David Centeno Valencia and Víctor Eduardo Jiménez Herrera. Manual consultivo de control neumático y electroneumático utilizando el software festo fluidsim. 2010.
3. Yana Maquera and C Juan. Automatización de máquinas industriales con la aplicación del plc simatic s7-200. *REVISTA CIENTÍFICA ELECTRÓNICA ETN Nº 1/2021*, page 17, 2021.
4. Operaciones SIMATIC. Sistema de automatización s7-200. *Memoria de la CPU: tipos de datos y direccionamiento. Sección*, pages 7–3, 1999.

Chapter 12
Mixed Sensitivity Control of Euler-Lagrange Models

R. Galindo

12.1 Introduction

Most of the control of rigid robots, mechanisms, and multi-domain systems are based on Euler-Lagrange (E-L) or Hamilton equations of motion. The E-L and Hamilton techniques are useful when there are many interconnection relations (laws of balance) since they do not require an explicit formulation of these relations. Also, robust control is a widely used control methodology for linearized Euler-Lagrange (E-L) models (see the book of [1]), and among the robust control techniques, mixed sensitivity control (MSC) ensures stability under high-frequency nonstructured uncertainties and low-frequency norm-bounded disturbances.

The recent results of an approximated analytic solution of the MSC proposed by [2] motivate the present work. It is reviewed and further developed. The results of [2] give conditions for strong stability and overcome the pole-zero cancellations between the plant and the controller of non-iterative solutions, keeping the low computational effort advantage of non-iterative solutions. The solution of [2] is based on analytic solutions of the parametrization of all stabilizing controllers or Youla-Jabr-Bongiorno-Kucera (YJBK) parametrization of [3] and [4] and on the formulae for the minimal value of the mixed sensitivity criterion of [5]. Strong stability, i.e., a stable controller, is desired for practical reasons, such as loop breaking and faults, or to minimize numerical errors. The criterion to be minimized of [5] uses the most common transfer functions and allows us to get analytically its minimum value. Low- and high-frequency approximations of these transfer functions are considered in the criterion of [2], and the minimum value is used for an optimal pole placement.

R. Galindo (✉)
Faculty of Mechanical Engineering, Autonomous University of Nuevo Leon, San Nicolas de los Garza, Mexico
e-mail: rene.galindoorz@uanl.edu.mx

E. Campos-Cantón et al. (eds.), *Complex Systems and Their Applications*,
https://doi.org/10.1007/978-3-031-51224-7_12

In Sect. 12.2, the problem statement is given, where fully actuated, with full-state information, linearized E-L models are considered. In Sect. 12.3, an additive uncertainty model is proposed based the state difference of the linearized E-L model with the nonlinear E-L model. When an input of norm 2 equal to one is applied, the norm 2 of difference of the states of these models gives an upper bound of the finite gain of the uncertainty that is used for the approximated optimal pole placement. Stability is guaranteed by the small-gain theorem when the MSC is applied to nonlinear model, and for the additive uncertainty level, an approximated optimal pole placement minimizes simultaneously the infinity norms of the output sensitivity function at low frequency and at high frequencies the transfer function from the output additive disturbance to the controller output, so assuring a desired output stationary state at low frequency and maximizing the set of admissible uncertainties at high frequencies. Also, an approximated optimal pole placement is proposed for input multiplicative uncertainty models. Also, in Sect. 12.3, based on quadratic forms, the Hamilton equations of motion are gotten, and a control law guaranteeing stability and assigning a desired stationary state for the closed-loop system is proposed based on Hamilton equations. Applications to a double pendulum on a plane and a Robot Maker 110 are given in Sect. 12.4, and the calculated torque is compared with the MSC. Finally, conclusions are given in Sect. 12.5.

Notation $\mathfrak{R}(s)$ is the set of all rational functions of the complex variable s with real coefficients. $\mathfrak{R}\mathcal{H}_\infty$ is the set of proper stable rational functions. $\mathfrak{R}$ is the set of real numbers. c_θ and s_θ are $\cos(\theta)$ and $\sin(\theta)$, respectively. For $A \in \mathfrak{R}^{m\times n}$ and $B \in \mathfrak{R}^{p\times q}$, $A \otimes B = \begin{bmatrix} a_{11}B & \cdots & a_{1n}B \\ \vdots & & \vdots \\ a_{m1}B & \cdots & a_{mn}B \end{bmatrix}$ is the Kronecker product. $\frac{\partial A}{\partial q} = \left[\left(\frac{\partial A}{\partial q_1}\right)^T \cdots \left(\frac{\partial A}{\partial q_n}\right)^T \right]^T$ is the partial derivative of a matrix $A \in \mathfrak{R}^{m\times n}$ with respect to $q \in \mathfrak{R}^{n\times 1}$. $x_{ss} := \lim_{t\to\infty} x(t)$. I_n is the identity matrix of dimensions $n \times n$. $A_l := \lim_{s\to 0} A(s)$ and $A_h := \lim_{s\to\infty} A(s)$ are the asymptotic approximations of a matrix $A(s) \in \mathfrak{R}(s)$ in low and high frequencies, respectively.

12.2 Problem Statement

Consider the fully actuated Euler-Lagrange (E-L) system,

$$M(q)\ddot{q} + V(q, \dot{q}) + G(q) + \mathcal{F}(\dot{q}) + \tau_d = \tau, \; q(t_0), \; \dot{q}(t_0), \tag{12.1}$$

where $q \in \mathfrak{R}^n$ is the vector of generalized coordinates; $q(t_0)$ and $\dot{q}(t_0)$ are the initial conditions of generalized position and velocity, respectively; $\tau_d \in \mathfrak{R}^n$ is an external disturbance; $\tau \in \mathfrak{R}^n$, $G(q) \in \mathfrak{R}^n$, and $\mathcal{F}(\dot{q}) \in \mathfrak{R}^n$ are the generalized, the gravity, and the friction forces, respectively; $M(q) = M^T(q) > 0$ is the inertia matrix; and

$V(q, \dot{q})$ is the vector associated with coriolis-centripetal effects. There is a coriolis $C(q, \dot{q})$ matrix that satisfies the skew-symmetric property:

$$\begin{aligned} C(q, \dot{q})\dot{q} &= V(q, \dot{q}), \\ \dot{M}(q) - 2C(q, \dot{q}) &= -\left(\dot{M}(q) - 2C(q, \dot{q})\right)^T. \end{aligned} \tag{12.2}$$

It is assumed that

Assumption 1 *The kinetic energy is a quadratic form, and the potential energy does not depend on* $\dot{q}$.

The kinetic energy of a system with holonomic constraints can be expressed as (see the book of [6]) $E_k = \frac{1}{2}\sum_{i=1}^{N} m_i (\dot{r}_i)^T \dot{r}_i$ where $r_1, \ldots, r_N$ are the positions of the centers of mass (CM) that are functions of q and of the time t. Hence, E_k can be decomposed in a quadratic term $\frac{1}{2}(\dot{q})^T M(q)\dot{q}$ and terms of the $\frac{\partial r_i}{\partial t}$. So, Assumption 1 is equivalent to assuming that r_i does not depend explicitly on t. The physical meaning of Assumption 1 is that when $\tau_d = \tau = 0$ and without friction, $H := E_k + E_p$ does not depend explicitly on time, and since $\frac{dH}{dt} = \frac{\partial H}{\partial t}$ (see [7]), then H is conserved.

Problem 1 Given an additive uncertainty level, design an MSC for a linearized E-L model, assuring stability for the MSC applied to the nonlinear E-L model and assigning a desired output stationary state for the closed-loop system, and compare with a proposed calculated torque control law.

In the next section, a solution to Problem 1 is given.

12.3 Mixed Sensitivity Control

First, the MSC of the work of [2] is reviewed and summarized. For simplicity, fully actuated E-L models, with full-state information, linearized around $(q_0, \dot{q}_0)$ are considered. Under Assumption 1, to linearize the E-L model (see the book of [7]), it is sufficient to replace E_k and E_p by $\frac{1}{2}\dot{q}^T M(q_0)\dot{q}$ and $\frac{1}{2}q^T \hat{B}(q_0)q$, respectively, where $\hat{B}(q_0) := \left[\frac{\partial G(q)}{\partial q}\right]_{q=q_0, \dot{q}=\dot{q}_0}$. From the works of [8] and [9], $V(q, \dot{q}) = \left(\dot{M}(q) - \frac{1}{2}U\right)\dot{q}$ where $U = \left(I_n \otimes \dot{q}^T\right)\frac{\partial M(q)}{\partial q}$, so, since $\dot{M}(q_0) = 0$ and $\frac{\partial M(q_0)}{\partial q} = 0$, then $V(q, \dot{q}) = 0$. Hence, from Eq. (12.1), the linearized E-L model is $M(q_0)\ddot{q} + \hat{B}(q_0)q + \mathcal{F}(\dot{q}) + \tau_d = \tau$, and a state-space description is

$$\begin{bmatrix} \dot{q} \\ \ddot{q} \end{bmatrix} = \begin{bmatrix} 0 & I_m \\ A_{21} & 0 \end{bmatrix}\begin{bmatrix} q \\ \dot{q} \end{bmatrix} + \begin{bmatrix} 0 \\ B_m \end{bmatrix}(\tau - \tau_d - \mathcal{F}(\dot{q})), \tag{12.3}$$

where $A_{21} := -M^{-1}(q_0)\hat{B}(q_0)$, $B_m := M^{-1}(q_0)$, the state dimension is $2n$, and n is the input dimension, as required for the mixed sensitivity control (MSC) of the work of [2]. Also, for the E-L model given by Eq. (12.3), the similarity transformations used by [2] are identity matrices.

The criterion to be minimized with the most common transfer functions of the work of [5] is

$$\gamma^* = \inf_{K(s)} \left\| \begin{bmatrix} S_o(s) & S_o(s)P(s) \\ K(s)S_o(s) & T_i(s) \end{bmatrix} \right\|_\infty, \tag{12.4}$$

where $P(s) = (sI_n - A)^{-1} B$ is the nominal plant, $K(s)$ is a stabilizing controller (see the book of [10]), $S_o(s) := (I + P(s)K(s))^{-1}$, and $T_i(s) := K(s)S_o(s)P(s)$. The criterion given by Eq. (12.4) is approximated in the work of [2] to

$$\min_{R(s),\tilde{R}(s)} \left\| \begin{bmatrix} G_l \\ G_h \end{bmatrix} \right\|_\infty \quad inf \left\{ \begin{array}{c} \min\limits_{R(s),\tilde{R}(s)} \|G_l\|_\infty \\ \text{subjet to} \\ \|G_l\|_\infty = \|G_h\|_\infty \end{array} \right. \tag{12.5}$$

where $R(s) \in \Re\mathcal{H}_\infty$ is a free control parameter of $K(s)$, $G_l := \begin{bmatrix} S_{ol} & S_{ol}P_l \end{bmatrix}$, and $G_h := \begin{bmatrix} K_h S_{oh} & T_{ih} \end{bmatrix}$. The stabilizing controller is implemented in the two-parameter feedback configuration of Fig. 12.1 (see the book of [10]), where $\tilde{P}(s)$ described by the dotted box is the generalized plant, d_1 is an external 2-norm bounded disturbance, d_2 is an input reference, z_i are the regulated outputs, y_m is the output measurement, u is the control output, and $P(s) = N(s)D^{-1}(s) = \tilde{D}^{-1}(s)\tilde{N}(s)$ are right and left coprime factorizations of $P(s)$, respectively,

$$K(s) = \tilde{D}_k^{-1}(s)\tilde{N}_k(s) \text{ and } K_r(s) = \tilde{D}_k^{-1}(s)Q(s), \tag{12.6}$$

being $\tilde{D}_k(s) = Y(s) - R(s)\tilde{N}(s)$ and $\tilde{N}_k(s) = X(s) + R(s)\tilde{D}(s)$ with $R(s) = \begin{bmatrix} R_1(s) & R_2(s) \end{bmatrix} \in \Re\mathcal{H}_\infty^{n\times 2n}$ and $Q(s) \in \Re\mathcal{H}_\infty^{n\times 2n}$ free control parameters and $X(s)$ and $Y(s)$ satisfying $X(s)N(s) + Y(s)D(s) = I_n$.

In the control scheme of Fig. 12.1, the closed-loop transfer functions from d_1 and d_2 to z_1 and z_2 are the transfer functions considered in the criterion of Eq. (12.4).

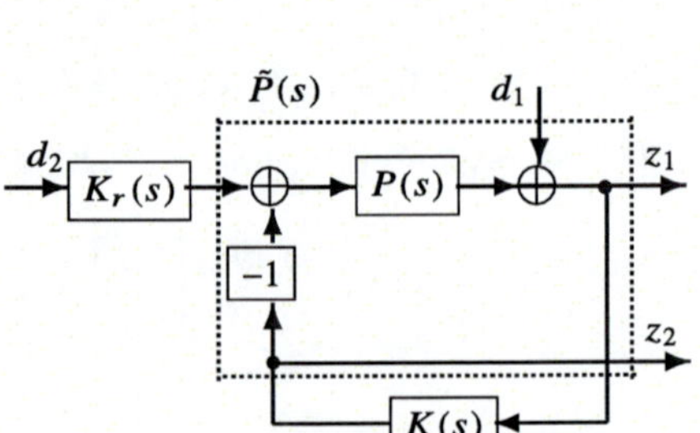

Fig. 12.1 General control scheme for the criterion (12.4)

Analytic solutions to the left and right factorizations of $P(s)$ over $\Re\mathcal{H}_\infty$ and solutions of $X(s)N(s) + Y(s)D(s) = I$ are given in the work of [4]

$$\tilde{D}(s) := \begin{bmatrix} \frac{s}{(s+a)} I_n & \frac{-1}{(s+a)} I_n \\ a\Gamma(s) & \Gamma(s) \end{bmatrix}, \tilde{N}(s) := \frac{1}{s+a} \begin{bmatrix} 0 \\ B_m \end{bmatrix}, \tag{12.7}$$

$$X(s) := \begin{bmatrix} a^2 I_n + A_{21} & 2a I_n \end{bmatrix}, Y(s) := B_m, \tag{12.8}$$

where $\Gamma(s) := \frac{1}{(s+a)^2}\left(s^2 I_n - A_{21}\right)$ and $0 < a \in \Re$. Alternative analytic solutions are given in the work of [3]. From the book of [10], the closed-loop transfer functions of Fig. 12.1 are affine functions of $N(s)$, $\tilde{N}(s)$, $D(s)$, $\tilde{D}(s)$, $R(s)$, $Q(s)$, $X(s)$, and $Y(s)$ that belong to $\Re\mathcal{H}_\infty$, since $s = -a$ is a stable pole. So, the closed-loop transfer functions also belong to $\Re\mathcal{H}_\infty$ as desired.

The following solution to the restriction equation $\|G_l\|_\infty = \|G_h\|_\infty$ of Eq. (12.4) that is part of Problem 1, when the velocity entries of the state input reference are zero and the state can be measured, estimated, or known, is proposed in the work of [2]. The solution is illustrated in Fig. 12.2.

Theorem 12.1 *Consider the plant $P(s) \in \Re^{2n\times n}$ having a controllable realization given by Eq. (12.3) in the two-parameter control configuration of Fig. 12.1. Let the state input reference be $x_d(t) = \begin{bmatrix} y_d(t) & 0 \end{bmatrix}^T$ and $R(s)$ be $\begin{bmatrix} 0 \; rI_m \end{bmatrix} \in \Re\mathcal{H}_\infty^{m\times n}$, where $r \in \Re$, and suppose that $K(s)$ is given by Eq. (12.6) and that A_{21} and B_m are non-singular matrices. Then, the value of r solving $\|G_l\|_\infty = \|G_h\|_\infty$ of the approximated criterion given by Eq. (12.5) is*

$$r(a) = \frac{a\left(\rho - a^2 b_3(a)\right)}{a^2(b_2(a) - b_3(a)) + \rho}, \tag{12.9}$$

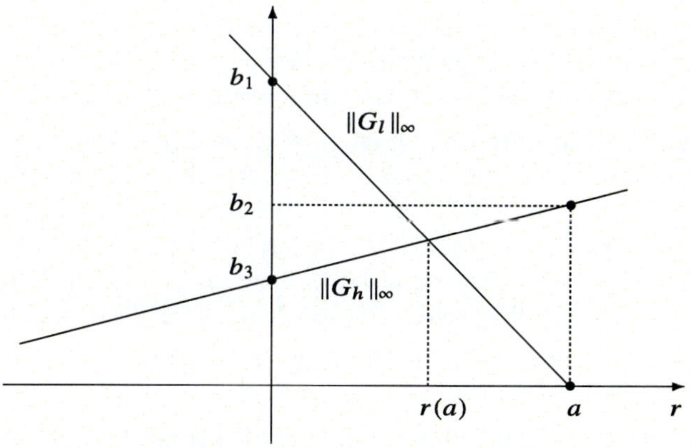

Fig. 12.2 Intersection function for one $d.o.f.$ feedback configuration

where $0 < a \in \Re$,

$$\begin{aligned} \rho &:= \left\| \left[-A_{21} \; B_m \right] \right\|_\infty, \\ b_2(a) &:= \left\| \left[B_m^{-1} \left(2a^2 I_n + A_{21}\right) \; \tfrac{1}{w_h} 3a I_n \right] \right\|_\infty, \\ b_3(a) &:= \left\| \left[B_m^{-1} \left(a^2 I_n + A_{21}\right) \; \tfrac{1}{w_h} 2a I_n \right] \right\|_\infty, \end{aligned} \tag{12.10}$$

being w_h *a fixed frequency in the high-frequency bandwidth of* $P(s)$*. Moreover,*

$$\|G_l\|_\infty = \|G_h\|_\infty = \tfrac{b_2 \rho}{a^2 (b_2 - b_3) + \rho}, \tag{12.11}$$

where $G_l := \left[S_{ol} \; S_{ol} P_l \right]$ *and* $G_h := \left[K_h S_{oh} \; T_{ih}. \right]$

Proof See [2] ■

Let $(A, B, I_n, 0)$ be a controllable realization of $P(s)$. From the work of [2], if $a^2 \gg \|A_{21}\|_\infty$ and $a^4 \|B_m\|_\infty \gg \rho$, then an approximately optimal pole placement is $s = -a^*$ with

$$a^* \cong \sqrt{\tfrac{2\rho}{\gamma^*}}, \tag{12.12}$$

where from the work of [5]

$$\gamma^* = \sqrt{1 + \lambda_{max}(Z X_r)}, \tag{12.13}$$

being $Z \geq 0$ and $X_r \geq 0$ the solutions of the algebraic Riccati equations $A_f Z + Z A_f^* - Z R_r^{-1} Z + B \left(I - E^* R_r^{-1} E\right)^{-1} B^* = 0$ and $X_r A_k + A_k^* X_r - X_r B S^{-1} B^* X_r + \left(I - E S^{-1} E^*\right) = 0$, being $R_r := I + E E^*$, $A_f := A - B E^* R_r^{-1}$, $A_k := A - B S^{-1} E^*$, and $S := I + E^* E$. The criterion of Eq. (12.4) is simplified using a normalized left coprime factorization proposed in the work of [5], and for it, γ^* is gotten, so, applying the small-gain theorem (see the book of [1]) to Lure's systems, coprime factor uncertainties whose ∞-norm is less than $\frac{1}{\gamma^*}$ is the set of admissible uncertainties. Given an uncertainty level $\frac{1}{\gamma} > \frac{1}{\gamma^*}$, since the ∞-norm of the closed-loop transfer functions decreases as a increases, this set increases as the value of a increases.

The closed-loop poles are located at $s = -a$, and if the plant satisfies the parity interlacing property (see the book of [10]), there exists a stable $K(s)$ stabilizing $P(s)$, that is, strong stability can be achieved. In this case, from $\tilde{D}_k(s)$ with $Y(s) = B_m$ and $R(s)$ being $\left[0 \; r I_n \right]$, the poles of $K(s)$ are located at $s = -(a - r(a)) = \frac{-a^3 b_2(a)}{a^2 (b_2(a) - b_3(a)) + \rho}$.

In the following, the approximated optimal pole placement given by Eq. (12.12) is improved, and $\|G_l\|_\infty = \|G_h\|_\infty$ takes the value of γ^* or γ for optimal or suboptimal control. The set of admissible uncertainties is $\left\{ \Delta(s) : \|\Delta(s)\|_\infty \leq \frac{1}{\|G_h\|_\infty} \right\}$, where for additive and input multiplicative uncertainty models $\|G_l\|_\infty = \|G_h\|_\infty$ is $\|K_h S_{oh}\|_\infty$ and $\|T_{ih}\|_\infty$, respectively.

Corollary 12.1 *Under the assumptions and definitions of Theorem 12.1. For additive uncertainty models, if*

$$\|K_h S_{oh}\|_\infty^2 - \theta \|K_h S_{oh}\|_\infty - \beta\rho_1 \leq 0, \tag{12.14}$$

where $\theta := \left\|B_m^{-1} A_{21}\right\|_\infty$, $\beta := \left\|B_m^{-1}\right\|_\infty$, *and* $\rho_1 := \left\|\left[A_{21}\right]\right\|_\infty$, *then select* $a > 0$ *from*

$$a^2 \cong \frac{\beta\rho_1 \pm \sqrt{\beta\rho_1 - \|K_h S_{oh}\|_\infty (\|K_h S_{oh}\|_\infty - \theta)}\sqrt{\beta\rho_1}}{\beta\|K_h S_{oh}\|_\infty}. \tag{12.15}$$

Moreover, selecting $\|K_h S_{oh}\|_\infty > 0$ *from*

$$\|K_h S_{oh}\|_\infty = \frac{1}{2}\left(\theta \pm \sqrt{\theta^2 + 4\beta\rho_1}\right), \tag{12.16}$$

then

$$a \cong \sqrt{\frac{\rho_1}{\|K_h S_{oh}\|_\infty}}. \tag{12.17}$$

For input multiplicative uncertainty models, if

$$\|T_{ih}\|_\infty \leq \sqrt[3]{\frac{4\rho_2}{w_h^2}}, \tag{12.18}$$

where $\rho_2 := \|B_m\|_\infty$ *and* w_h *is a fixed frequency in the high-frequency bandwidth of* $P(s)$, *then select* a_1, a_2, *or* a_3 *real and positive from*

$$\begin{aligned} a_1 &= 2\sqrt{\frac{\rho_2}{\|T_{ih}\|_\infty}} cos(\tfrac{\theta}{3}), \\ a_2 &= 2\sqrt{\frac{\rho_2}{\|T_{ih}\|_\infty}} cos(\tfrac{\theta}{3} + \tfrac{2\pi}{3}), \\ a_3 &= 2\sqrt{\frac{\rho_2}{\|T_{ih}\|_\infty}} cos(\tfrac{\theta}{3} + \tfrac{4\pi}{3}), \end{aligned} \tag{12.19}$$

where $\theta = cos^{-1}\left(\frac{w_h\|T_{ih}\|_\infty}{2}\right)\sqrt{\frac{\|T_{ih}\|_\infty}{\rho_2}}$.

Proof For additive uncertainty models, from Eq. (12.10), ρ is ρ_1, and applying triangle inequality,

$$\begin{aligned} b_2(a) &\leq 2a^2\beta + \theta, \\ b_3(a) &\leq a^2\beta + \theta. \end{aligned} \tag{12.20}$$

Taken the upper bounds of these inequalities, from Eq. (12.11),

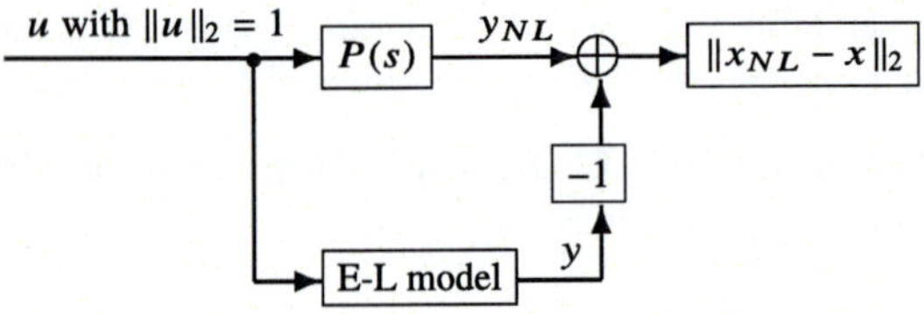

Fig. 12.3 Finite gain of the additive uncertainty model

$$\left\| K_h S_{oh} \right\|_\infty \cong \frac{(2a^2\beta+\theta)\rho_1}{\beta a^4+\rho_1}, \tag{12.21}$$

that is, $\beta \left\| K_h S_{oh} \right\|_\infty a^4 - 2\beta\rho_1 a^2 + \rho_1 (\left\| K_h S_{oh} \right\|_\infty - \theta) \cong 0$ that has real roots under condition (12.14), and the result of Eq. (12.15) follows. When condition (12.14) is an equality, its solution given by Eq. (12.16), then Eq. (12.15) simplifies to the result of Eq. (12.17).

For input multiplicative uncertainty models, from Eq. (12.10), ρ is ρ_2, and applying triangle inequality,

$$\begin{aligned} b_2(a) &\leq \tfrac{3a}{w_h} + \tfrac{1}{w_h}\psi, \\ b_3(a) &\leq \tfrac{2a}{w_h} + \tfrac{1}{w_h}\psi. \end{aligned} \tag{12.22}$$

Taken the upper bounds of these inequalities, from Eq. (12.11),

$$\left\| T_{ih} \right\|_\infty \cong \frac{3a\rho_2}{a^3+\rho_2 w_h}, \tag{12.23}$$

that is, $\left\| T_{ih} \right\|_\infty a^3 - 3\rho_2 a + \rho_2 w_h \left\| T_{ih} \right\|_\infty \cong 0$. This cubic equation has a discriminant $D = \rho_2^2 \left(\frac{w_h^2}{4} - \frac{\rho_2}{\left\| T_{ih} \right\|_\infty^3} \right)$ that has real roots if $D \leq 0$, i.e., inequality (12.18), and its solution is the result of Eq. (12.19). ∎

As shown in Fig. 12.3, an additive uncertainty model is proposed based on the difference of the linearized E-L model with the nonlinear E-L model. Usually, $\|x_{NL} - x\|_2$ increases with time, and for Lure's systems, the output of the additive uncertainty satisfies $\|y_\Delta\|_2 \leq K_1\|x\|_2 + K_2\|u\|_2$ (see [11]). So, the following is proposed:

Proposition 12.1 *Select the higher uncertainty of (i) u such that* $\|u\|_2 = 1$ *and* $x(0) = 0$ *or (ii)* $u = 0$ *and* $x(0) \neq 0$*. Let an approximated upper bound of* $\|x_{NL} - x\|_2$ *be of the form kt for* $t > t_1$ *where k is the biggest slope of* $\|x_{NL} - x\|_2$ *or equivalently* $\Delta(s) \cong k\frac{1}{s^2}$*. Hence, an upper bound of the finite gain is,*

$$\|\Delta(s)\|_\infty \leq k. \tag{12.24}$$

So,

$$\|K_h S_{oh}\|_\infty \cong \tfrac{1}{k}, \tag{12.25}$$

is used for the approximated optimal pole placement proposed in Corollary 12.1.

Under Proposition 12.1, stability of the uncertainty system, that is, of the MSC applied to the nonlinear E-L model in closed loop, is guaranteed by the small-gain theorem (see [12]).

In the next section, a calculated torque is proposed based on Hamilton's equations and quadratic forms.

12.4 Calculated Torque Control

A useful and well-known property of quadratic forms of symmetric matrices is used. To highlight it, letting $A = A^T \in \Re^{2\times 2}$ and $x \in \Re^2$, then the quadratic form of A is $x^T Ax = a_{11}x_1^2 + 2a_{12}x_1x_2 + a_{22}x_2^2$ that is used either to obtain the quadratic form given A or to get the coefficients of A given the quadratic form. Based on quadratic forms, the following is proposed:

Proposition 12.2 *Assuming that E_k is a quadratic form and E_p does not depend on $\dot{q}$, the Hamiltonian equations are*

$$\begin{aligned} \dot{p} &= \tfrac{-1}{2}\begin{bmatrix} p^T Q_1(q)p \\ \vdots \\ p^T Q_n(q)p \end{bmatrix} - G(q) + B\tau - \tau_d - F(\dot{q}), \\ \dot{q} &= M^{-1}(q)p, \end{aligned} \tag{12.26}$$

where $Q_i(q) := \frac{\partial M^{-1}(q)}{\partial q_i}$, $i = 1, \ldots, n$, $p \in \Re^n$ is the generalized momentum, and the terms of $\frac{1}{2}p^T Q_i(q)p$ are the (i, i) and (i, j) with $i \neq j$ elements of $Q_i(q)$ multiplied by $\frac{1}{2}p_i^2$ and $p_i p_j$, respectively, for $i = 1, \ldots, n$ and $j = 1, \ldots, n$.

Proof Under Assumption 1, the generalized momentum (see the book of [6] is $p = \frac{\partial L}{\partial \dot{q}} = M(q)\dot{q}$; hence, the Hamiltonian function is $H(q, p) = \frac{1}{2}p^T M^{-1}(q)p + E_p$. So, from the works of [8] and [9],

$$\frac{\partial H(q,p)}{\partial q} = \frac{1}{2}\left(I_n \otimes p^T\right)\frac{\partial M^{-1}(q)}{\partial q}p + G(q), \tag{12.27}$$

and the result follows from the Hamilton equations of motion, $\dot{p} = \frac{-\partial H(q,)}{\partial q} + B\tau - \tau_d - F(\dot{q})$, $\dot{q} = \frac{\partial H}{\partial p}$, and the definition of the Kronecker product. ∎

In case for certain i, q_i is a cyclic coordinate (see the book of [6]), then $Q_i(q) = 0$, simplifying Eq. (12.26), and when $\tau_d = \tau = 0$ and without friction, the generalized momentum $\frac{\partial L}{\partial \dot{q}_i}$ is conserved, where $L := E_k - E_p$ is the Lagrangian function. Physically, it is due to certain symmetry of the rigid body.

The Hamilton equations and the output $y := B^T M^{-1}(q)p$ that satisfies $\frac{dH}{dt} = y^T \tau$ are important in energy conservation and passivity analysis (see for instance the work of [13]) and passivity-based control (see, for instance, the work of [14]). The results of [14] can be applied to the Hamilton equations noting that the state is $x = \left[p^T \; q^T\right]^T$, the co-energy is $z = \left[\left(\frac{\partial H(q,p)}{\partial q}\right)^T \; \left(\frac{\partial H(q,p)}{\partial p}\right)^T\right]^T$, and the dissipative input and output are $D_i = \dot{q}$ and $D_o = F(\dot{q})$, respectively. Instead, Proposition 12.1 arrives at an explicit first-order state equation that can be used to control analysis and design. In the works of [15] and [14], it is proposed that the controller has a structure similar to the plant. A calculated torque control law for fully actuated systems is proposed in the following:

Lemma 12.1 *Under Assumption 1, suppose that the fully actuated plant is described by Eq. (12.26) of Proposition 12.2 and is interconnected with a controller in feedback with no loading effect. Then, the control law*

$$\begin{aligned} \tau &= -K_a p + \frac{1}{2} \begin{bmatrix} p^T Q_1(q) p \\ \vdots \\ p^T Q_n(q) p \end{bmatrix} + G(q) + \tau_d + F(\dot{q}) + M(q)\tilde{\tau}, \\ \tilde{\tau} &= K_p \left(q_d - q\right) + K_d \left(\dot{q}_d - \dot{q}\right), \end{aligned} \tag{12.28}$$

where $q_d \in \Re^n$ is a desired reference input and $K_a > 0$, $K_p > 0$, and $K_d > 0$ are control gains. This control law arrives at the stable closed-loop system

$$\begin{aligned} \dot{p} &= -K_a p + M(q)\tilde{\tau}, \\ \dot{q} &= M^{-1}(q)p, \end{aligned} \tag{12.29}$$

and in the stationary state

$$q_{ss} = q_{dss} \text{ and } p_{ss} = 0. \tag{12.30}$$

Proof The closed-loop system follows directly substituting τ from Eq. (12.28) in Eq. (12.26) of the plant. Also, from Eqs. (12.28) and (12.29), the stationary state (SS) of the closed-loop system is $p_{ss} = K_a^{-1} M(q_{ss})\tilde{\tau}_{ss}$ and $\tilde{\tau}_{ss} = K_p\left(q_{dss} - q_{ss}\right)$, implying from Eq. (12.29) that $q_{ss} = q_{dss}$ and $p_{ss} = 0$. ■

Alternatively, a calculated torque control can be designed for the E-L model given by Eq. (12.1), and K_a is not required; however, the approach of Lemma 12.1 is in the control of Hamiltonian systems. Also, other controls for linear systems such as optimal or robust control can be used instead of the proportional derivative control $\tilde{\tau}$ of Eq. (12.28). The control gain K_a allows tuning the dynamics of p and hence indirectly the dynamics of q.

In the following section, the MSC is compared with the calculated torque of Lemma 12.1. The MSC and the calculated torque are applied to a double pendulum on a plane and a Robot Maker 110.

12.5 Application to the Control of E-L Models

Example 12.1 A double pendulum in a plane or planar rotational robot is shown in Fig. 12.4, where q_1 and q_2 are the joint variables, l_1 and l_2 are the lengths of the links, m_1 and m_2 are the masses concentrated on the CM, and for simplicity, the inertias are neglected. In the worst case for control and design, the CMs are considered at the end of the links.

The position vectors of the CM are

$$r_1 = \begin{bmatrix} l_1 c_{q_1} \\ l_1 s_{q_1} \\ 0 \end{bmatrix}, r_2 = \begin{bmatrix} l_1 c_{q_1} + l_2 c_{(q_1+q_2)} \\ l_1 s_{q_1} + l_2 s_{(q_1+q_2)} \\ 0 \end{bmatrix}. \tag{12.31}$$

So, the translational kinetic energy has the quadratic form $E_k = \frac{1}{2}\dot{q}^T M(q)\dot{q}$, where

$$M(q) = \begin{bmatrix} \theta + 2m_2 l_1 l_2 c_{q_2} & m_2 l_2^2 + m_2 l_1 l_2 c_{q_2} \\ m_2 l_2^2 + m_2 l_1 l_2 c_{q_2} & m_2 l_2^2 \end{bmatrix}, \tag{12.32}$$

being $\theta := (m_1 + m_2) l_1^2 + m_2 l_2^2$. Since q_1 is a cyclic coordinate, then $Q_1(q) = 0$. From the works of [8] and [9], $V(q, \dot{q}) = \left(\dot{M}(q) - \frac{1}{2}U\right)\dot{q}$, and $C(q, \dot{q}) = \frac{1}{2}\left(\dot{M}(q) + U^T - U\right)$ where $U = \left(I_n \otimes \dot{q}^T\right)\frac{\partial M(q)}{\partial q}$. So,

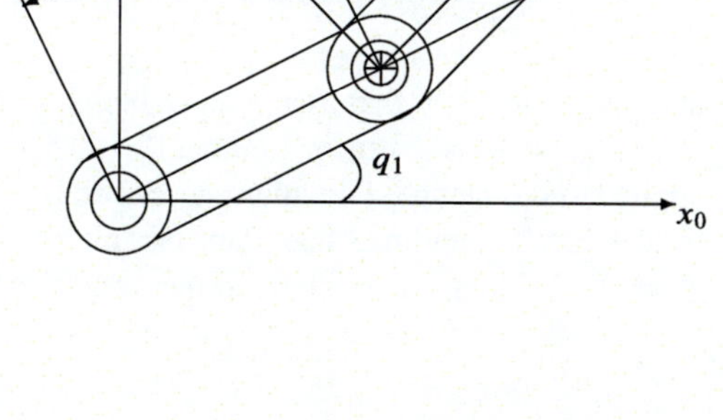

Fig. 12.4 Double pendulum in a plane

$$V\left(q,\dot{q}\right)=\begin{bmatrix} w_{12}\dot{q}_2 \\ m_2 l_1 l_2 \dot{q}_1^2 s_{q_2} \end{bmatrix}, \tag{12.33}$$

where $w_{12} := -m_2 l_1 l_2 \left(2\dot{q}_1+\dot{q}_2\right) s_{q_2}$. The potential energy is $E_p = m_1 g l_1 s_{q_1} + m_2 g \left(l_1 s_{q_1} + l_2 s_{(q_1+q_2)}\right)$, so,

$$G(q)=\begin{bmatrix} m_1 g l_1 c_{q_1} + m_2 g \left(l_1 c_{q_1} + l_2 c_{(q_1+q_2)}\right) \\ m_2 g l_2 c_{(q_1+q_2)} \end{bmatrix}, \tag{12.34}$$

arriving at the E-L equations given by Eq. (12.1). If $C\left(q,\dot{q}\right)$ is required,

$$U^T - U = \begin{bmatrix} 0 & w_{12} \\ -w_{12} & 0 \end{bmatrix}, \tag{12.35}$$

and

$$C\left(q,\dot{q}\right)=\begin{bmatrix} -m_2 l_1 l_2 \dot{q}_2 s_{q_2} & -m_2 l_1 l_2 \left(\dot{q}_1+\dot{q}_2\right) s_{q_2} \\ m_2 l_1 l_2 \dot{q}_1 s_{q_2} & 0 \end{bmatrix}. \tag{12.36}$$

It can be verified that $\frac{1}{2}\left(U^T-U\right)\dot{q} = V\left(q,\dot{q}\right) - \frac{1}{2}\dot{M}(q)\dot{q}$.

Applying Proposition 12.2 and Lemma 12.1,

$$M^{-1}(q)=\frac{1}{l_1^2\psi(q_2)}\begin{bmatrix} 1 & -\left(1+\phi_1 c_{q_2}\right) \\ -\left(1+\phi_1 c_{q_2}\right) & \phi_2 + 2\phi_1 c_{q_2} \end{bmatrix}, \tag{12.37}$$

where $\psi(q_2) := m_1 + m_2 s_{q_2}^2$, $\phi_1 := \frac{l_1}{l_2}$ and $\phi_2 := \frac{\theta}{m_2 l_2^2}$, and $p^T Q_1(q) p = 0$,

$$\begin{aligned} Q_2(q) = &\frac{-2m_2 s_{q_2} c_{q_2}}{l_1^2\psi^2(q_2)}\begin{bmatrix} 1 & -\left(1+\phi_1 c_{q_2}\right) \\ -\left(1+\phi_1 c_{q_2}\right) & \phi_2 + 2\phi_1 c_{q_2} \end{bmatrix} \\ &+\frac{1}{l_1^2\psi(q_2)}\begin{bmatrix} 0 & \phi_1 s_{q_2} \\ \phi_1 s_{q_2} & -2\phi_1 s_{q_2} \end{bmatrix}. \end{aligned} \tag{12.38}$$

So,

$$\begin{aligned} \tfrac{1}{2}p^T Q_2(q) p = &\frac{-m_2 s_{q_2} c_{q_2}}{l_1^2\psi^2(q_2)} p_1^2 + \frac{2m_2 s_{q_2} c_{q_2}\left(1+\phi_1 c_{q_2}\right)}{l_1^2\psi^2(q_2)} p_1 p_2 + \\ &\frac{\phi_1 s_{q_2}}{l_1^2\psi(q_2)} p_1 p_2 - \frac{m_2 s_{q_2} c_{q_2}\left(\phi_2 + 2\phi_1 c_{q_2}\right)}{l_1^2\psi^2(q_2)} p_2^2 - \frac{\phi_1 s_{q_2}}{l_1^2\psi(q_2)} p_2^2, \end{aligned} \tag{12.39}$$

and the control law that assures a stable closed-loop system and assigns a desired stationary state is given in Lemma 12.1. The control law given by Eq. (12.28) is applied to the Double Pendulum described by Proposition 12.2 and is implemented in MATLAB Simulink. The plant parameters are $m_1 = 2$, $m_2 = 1$, $l_1 = 2$, and $l_2 = 1$. It is considered zero initial conditions $q(t_0) = 0$, $\dot{q}(t_0) = 0$, a desired

Table 12.1 MSC design parameters

	Eq. (12.13)	Eq. (12.12)	Eq. (12.11)
Double Pendulum	$\gamma^* = 124.943$	$a^* = 0.627$	$\|K_h S_{oh}\|_\infty = 83.676$
Robot Maker 110	$\gamma^* = 2.568$	$a^* = 6.192$	$\|K_h S_{oh}\|_\infty = 2.564$
Corollary 12.1	Eq. (12.16)	Eq. (12.15) or Eq. (12.17)	Eq. (12.11)
Double Pendulum	$\|K_h S_{oh}\|_\infty = 84.309$	$a = 0.539$	$\|K_h S_{oh}\|_\infty = 84.309$
Robot Maker 110	$\|K_h S_{oh}\|_\infty = 34.702$	$a = 1.191$	$\|K_h S_{oh}\|_\infty = 32.063$
Proposition 12.1	Eq. (12.25)	Eq. (12.15) or Eq. (12.17)	Eq. (12.11)
Double Pendulum	$\|K_h S_{oh}\|_\infty = \frac{1}{15}$	$a = 27.186$	$\|K_h S_{oh}\|_\infty = 0.066$
Robot Maker 110	$\|K_h S_{oh}\|_\infty = \frac{1}{32.5}$	$a = 56.574$	$\|K_h S_{oh}\|_\infty = 0.030$

reference input $q_d = \begin{bmatrix} 40 \ 20 \end{bmatrix}^T$, $\dot{q}_d = 0$, viscous friction $F(\dot{q}) = 0.5\dot{q}$, without external disturbance, i.e., $\tau_d = 0$, and the constant of gravity $g = 9.81$. If $\tau = 0$, an equilibrium point of the double pendulum is $q_0 = \begin{bmatrix} \frac{\pi}{2} \ 0 \end{bmatrix}^T$. So,

$$M(q_0) = \begin{bmatrix} \theta + 2m_2 l_1 l_2 & m_2 l_2^2 + m_2 l_1 l_2 \\ m_2 l_2^2 + m_2 l_1 l_2 & m_2 l_2^2 \end{bmatrix}, \tag{12.40}$$

$$\hat{B} = \begin{bmatrix} -m_1 g l_1 - m_2 g (l_1 + l_2) & -m_2 g l_2 \\ -m_2 g l_2 & -m_2 g l_2 \end{bmatrix}, \tag{12.41}$$

and the linearized plant is given by Eq. (12.3), which is a minimal realization and satisfies the parity interlacing property since it does not have transmission zeros. Then, the MSC is designed for the linearized E-L model, and the relevant parameters are shown in Table 12.1.

Using a^*, the stabilizing controller has a "good" performance when it is applied to the linear model of the plant; however, due to the uncertainty, the output has significant oscillations when it is applied to the nonlinear model given by Eq. (12.26). As expected, applying Corollary 12.1, a from Eq. (12.15) or Eq. (12.17), arrives to similar values of $\|K_h S_{oh}\|_\infty$ than a^*, as shown in Table 12.1. To increase the size of the admissible uncertainties, an approximated uncertainty level is obtained from Fig. 12.3 and Proposition 12.1, as shown by Fig. 12.5. It is proposed the approximation $\|x_{NL} - x\|_2 = 15t$ for $t > 0.6$ sec., so, $k = 15$ is used on Proposition 12.2, and Corollary 12.1 gives the approximated pole placement $s = -a$ that achieves the desired value of $\|K_h S_{oh}\|_\infty = \frac{1}{15}$ as shown in Table 12.1. This value of a is selected for $K(s)$, as shown in Figs. 12.6, 12.7, 12.8, 12.9. So, from Eq. (12.9), $r = -27.331$. Also, the control gains $K_a = K_d = aI_2$ and $K_p = a^2 I_2$, $R(s)$ being $\begin{bmatrix} 0 \ rI_2 \end{bmatrix}$ and $Q(s)$ being $\begin{bmatrix} a^2 I_2 \ 0 \end{bmatrix}$, are selected. The overshoot of the output increases, and the time response decreases as the value of K_a decreases. Also, the state of the linearized E-L model used for $K(s)$ is related to the state of the nonlinear model by $\dot{q} = M^{-1}(q)p$.

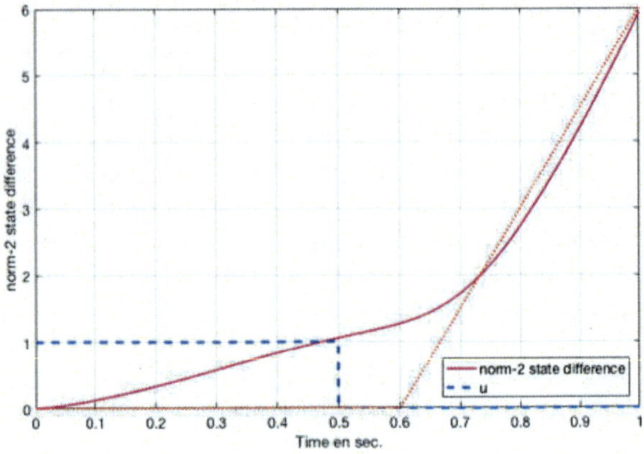

Fig. 12.5 $\|x_{NL} - x\|_2$ of the double pendulum

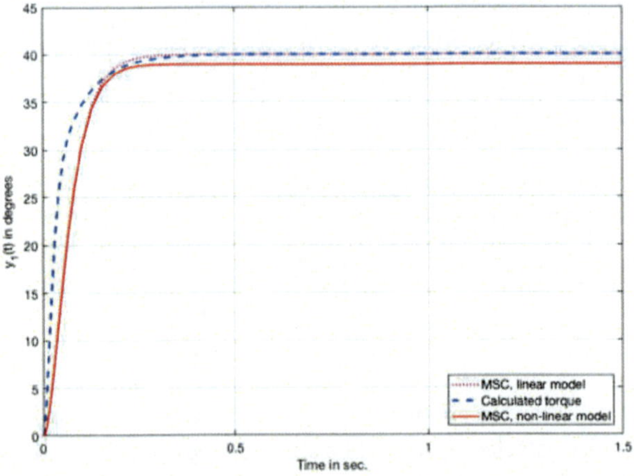

Fig. 12.6 Angular position y_1 of the double pendulum

Figures 12.6, 12.7, 12.8, 12.9 show a stable and "good" performance of the closed-loop system, where q converges to q_d in the stationary state, except when the MSC is applied to the nonlinear system that the output has a "small" stationary state error. As shown in Figs. 12.6 and 12.7, the time response of the output of the calculated torque applied to the nonlinear system is "small" than other control strategies, and so, a "bigger" control magnitude is required as shown in Figs. 12.8 and 12.9. □

Example 12.2 A three-degrees-of-freedom robot is shown in Fig. 12.11, where a, b, and f are constants, q_1 and q_3 are revolute, and q_2 is a prismatic joint-variables, m_1, m_2, and m_3 are the masses concentrated on the CM, and for simplicity, the inertias are neglected. Considering the worst case, the CMs are considered at the end of the links.

From Fig. 12.10, the position vectors of the CM are

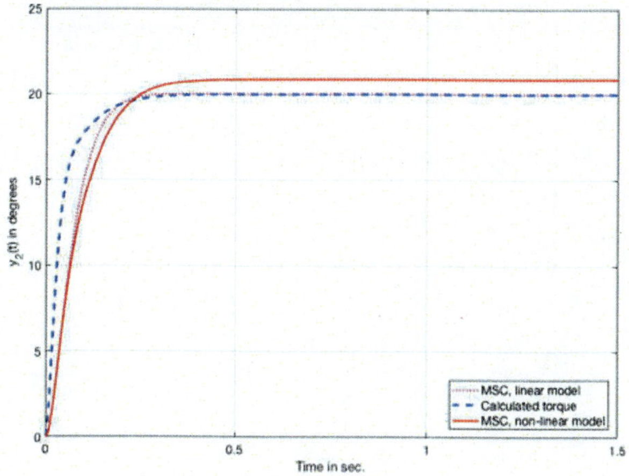

Fig. 12.7 Angular position y_2 of the double pendulum

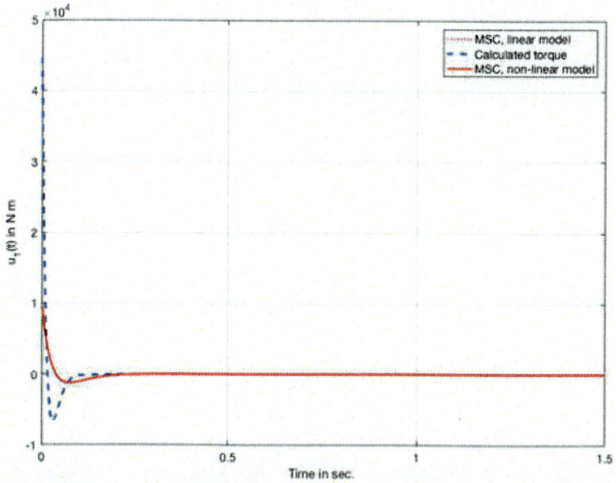

Fig. 12.8 Torque applied u_1 to the double pendulum

$$r_1 = \begin{bmatrix} ac_{q_1} \\ as_{q_1} \\ b \end{bmatrix}, r_2 = \begin{bmatrix} ac_{q_1} - q_2 s_{q_1} \\ as_{q_1} + q_2 c_{q_1} \\ b \end{bmatrix}, r_3 = \begin{bmatrix} ac_{q_1} - \left(q_2 + f c_{q_3}\right) s_{q_1} \\ as_{q_1} + \left(q_2 + f c_{q_3}\right) c_{q_1} \\ b - f s_{q_3} \end{bmatrix}. \tag{12.42}$$

An attempt to reduce the CM does not have physical meaning, and hence, the properties of the E-L model can be loss. So, the translational kinetic energy has the quadratic form $E_k = \frac{1}{2}\dot{q}^T M(q)\dot{q}$, where

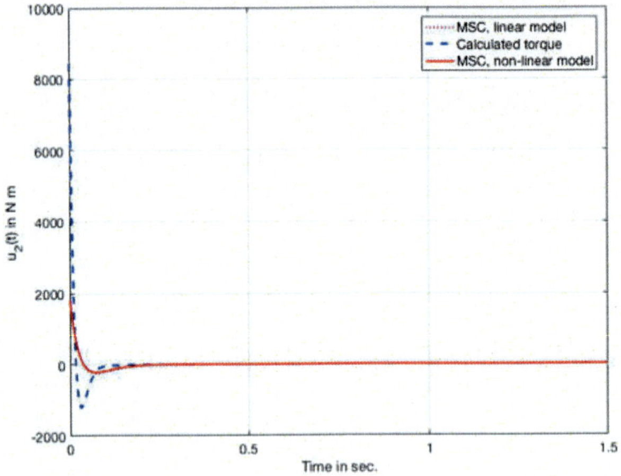

Fig. 12.9 Torque applied u_2 to the double pendulum

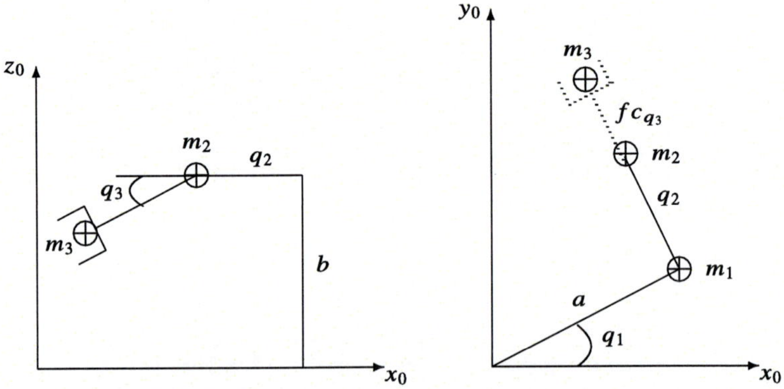

Fig. 12.10 From left to right, side, and top views of the Robot Maker 110

$$M(q) = \begin{bmatrix} \theta_1 + m_2\left(q_2 + f c_{q_3}\right)^2 + m_2 q_2^2 & \theta_4 & -\theta_3 s_{q_3} \\ \theta_4 & m_2 + m_3 & -\theta_2 s_{q_3} \\ -\theta_3 s_{q_3} & -\theta_2 s_{q_3} & \theta_2 f s_{q_3}^2 \end{bmatrix}, \tag{12.43}$$

being $\theta_1 := (m_1 + m_2 + m_3)\, a^2$, $\theta_2 := m_3 f$, $\theta_3 := m_3 a f$, and $\theta_4 := (m_2 + m_3)\, a$.

Since q_1 is a cyclic coordinate, then $Q_1(q) = 0$. So,

$$\dot{M}(q) = \begin{bmatrix} 2m_2\left(q_2 + f c_{q_3}\right)\left(\dot{q}_2 - f \dot{q}_3 s_{q_3}\right) + 2m_2 q_2 \dot{q}_2 & 0 & -\theta_3 \dot{q}_3 c_{q_3} \\ 0 & 0 & -\theta_2 \dot{q}_3 c_{q_3} \\ -\theta_3 \dot{q}_3 c_{q_3} & -\theta_2 \dot{q}_3 c_{q_3} & 2\theta_2 f \dot{q}_3 s_{q_3} c_{q_3} \end{bmatrix}. \tag{12.44}$$

Fig. 12.11 Robot Maker 110

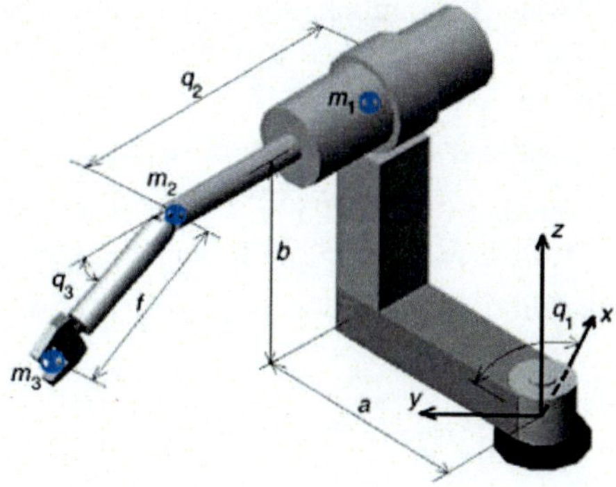

Also, $V(q, \dot{q}) = \begin{bmatrix} (w_{12} + 2m_2 q_2 \dot{q}_1)\, \dot{q}_2 + w_{13}\dot{q}_3 \\ w_{23}\dot{q}_3 - m_2\left(2q_2 + f c_{q_3}\right)\dot{q}_1^2 \\ m_2 f\left(q_2 + f c_{q_3}\right)\dot{q}_1^2 s_{q_3} - w_{23} f \dot{q}_3 s_{q_3} \end{bmatrix}$ where

$$\begin{aligned} w_{12} &:= 2m_2\left(2q_2 + f c_{q_3}\right)\dot{q}_1, \\ w_{13} &:= -2m_2 f\left(q_2 + f c_{q_3}\right)\dot{q}_1 s_{q_3} - \theta_3 \dot{q}_3 c_{q_3}, \\ w_{23} &:= -\theta_2 \dot{q}_3 c_{q_3}. \end{aligned} \tag{12.45}$$

The potential energy is $E_p = (m_1 + m_2)\, gb + m_3 g\left(b - f s_{q_3}\right)$, so,

$$G(q) = \begin{bmatrix} 0 \\ 0 \\ -m_3 g f c_{q_3} \end{bmatrix}, \tag{12.46}$$

arriving at the E-L equations given by Eq. (12.1). If $C(q, \dot{q})$ is required,

$$U^T - U = \begin{bmatrix} 0 & w_{12} & w_{13} \\ -w_{12} & 0 & w_{23} \\ -w_{13} & -w_{23} & 0 \end{bmatrix}, \tag{12.47}$$

and

$$C(q, \dot{q}) = \begin{bmatrix} c_{11} & c_{12} & -\theta_3 \dot{q}_3 c_{q_3} - c_{31} \\ -c_{12} & 0 & -\theta_2 \dot{q}_3 c_{q_3} \\ c_{31} & 0 & \theta_2 f \dot{q}_3 s_{q_3} c_{q_3} \end{bmatrix}, \tag{12.48}$$

where $c_{11} := m_2\left(q_2 + f c_{q_3}\right)\left(\dot{q}_2 - f \dot{q}_3 s_{q_3}\right) + m_2 q_2 \dot{q}_2$, $c_{12} := m_2\left(2q_2 + f c_{q_3}\right)\dot{q}_1$ and $c_{31} := m_2 f\left(q_2 + f c_{q_3}\right)\dot{q}_1 s_{q_3}$.

The MSC is compared with a calculated torque $\tau = M(q)\bar{\tau} + C(q, \dot{q})\,\dot{q} + G(q) + F(\dot{q})$ where $\bar{\tau} = K_p(q_d - q) + K_d(\dot{q}_d - \dot{q})$ and is implemented in

MATLAB Simulink. A comparison of the MSC with the iterative function *hinfsyn* of MATLAB applied to a two-cart system is given in the work of [2]. The plant parameters are $m_1 = 2$, $m_2 = m_3 = 1$, $a = 2$, $b = 1$, and $f = 0.5$. It is considered initial conditions $q(t_0) = \begin{bmatrix} 0 \; 0 \; \frac{\pi}{2} \end{bmatrix}^T$, $\dot{q}(t_0) = 0$, a desired reference input $q_d = \begin{bmatrix} 30 \; 0.5 \; 50 \end{bmatrix}^T$, $\dot{q}_d = 0$, viscous friction $F(\dot{q}) = 0.5\dot{q}$, without external disturbance, i.e., $\tau_d = 0$, and the constant of gravity $g = 9.81$. If $\tau = 0$, an equilibrium point is $q_0 = \begin{bmatrix} 0 \; 0 \; \frac{\pi}{2} \end{bmatrix}^T$. So,

$$M(q_0) = \begin{bmatrix} \theta_1 & \theta_4 & -\theta_3 \\ \theta_4 & m_2 + m_3 & -\theta_2 \\ -\theta_3 & -\theta_2 & \theta_2 f \end{bmatrix}, \tag{12.49}$$

$$\hat{B} = \begin{bmatrix} 0 \; 0 & 0 \\ 0 \; 0 & 0 \\ 0 \; 0 & \theta_2 f \end{bmatrix}, \tag{12.50}$$

and the linearized plant is given by Eq. (12.3), which is a minimal realization and satisfies the parity interlacing property since it does not have transmission zeros. Using a^*, the stabilizing controller has a "good" performance with smooth trajectories when it is applied to the linear model of the plant; however, due to the uncertainty, the stability is not guaranteed, the output has significant oscillations, and the output has significant stationary state error when it is applied to the E-L nonlinear model. To increase the size of the admissible uncertainties, an approximated uncertainty level is obtained from Fig. 12.3 and Proposition 12.1, as shown by Fig. 12.12. It is proposed the approximation $\|x_{NL} - x\|_2 = 32.5t$ for $t > 0$ sec., so, $k = 32.5$ is used on Proposition 12.2, and Corollary 12.1 gives the approximated pole placement $s = -a$ that achieves the desired value of $\|K_h S_{oh}\|_\infty = \frac{1}{32.5}$ as shown in Table 12.1. This value of a is selected for $K(s)$, as shown in Figs. 12.13, 12.14, 12.15, 12.16, 12.17, 12.18. So, from Eq. (12.9), $r = -56.574$. Also, the control gains $K_d = 2aI_3$ and $K_p = aI_3$, $R(s)$ being $\begin{bmatrix} 0 \; rI_3 \end{bmatrix}$ and $Q(s)$ being $\begin{bmatrix} a^2 I_3 \; 0 \end{bmatrix}$, are selected.

Figures 12.13, 12.14, 12.15, 12.16, 12.17, 12.18 show a stable and "good" performance of the closed-loop system, where q converges to q_d in the stationary state, except when the MSC is applied to the nonlinear system that the output has a "small" stationary state error. As shown in Figs. 12.13 and 12.15, the time response of the output of the calculated torque applied to the nonlinear system is "small" than other control strategies, and so, a "bigger" control magnitude is required as shown in Figs. 12.16 and 12.18. Also, all the controllers require the same amount of energy at stationary state. Since the closed-loop poles are located at $-a$, the time response decreases, the closed-loop low-frequency bandwidth increases, and the overshoots of $u_1(t)$, $u_2(t)$, and $u_3(t)$ are "bigger" as a increases, as required for the MSC applied to the nonlinear model. □

Finally, conclusions are given in the next section.

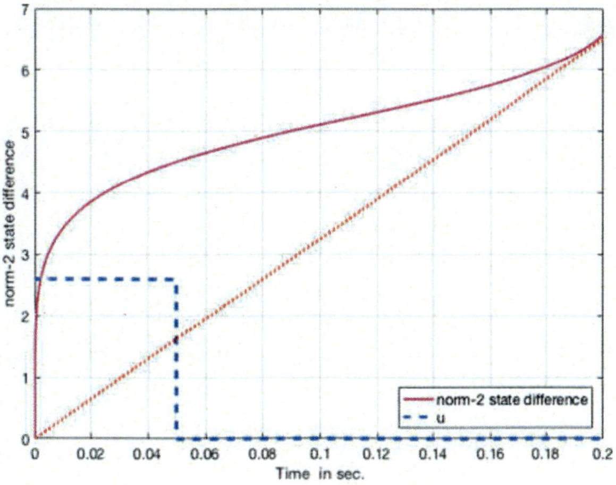

Fig. 12.12 $\|x_{NL} - x\|_2$ of the Robot Maker

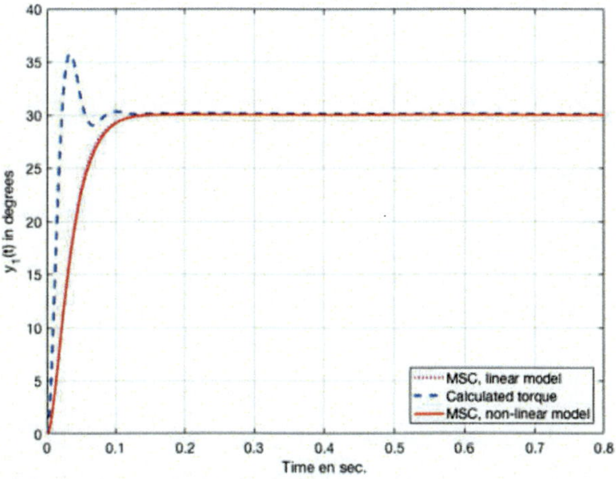

Fig. 12.13 Angular position y_1 of the Robot Maker

12.6 Conclusions

The mixed sensitivity control applied to linearized Euler-Lagrange (E-L) models has less computational effort than iterative methodologies. The examples show "good" output performance in closed loop for the calculated torque proposed to control Hamiltonian systems and the mixed sensitivity control applied to the nonlinear model. Stability is guaranteed by the small-gain theorem when the pole placement is gotten from a given approximated additive uncertainty level. Also, taking advantage of the involved quadratic forms arising in the E-L equations based on Kronecker

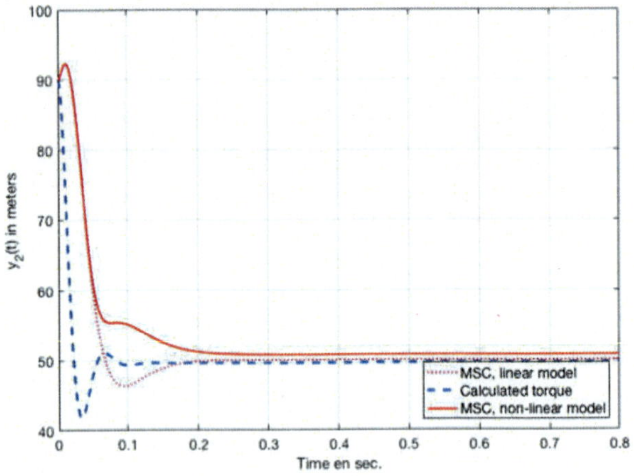

Fig. 12.14 Position y_2 of the Robot Maker

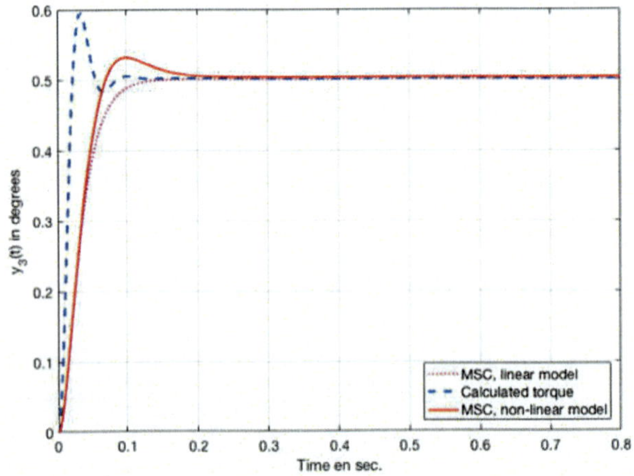

Fig. 12.15 Angular position y_3 of the Robot Maker

products, the Hamilton equations of motion are gotten. In particular, simplifications can easily be seen when there are cyclic coordinates or when the elements of the inertia matrix are constant. The applications to a double pendulum on a plane and a Robot Maker 110 clearly show less work and complexity when the MSC or the calculated torque is applied. Moreover, the proposed methodology has implications for the design and control of systems.

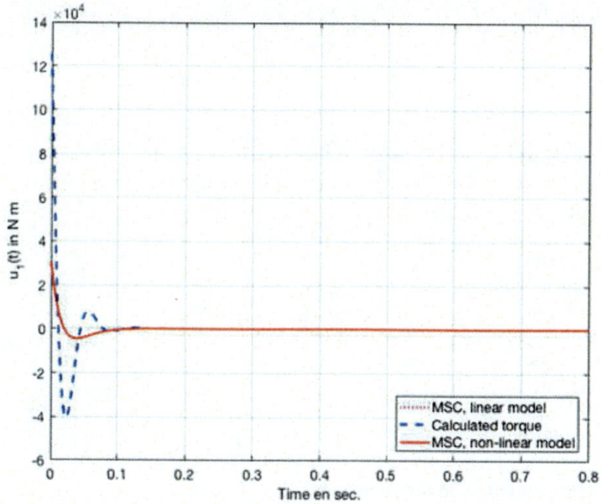

Fig. 12.16 Torque applied u_1 to the Robot Maker

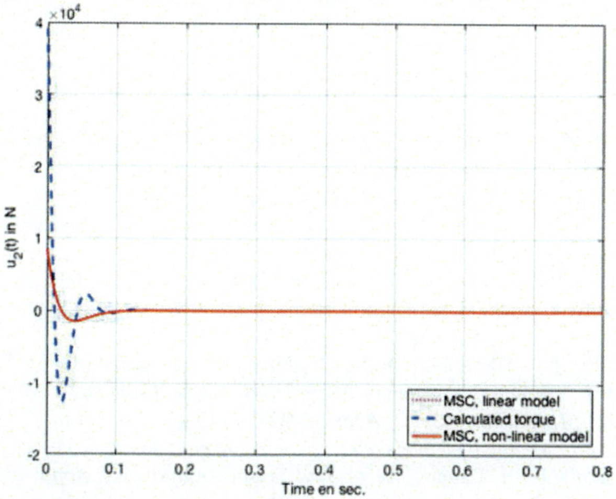

Fig. 12.17 Force applied u_2 to the Robot Maker

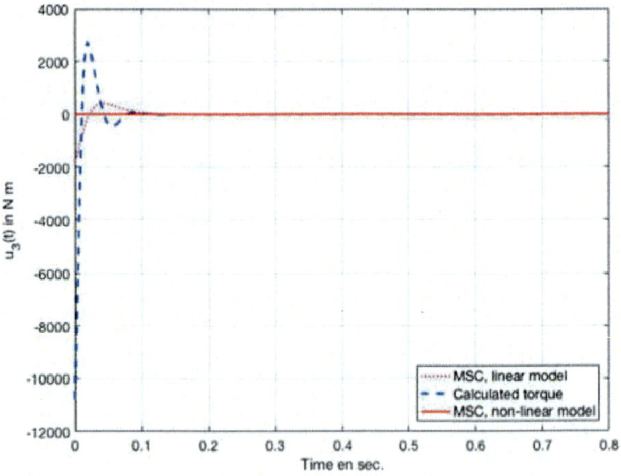

Fig. 12.18 Torque applied u_3 to the Robot Maker

Competing Interests The author has no conflicts of interest to declare that are relevant to the content of this chapter.

References

1. K. Zhou, J. C. Doyle, and K. Glover. *Robust and Optimal Control*. Prentice Hall, 1996.
2. R. Galindo. Mixed sensitivity control: A non-iterative approach. *System Science and Control Engineering: An Open Access Journal*, 8(1):442–453, 2020.
3. R. Galindo. Input/output decoupling of square linear systems by dynamic two-parameters stabilizing control. *Asian J. of Control*, 18 (6):2310–2316, 2016.
4. R. Galindo and C. Conejo. A parametrization of all one parameter stabilizing controllers and a mixed sensitivity problem, for square systems. pages 171–176. Int. Conference on Electrical Engineering, Computing Science and automatic Control, 2012.
5. K. Glover and D. McFarlane. Robust stabilization of normalized coprime factor plant descriptions with bounded uncertainty. *IEEE Trans. Autom. Control*, 34:821–830, 1989.
6. H. Goldstein. *Classical mechanics*. Addison-Wesley, 1980.
7. V.I. Arnold. *Mathematical methods of classical mechanics*. Springer-Verlag, 1989.
8. A.I. Borisenko and I.E. Tarapov. *Vector and Tensor Analysis with Applications*. Englewood Cliffs, New York, Prentice Hall, 1968.
9. F.L. Lewis, D.M. Dawson, and Ch.T. Abdallah. *Robot Manipulator Control Theory and Practice*. Marcel Dekker Inc., New York, 2004.
10. M. Vidyasagar. *Control System Synthesis: A Factorization Approach*. The MIT Press Cambridge, 1985.
11. G.E. Dullerud and F. Paganini. *A Course in Robust Control Theory: A Convex Approach*. Springer, 1999.
12. H. K. Khalil. *Non-Linear Systems*. Prentice Hall, 1996.
13. A. van der Schaft. Port-Hamiltonian systems: an introductory survey. Proc. of the Int. Congress of Mathematicians, 2006.
14. R. Galindo and R. F. Ngwompo. Passivity analysis and control of nonlinear systems modelled by bond graphs. *International Journal of Control*, 0(0):1–11, 2022.
15. R. Galindo and R. F. Ngwompo. Passivity-based control of linear time-invariant systems modelled by bond graph. *Int. J. of Control*, pages 1–17, 2017.

Part IV
Applications of Chaotic and Complex Systems

Chapter 13
A New 4-D Four-Scroll Hyperchaotic System with Multistability, Coexisting Attractors and Its Circuit Realization

Sundarapandian Vaidyanathan, Fareh Hannachi, and Aceng Sambas

13.1 Introduction

Chaotic systems and control techniques have numerous applications in engineering and science [1]. Chaotic dynamical systems have a wide range of applications in fields such as oscillators [2, 3], lasers [4, 5], robotics [6, 7], biology [8, 9], economics [10, 11], deep learning [12, 13], memristors [14, 15], etc. To improve the security of chaotic secure communication and information encryption, many research studies have used chaotic systems with complex dynamic behaviors [16–18].

Recently, significant attention has been made in the chaos literature on the modeling, and applications of multi-scroll hyperchaotic systems [19–22]. Yan et al. [19] derived an image encryption algorithm based on a multi-scroll hyperchaotic system. Jia et al. [20] proposed a new memristor-based multi-scroll hyperchaotic system using a modulating sine nonlinear function and a memristor. Pan et al. [21] designed a fractional-order multi-scroll hyperchaotic system and implemented it

S. Vaidyanathan (✉)
Centre for Control Systems, Vel Tech University, Chennai, Tamil Nadu, India

Centre of Excellence for Research, Value Innovation & Entrepreneurship (CERVIE), UCSI University, Kuala Lumpur, Malaysia
e-mail: sundar@veltech.edu.in

F. Hannachi
Larbi Tebessi University - Tebessi, Tebessa, Algeria
e-mail: Fareh.hannachi@univ-tebessa.dz

A. Sambas
Faculty of Informatics and Computing, Universiti Sultan Zainal Abidin, Terengganu, Malaysia

Department of Mechanical Engineering, Universitas Muhammadiyah Tasikmalaya, Jawa Barat, Indonesia
e-mail: acengsambas@unisza.edu.my; acengs@umtas.ac.id

E. Campos-Cantón et al. (eds.), *Complex Systems and Their Applications*,
https://doi.org/10.1007/978-3-031-51224-7_13

with an electronic circuit. Wang et al. [22] proposed a multi-scroll 4-D hyperchaotic system using a jerk system and designed an electronic circuit.

In this work, we propose a new four-scroll hyperchaotic system by adding a feedback control to the Zhang chaotic system with a cubic nonlinearity [23]. Using bifurcation analysis, we investigate the novel features of new four-scroll hyperchaotic system. We carry out a detailed bifurcation analysis of the new 4-D four-scroll hyperchaotic system and describe the changes in the qualitative properties of the new system by varying the parameter values.

Recently, there is significant interest in reporting chaotic and hyperchaotic systems having multistability property [24, 25]. In this research work, we exhibit that the new 4-D four-scroll hyperchaotic system has multistability with coexisting chaotic attractors, and we discuss the offset boosting control [26, 27] for the new four-scroll hyperchaotic system. For real-world applications, we have designed an electronic circuit of the new 4-D hyperchaotic four-scroll system using MultiSim. Circuit designs of chaotic and hyperchaotic systems are useful in practical applications of the systems [28–30].

13.2 Description of the New Four-Scroll Hyperchaotic System

In 2016, Zhang et al. [23] presented the 3-D four-scroll chaotic system:

$$\begin{cases} \dot{x} = -ax + cyz, \\ \dot{y} = -dy^3 + bxz, \\ \dot{z} = bz - xy. \end{cases} \tag{13.1}$$

Here, $a, b, c, d,$ are positive parameters. The 3-D Zhang system (13.1) exhibits a four-scroll chaotic attractors for the parameter values taken as follows:

$$a = 2, \ b = 3, \ c = 10, \ d = 6. \tag{13.2}$$

For the above parameter values and the initial conditions $x_0 = 1, y_0 = -1, z_0 = 1$, the Lyapunov exponents were computed using MATLAB for $T = 1E4$ seconds as follows:

$$L_1 = 0.7290, \ L_2 = 0, L_3 = -12.6221. \tag{13.3}$$

The Zhang chaotic system (13.1) is dissipative for the parameter values (13.2), and it exhibits a four-scroll chaotic attractor.

In this work, we present a new 4-D hyperchaotic four-scroll system obtained by introducing a new state variable w to the Zhang system (13.1). The new 4-D system dynamics is described as follows:

$$\begin{cases} \dot{x} &= -ax + cyz, \\ \dot{y} &= -dy^3 + bxz - w, \\ \dot{z} &= bz - xy, \\ \dot{w} &= py. \end{cases} \tag{13.4}$$

where x, y, z, w are the state variables and a, b, c, d, p are positive parameters. There are eight terms on the right-hand side, but it mainly relies on four nonlinearities, namely, y^3, xy, xz, yz, respectively.

We shall show that the new 4-D system (13.4) exhibits a four-scroll hyperchaotic attractor for the parameter values taken as follows:

$$a = 6, b = 6, c = 10, d = 5, p = 0.3. \tag{13.5}$$

We take the initial values as follows:

$$x_0 = 1, y_0 = -1, z_0 = 1, w_0 = -1. \tag{13.6}$$

The Lyapunov exponents (LEs) were computed using MATLAB as follows:

$$L_1 = 0.91781, L_2 = 0.01537, L_3 = 0, L_4 = -45.21791. \tag{13.7}$$

From (7), the 4-D system (13.4) has two positive LE values with $L_1 + L_2 + L_3 + L_4 < 0$; thus, the new hyperchaotic system (13.4) is dissipative for the parameter values (13.6), and it exhibits a four-scroll hyperchaotic attractor.

The maximum Lyapunov exponent (MLE) of the new hyperchaotic system is $L_1 = 0.917811$, which is greater than the maximum Lyapunov exponent (MLE) of the Zhang system given by $L_1 = 0.7290$.

We also note that the new hyperchaotic four-scroll system (13.4) has rotation symmetry about the z-axis.

Also, The Kaplan-Yorke dimension of system (13.4) is calculated as:

$$D_{KL} = 3 + \frac{L_1 + L_2 + L_3}{|L_4|} = 3.01970. \tag{13.8}$$

Next, we calculate the balance points of the system (13.4) and check their stability.

To find the balance points of the system (13.4), we solve the equations

$$-ax + cyz = 0, \tag{13.9a}$$

$$-dy^3 + bxz - w = 0, \tag{13.9b}$$

$$bz - xy = 0, \tag{13.9c}$$

$$py = \quad 0. \tag{13.9d}$$

Since $p > 0$, we derive from Eq. (13.9d) that $y = 0$.

Substituting $y = 0$ into Eq. (13.9a) and (13.9c), we get $x = 0$ and $z = 0$.

Substituting $x = y = 0$ in Eq. (13.9b), we get $y = 0$.

Thus, we have shown that $E_0 = (0, 0, 0, 0)$) is the unique balance point of the 4-D system (13.4).

In order to check the stability of E_0, we derive the Jacobian matrix at E_0. The Jacobian matrix of the system (13.4) at E_0 has the eigenvalues:

$$\lambda_1 = 6,\ \lambda_2 = -6,\ \lambda_3 = -0.54772i,\ \lambda_4 = 0.54772i. \tag{13.10}$$

Hence, E_0 is a saddle-focus equilibrium point of the new four-scroll hyperchaotic system (13.4), which is unstable.

MATLAB simulations of the new four-scroll hyperchaotic system (13.4) in various 2D planes for the initial state $(1, -1, 1, -1)$ and parameter set $(a, b, c, d, p) = (6, 6, 10, 5, 0.3)$ are depicted in Fig. 13.1.

13.3 Dynamic Analysis of the New Four-Scroll Hyperchaotic System

In this section, we investigate numerically the dynamical behaviors of the new four-scroll hyperchaotic system (13.4) using the Lyapunov exponents spectrum and bifurcation diagrams.

13.3.1 Varying the Parameter a

Figure 13.2 shows the Lyapunov exponents spectrum and the bifurcation diagram of the new four-scroll hyperchaotic system (13.4) with respect to parameter a.

We fix the values of the parameters b, c, d and p as $(b, c, d, p) = (6, 10, 5, 0.3)$. We vary the parameter a in the range $[5.5, 6.5]$.

We can identify the dynamic behavior of the system (13.4) when the parameter a varies in the range $[5.5, 6.5]$ as follows:

Let $a \in [5.5, 6.5]$. We define:

$A = [5.52, 5.5.667] \cup [5.715, 5.782] \cup [5.84, 5.926] \cup [6.076, 6.177]$.
$B = [6.236, 6.28] \cup [6.307, 6.439] \cup [6.481, 6.5]$.
$D = [5.5, 5.52) \cup (5.667, 5, 715) \cup (5.782, 5.84) \cup (5.926, 6.076)$.
$E = (6.177, 6.236) \cup (6.28, 6.307) \cup (6.439, 6.481)$.

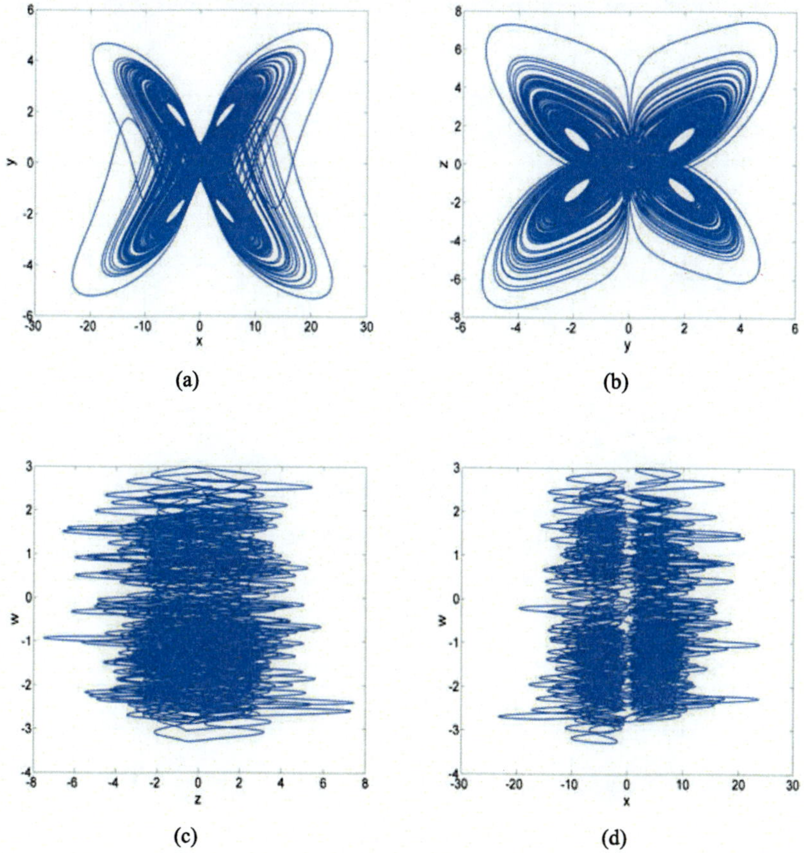

Fig. 13.1 MATLAB simulations of the new four-scroll hyperchaotic system (13.4) in various 2D planes for the initial state $(1, -1, 1, -1)$ and parameter set $(a, b, c, d, p) = (6, 6, 10, 5, 0.3)$

When $a \in A \cup B$, we can see from Figs. 13.9 and 13.10 that the system (4) has only one positive Lyapunov exponent ($L_1 > 0$) and the other Lyapunov exponents are negative ($L_{2,3,4} < 0$). Thus, the system (4) is chaotic and generate chaotic attractor. The values of Lyapunov exponents obtained when $a = 5.75$ are:

$$L_1 = 0.9925, L_2 = -0.006968, L_3 = -0.05346, L_4 = -44.17, \tag{13.11}$$

and the values of Lyapunov exponents when $a = 6.4$ are:

$$L_1 = 0.9933, L_2 = -0.02838, L_3 = -0.0336, L_4 = -48.27. \tag{13.12}$$

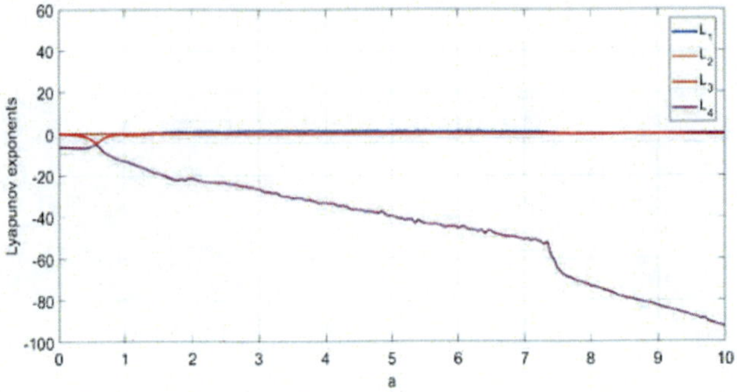

(a)

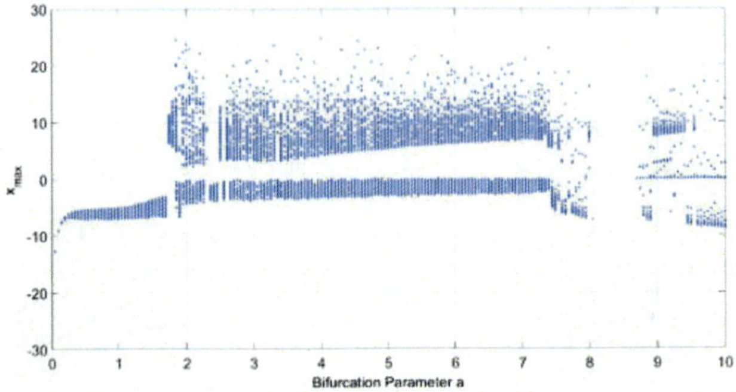

(b)

Fig. 13.2 Lyapunov exponents spectrum and bifurcation diagram of the 4-D system (13.4) for $a \in [0, 10]$

When $a \in D \cup E$, the system (4) has two positive Lyapunov exponents ($L_{1,2} > 0$) and two negative Lyapunov exponent ($L_{3,4} < 0$). Thus the system(4) is hyperchaotic in this range.

The values of Lyapunov exponents when $a = 5.5$ are:

$$L_1 = 0.8276, L_2 = 0.03422, L_3 = -0.4854, L_4 = -43.25, \tag{13.13}$$

and the values of Lyapunov exponents when $a = 5.8$ are:

$$L_1 = 0.8069, L_2 = 0.0295, L_3 = -0.01223, L_4 = -45.16, \tag{13.14}$$

and the values of Lyapunov exponents when $a = 6.2$ are:

$$L_1 = 0.9665, L_2 = 0.009683, L_3 = -0.03279, L_4 = -46.15, \tag{13.15}$$

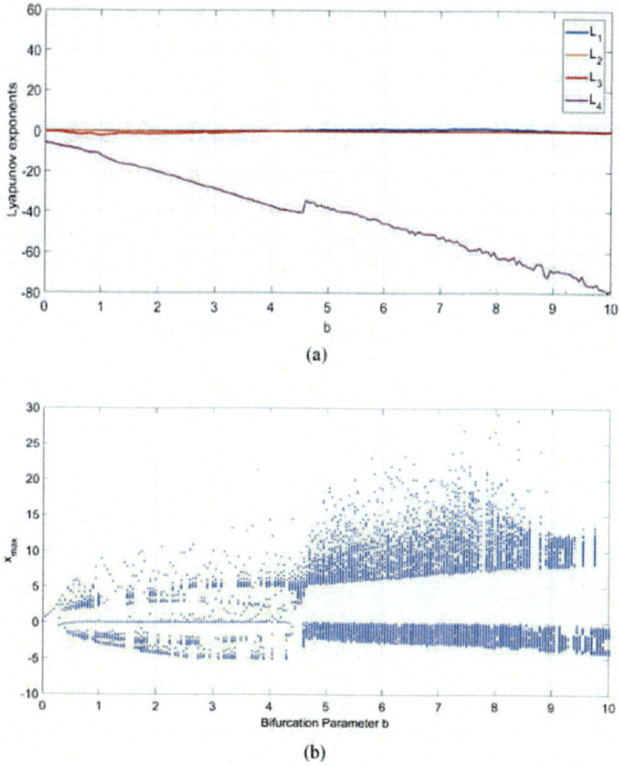

Fig. 13.3 Lyapunov exponents spectrum and bifurcation diagram of the 4-D system (13.4) for $b \in [0, 10]$

13.3.2 *Varying the Parameter b*

Figure 13.3 shows the Lyapunov exponents spectrum and the bifurcation diagram of the new four-scroll hyperchaotic system (13.4) with respect to parameter b.

We fix the values of the parameters a, c, d and p as $(a, c, d, p) = (6, 10, 5, 0.3)$. We vary the parameter b in the range $[5.5, 6.5]$.

We can identify the behavior of the system (13.4) when the parameter b varies in the range $[5.5, 6.5]$ as follows:

Let $b \in [5.5, 6.5]$. We define:

$E = [5.5, 5.58] \cup [5.168, 5.668] \cup [5.895, 5.967] \cup [6.026, 6.241] \cup [6.308, 6.5]$.

$F = (5.58, 5.618) \cup (5.668, 5.8) \cup (5.8, 5.895) \cup (5.967, 6.026) \cup (6.241, 6.308)$.

When $b \in E$, we can see from Figs. 13.9 and 13.10 that the system (13.4) has only one positive Lyapunov exponent ($L_1 > 0$) and the other Lyapunov exponents are negative ($L_{2,3,4} < 0$). Thus, the system (13.4) is chaotic and generate chaotic attractor. The values of Lyapunov exponents obtained when $b = 5.5$ are:

$$L_1 = 0.8139, L_2 = -0.0244, L_3 = -0.0566, L_4 = -40.85, \tag{13.16}$$

and the values of Lyapunov exponents when $b = 6.4$ are:

$$L_1 = 1.0423, L_2 = -0.01333, L_3 = -0.04437, L_4 = -48.23. \tag{13.17}$$

When $b \in F$, the system (13.4) has two positive Lyapunov exponents ($L_{1,2} > 0$) and two negative Lyapunov exponent ($L_{3,4} < 0$). Thus the system (13.4) is hyperchaotic in this range. The values of Lyapunov exponents when $b = 5.6$ are:

$$L_1 = 0.9181, L_2 = 0.008483, L_3 = -0.04265, L_4 = -42.13, \tag{13.18}$$

and the values of Lyapunov exponents when $b = 6.3$ are:

$$L_1 = 0.9069, L_2 = 0.003911, L_3 = -0.07449, L_4 = -48.11. \tag{13.19}$$

We also note that the new system (13.4) exhibits periodic behavior for some small ranges for the parameter b. The values of Lyapunov exponents when $b = 0.5$ are:

$$L_1 = 0.00, L_2 = -0.09272, L_3 = -1.252, L_4 = -8.561. \tag{13.20}$$

Also, the values of Lyapunov exponents when $b = 2$ are:

$$L_1 = 0.00, L_2 = -0.1531, L_3 = -1.365, L_4 = -20.1. \tag{13.21}$$

13.3.3 Varying the Parameter c

Figure 13.4 shows the Lyapunov exponents spectrum and the bifurcation diagram of the new four-scroll hyperchaotic system (13.4) with respect to parameter c.

We fix the values of the parameters a, b, d and p as $(a, b, d, p) = (6, 6, 5, 0.3)$. We vary the parameter c in the range [9.5, 10.5].

We can identify the behavior of the system (13.4) when the parameter c varies in the range [9.5, 10.5] as follows:

Let $c \in [9.5, 10.5]$. We define:

$$\begin{aligned}
A &= [9.5, 9.565) \cup (9.7825, 9.845) \cup (9.855, 9.918) \cup (9.982, 10.019)\,. \\
B &= (10092, 10.107) \cup (10.18325, 10.2215) \cup (10.394, 10.405) \cup (10.48, 10.5]\,. \\
C &= (9.565, 9.7825) \cup (9.845, 9.855) \cup (9.918, 9.982) \cup (10.019, 10.092)\,. \\
D &= (10.107, 10.18325) \cup (10.2215, 10.394) \cup (10.405, 10.48)\,.
\end{aligned}$$

When $c \in A \cup B$, the system (13.4) has two positive Lyapunov exponents ($L_{1,2} > 0$) and two negative Lyapunov exponents ($L_{3,4} < 0$). This implies the existence of

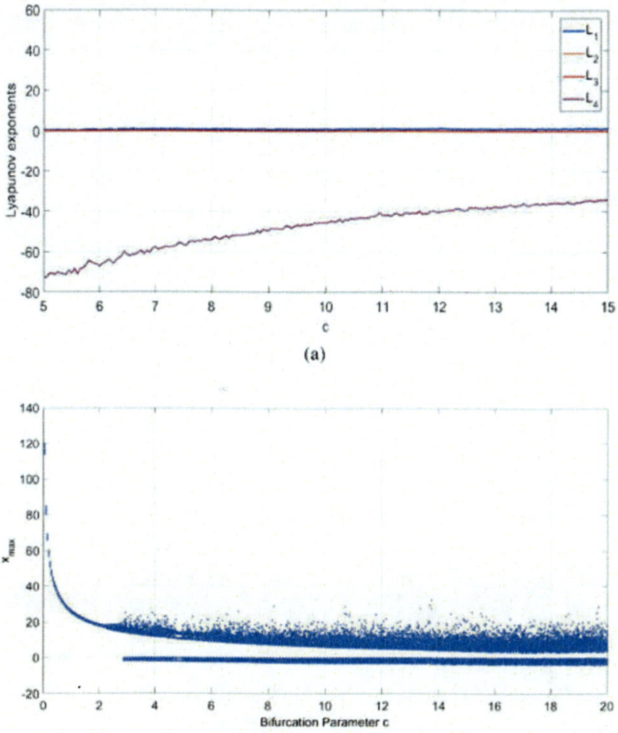

(a) (b)

Fig. 13.4 Lyapunov exponents spectrum and bifurcation diagram of the 4-D system (13.4) for $c \in [5, 15]$

hyperchaos for the system (4). The values of Lyapunov exponents when $c = 10.2$ are:

$$L_1 = 0.8014, L_2 = 0.01578, L_3 = -0.03099, L_4 = -44.74. \tag{13.22}$$

Also, the values of Lyapunov exponents when $c = 10.2$ are:

$$L_1 = 0.8227, L_2 = 0.0156, L_3 = -0.05183, L_4 = -45.61. \tag{13.23}$$

When $c \in C \cup D$, the system (13.4) has only one positive Lyapunov exponent ($L_1 > 0$), and the other Lyapunov exponents are negative ($L_{2,3,4} < 0$).

This implies the existence of chaos for the system (13.4). The values of Lyapunov exponents when $c = 10.25$ are:

$$L_1 = 0.9453, L_2 = -0.02142, L_3 = -0.07179, L_4 = -43.78. \tag{13.24}$$

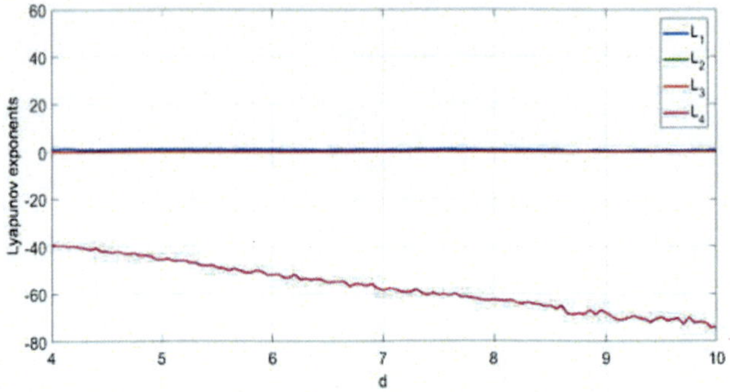

(a)

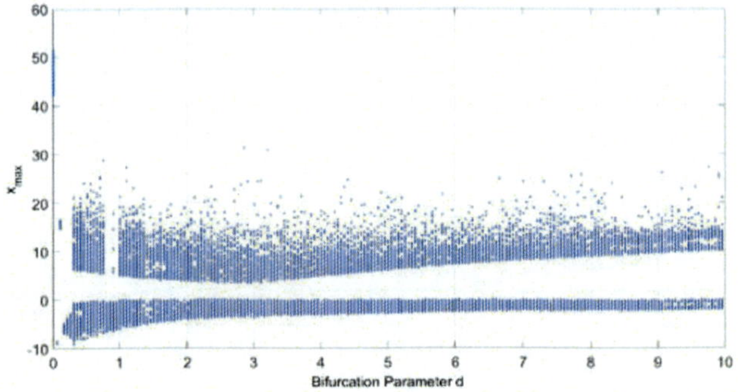

(b)

Fig. 13.5 Lyapunov exponents spectrum and bifurcation diagram of the 4-D system (13.4) for $d \in [0, 10]$

Also, the values of Lyapunov exponents when $c = 9.6$ are:

$$L_1 = 0.9091, L_2 = -0.03205, L_3 = -0.0237, L_4 = -46.35. \tag{13.25}$$

13.3.4 Varying the Parameter d

Figure 13.5 shows the Lyapunov exponents spectrum and the bifurcation diagram of the new four-scroll hyperchaotic system (13.4) with respect to parameter d.

We fix the values of the parameters a, b, c and p as $(a, b, c, p) = (6, 6, 10, 0.3)$. We vary the parameter d in the range [4.5, 5.5].

We can identify the behavior of the system (13.4) when the parameter d varies in the range [4.5, 5.5] as follows:

Let $d \in [4.5, 5.5]$. We define:

$$A = [4.5, 4.6] \cup [4.66, 4.7287] \cup [4.7725, 4.98].$$

$B = [5.0425, 5054] \cup [5.21025, 5.39] \cup [5.429, 5.5]$,
$C = (4.6, 4.66) \cup (4.7287, 4.7725) \cup (4.98, 5.0425)$,
$D = (5.054, 5.21025) \cup (5.39, 5.429)$.

When $d \in A \cup B$, we note that the system (13.4) has only one positive Lyapunov exponent ($L_1 > 0$), and the other Lyapunov exponents are negative ($L_{2,3,4} < 0$). Thus, the system (13.4) evolves into a chaotic attractor. The values of Lyapunov exponents obtained when $d = 4.8$ are:

$$L_1 = 1.0001, L_2 = -0.03751, L_3 = -0.01377, L_4 = -43.84. \qquad (13.26)$$

Also, the values of Lyapunov exponents when $d = 5.3$ are:

$$L_1 = 0.9624, L_2 = -0.01591, L_3 = -0.1065, L_4 = -46.88. \qquad (13.27)$$

When $d \in C \cup D$, the system (13.4) has two positive Lyapunov exponents ($L_{1,2} > 0$) and two negative Lyapunov exponent ($L_{3,4} < 0$). Thus the system (13.4) is hyperchaotic in this range. The values of Lyapunov exponents when $d = 4.75$ are:

$$L_1 = 1.0104, L_2 = 0.03047, L_3 = -0.03457, L_4 = -42.74. \qquad (13.28)$$

Also, the values of Lyapunov exponents when $d = 5.4$ are:

$$L_1 = 0.9071, L_2 = 0.01308, L_3 = -0.02481, L_4 = -47.98. \qquad (13.29)$$

13.3.5 Varying the Parameter *p*

Figure 13.6 shows the Lyapunov exponent spectrum and the bifurcation diagram of the new four-scroll hyperchaotic system (13.4) with respect to parameter p.

We fix the values of the parameters a, b, c and d as $(a, b, c, d) = (6, 6, 10, 5)$. We vary the parameter p in the range $[0, 0.5]$.

Next, we identify the behavior of the system (13.4) when the parameter p varies in the range $[0, 0.5]$ as follows:

Let $p \in [0, 0.5]$. We define:

$G = [0, 0.2673] \cup [0.3415, 0.45]$.
$H = (0.2673, 0.3415) \cup (0.45, 0.50)$.

When $p \in G$, we find that the system has only one positive Lyapunov exponent ($L_1 > 0$), and the other Lyapunov exponents are negative ($L_{2,3,4} < 0$). Thus, the system (13.4) is chaotic and evolves into a chaotic attractor. The values of Lyapunov exponents obtained when $p = 0.2$ are:

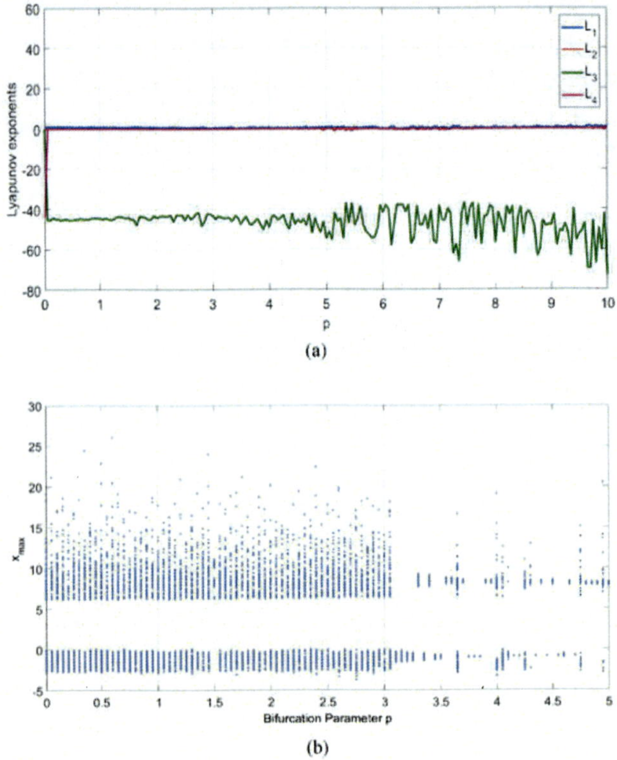

Fig. 13.6 Lyapunov exponents spectrum and bifurcation diagram of the 4-D system (13.4) for $p \in [0, 10]$

$$L_1 = 0.8139, L_2 = -0.0244, L_3 = -0.0566, L_4 = -40.85, \tag{13.30}$$

and the values of Lyapunov exponents when $p = 0.4$ are:

$$L_1 = 1.0423, L_2 = -0.01333, L_3 = -0.04437, L_4 = -48.23. \tag{13.31}$$

When $p \in H$, the system (13.4) has two positive Lyapunov exponents ($L_{1,2} > 0$) and two negative Lyapunov exponents ($L_{3,4} < 0$). Thus the system (13.4) is hyperchaotic in this range of paramater p. The values of the Lyapunov exponents when $p = 0.5$ are:

$$L_1 = 0.77763, L_2 = 0.006636, L_3 = -0.03073, L_4 = -45.43. \tag{13.32}$$

13.4 Multistability in the New 4D Four-Scroll Hyperchaotic System

In order to study the coexistence attractors and other characteristics of the system better, it is necessary to give some disturbance to the initial conditions under the condition of keeping the system parameters constant.

We take the parameters of the system (13.4) as in the hyperchaotic case, viz. $(a, b, c, d, p) = (6, 6, 10, 5, 0.3)$.

Figure 13.7 shows the dynamic behavior with coexistence bifurcation for the four-scroll hyperchaotic system (13.4) in which the initial conditions of blue trajectory and red trajectory are $X_0 = (1, -1, 1, -1)$ and $Y_0 = (-1, 1, 1, 1)$, respectively. Figure 13.8 depicts the coexistence of two hyperchaotic attractors with different initial states and same parameter values.

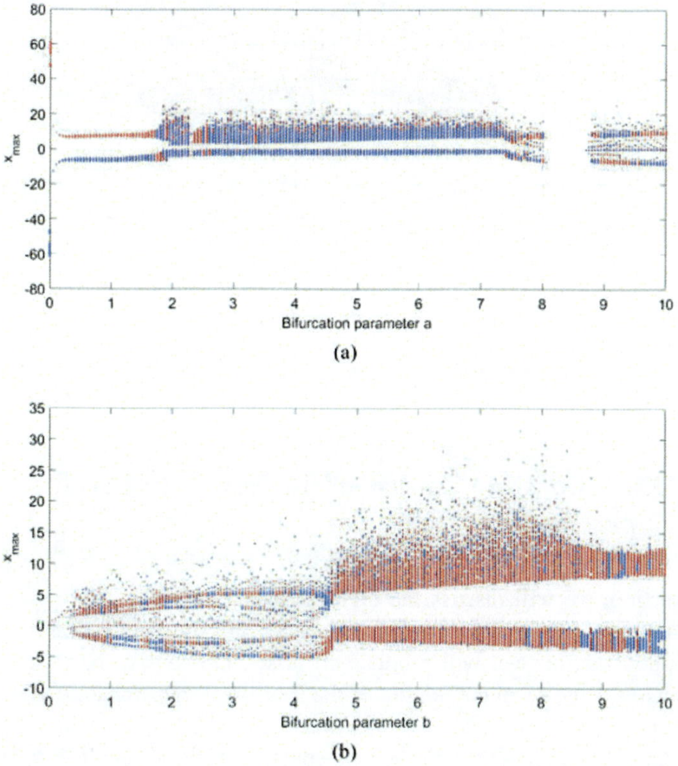

Fig. 13.7 Bifurcation diagram versus parameters a and b with different initial values for the four-scroll hyperchaotic system (13.4), where the blue color corresponds to the initial state $X_0 = (1, -1, 1, -1)$ and the red color corresponds to the initial state $Y_0 = (-1, 1, 1, 1)$

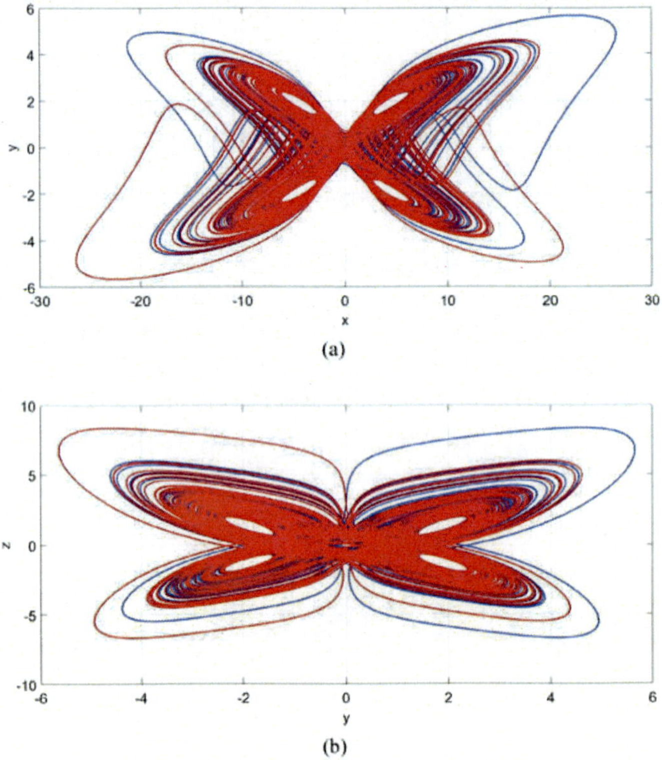

Fig. 13.8 Coexistence of two hyperchaotic attractors for the system (13.4) for $(a, b, c, d, p) = (6, 6, 10, 5, 0.3)$ with different initial states, viz. $X_0 = (1, -1, 1, -1)$ and $Y_0 = (-1, 1, 1, 1)$, where the blue trajectory corresponds to X_0 and the red trajectory corresponds to Y_0: **(a)** (x, y)-plot and **(b)** (y, z)-plot

13.5 Offset Boosting Control of the New Four-Scroll Hyperchaotic System

In this section, we will discuss the offset boosting control of the new four-scroll hyperchaotic system. When a variable appears only once in a nonlinear system, adding a constant m to it will produce an offset. Obviously, the state variable w only emerges in the second equation of the four-scroll hyperchaotic system (13.4). Thus, we can obtain an offset-boosted system from the hyperchaotic system (13.4) by replacing w with $w + m$ in the second equation of the system (13.4) as follows:

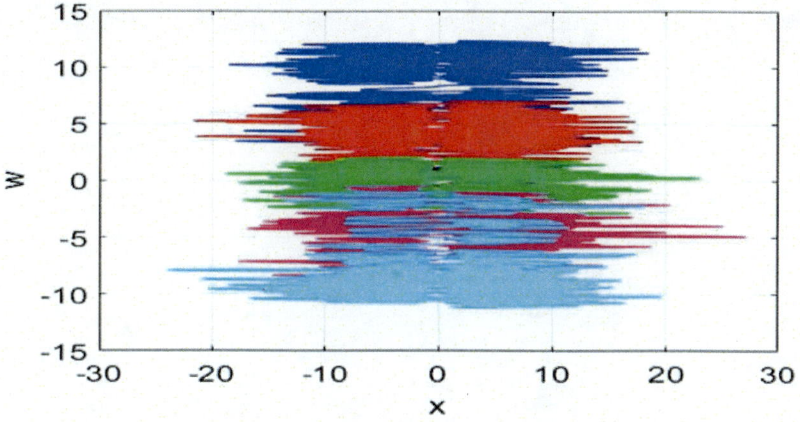

Fig. 13.9 Offset boosting plots: (x, w) phase portraits of the system (13.33) with different values of offset boosting controller: $m = -10$ (blue), $m = -5$ (red), $m = 0$ (green), $m = 5$ (violet), and $m = 10$ (cyan)

$$\begin{cases} \dot{x} &= -ax + cyz, \\ \dot{y} &= -dy^3 + bxz - (w + m), \\ \dot{z} &= bz - xy, \\ \dot{w} &= py. \end{cases} \tag{13.33}$$

Figure 13.9 shows the offset boosting plots in (x, w) plane of the system (13.33) with different values of offset boosting controller: $m = -10$ (blue), $m = -5$ (red), $m = 0$ (green), $m = 5$ (violet), and $m = 10$ (citron)

13.6 Circuit Simulation of the New 4D Hyperchaotic System

In this section, the new 4D hyperchaotic four-scroll system (13.4) is realized by the NI Multisim 14.1 platform. The electronic circuit design of the hyperchaotic four scroll system (13.4) is shown in Fig. 13.10 in which TLO84ACN is selected as OPAMP and the multipliers are of type AD633 with an output coefficient of 10. Applying the Kirchhoff's laws, the circuit presented in Fig. 13.10 is described by the following equations:

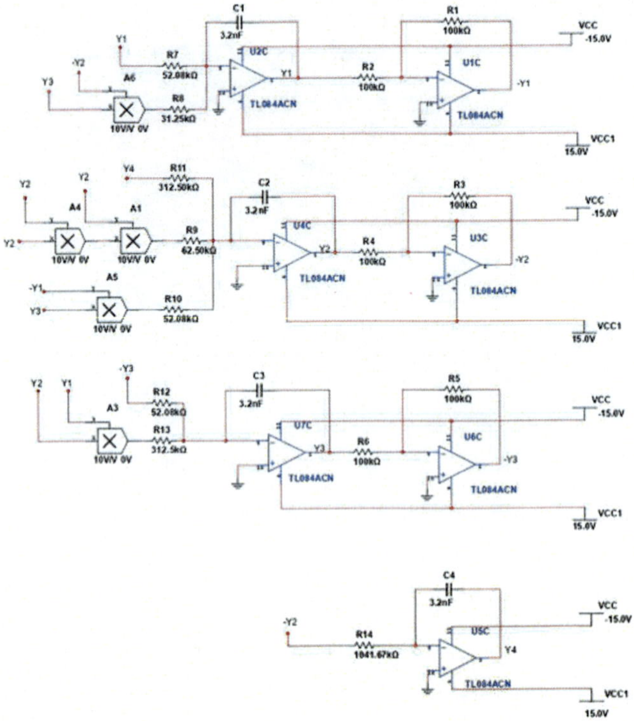

Fig. 13.10 Circuit design of the new 4-D four-scroll hyperchaotic system (13.4) (high resolution figure is made available online)

$$\begin{cases} \dot{y}_1 = -\frac{1}{R_7C_1}y_1 + \frac{1}{10R_8C_1}y_2y_3, \\ \dot{y}_2 = -\frac{1}{100R_9C_2}y_2^3 + \frac{1}{10R_{10}C_2}y_1y_3 - \frac{1}{R_{11}C_2}y_4, \\ \dot{y}_3 = \frac{1}{R_{12}C_3}y_3 - \frac{1}{10R_{13}C_3}y_1y_2, \\ \dot{y}_4 = \frac{1}{10R_{14}C_4}y_2. \end{cases} \tag{13.34}$$

Here y_1, y_2, y_3, y_4 correspond to the voltages on the integrators U2C, U4C, U7C, and U5C, respectively. The values of components in the circuit are selected as follows: $R_1 = R_2 = R_3 = R_4 = R_5 = R_6 = 100\,k\Omega$, $R_7 = R_{10} = R_{12} = 52.08\,k\Omega$, $R_8 = 31.25\,k\Omega$, $R_9 = 62.50\,k\Omega$, $R_{14} = 1041.67\,k\Omega$, $R_{11} = R_{13} = 312.50\,k\Omega$, $C_1 = C_2 = C_3 = C_4 = 3.2\,nF$.

MultiSim outputs of the circuit (13.34) are presented in Fig. 13.11. These results are consistent with the MATLAB simulation results for the new four-scroll hyperchaotic system (13.4) depicted in Fig. 13.1.

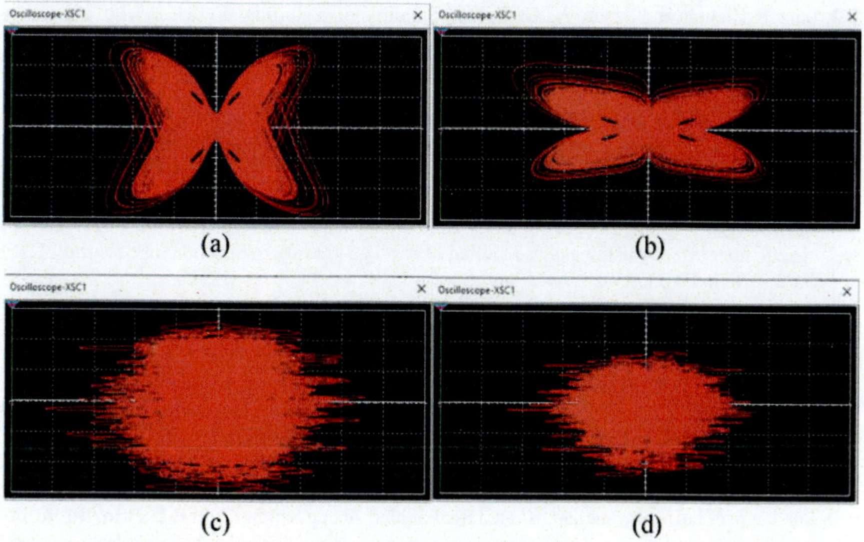

Fig. 13.11 MultiSim Outputs of the new 4D four-scroll hyperchaotic system (13.34) via oscilloscope (high resolution figure is made available online)

13.7 Conclusions

In this work, we described a new 4-D four-scroll hyperchaotic system with two positive Lyapunov exponents. Dynamic properties of the new hyperchaotic system have been presented in detail such as its Lyapunov exponents, Kaplan-Yorke dimension, phase portraits, equilibrium points, and their stability. Next, we carried out a bifurcation analysis of the new four-scroll hyperchaotic dynamics using the Lyapunov exponent spectrums and bifurcation diagrams and described offset boosting control. Moreover, we showed that the new 4-D four-scroll hyperchaotic system has multistability with coexisting hyperchaotic attractors. Finally, we designed an electronic circuit of the new four-scroll hyperchaotic system using MultiSim 14.1 for practical applications.

Competing Interests The authors have no conflicts of interest to declare that are relevant to the content of this chapter.

References

1. Sundarapandian Vaidyanathan and Ahmad Taher Azar. *Backstepping Control of Nonlinear Dynamical Systems*. Advances in Nonlinear Dynamics and Chaos (ANDC). Academic Press, Cambridge, Massachusetts, USA, 2021.

2. Luiz F. Ziebell and Jason A. C. Gallas. Stability of a duffing oscillator with a position-dependent mass. *European Physical Journal Plus*, 138(10), 2023. Article ID 930.
3. Godwin Sani, Jan Awrejcewicz, and Zeric Njitacke Tabekoueng. Modeling, analysis and control of parametrically coupled electromechanical oscillators. *Mechanism and Machine Theory*, 191, 2024. Article ID 105514.
4. Youjian Song, Defeng Zou, Omri Gat, Minglie Hu, and Philippe Grelu. Chaotic internal dynamics of dissipative optical soliton molecules. *Laser and Photonics Reviews*, 17(8), 2023. Article ID 2300066.
5. Zhanwu Xie, Wenlong Zeng, Pengfei Li, Haitao Yan, and Daofu Han. Thermal modulation of a chaotic fiber laser by using a phase-shifted fiber bragg grating. *IEEE Photonics Journal*, 15(5), 2023. Article ID 3302302.
6. Zeyi Liu, Jianshe Gao, Xiaobo Rao, Shunliang Ding, and Deping Liu. Complex dynamics of the passive biped robot with flat feet: Gait bifurcation, intermittency and crisis. *Mechanism and Machine Theory*, 191, 2024. Article ID 105500.
7. XiaoBo Rao, JianShe Gao, ShunLiang Ding, Jie Liang, and Jiangang Zhang. Multistability of gaits, the basin of attraction and its external topology in the simplest passive walking model on stairs. *Chaos, Solitons and Fractals*, 172, 2023. Article ID 113592.
8. Feng Li. Incorporating fractional operators into interaction dynamics of a chaotic biological model. *Results in Physics*, 54, 2023. Article ID 107052.
9. Sundarapandian Vaidyanathan, Khaled Benkouider, Aceng Sambas, and P. Darwin. Bifurcation analysis, circuit design and sliding mode control of a new multistable chaotic population model with one prey and two predators. *Archives of Control Sciences*, 33(1):127 â 153, 2023.
10. Aoqi Xu, Man-Wen Tian, Nasreen Kausar, Ardashir Mohammadzadeh, Dragan Pamucar, and Ebru Ozbilge. Optimal type-3 fuzzy control and analysis of complicated financial systems. *Journal of Intelligent and Fuzzy Systems*, 44(5):7121 â 7134, 2023.
11. Anam Azam, Danish Ali Sunny, and Muhammad Aqeel. Generation of multiscroll chaotic attractors of a finance system with mirror symmetry. *Soft Computing*, 27(6):2769 â 2782, 2023.
12. Jerol Soibam, Valentin Scheiff, Ioanna Aslanidou, Konstantinos Kyprianidis, and Rebei Bel Fdhila. Application of deep learning for segmentation of bubble dynamics in subcooled boiling. *International Journal of Multiphase Flow*, 169, 2023. Article ID 104589.
13. Juntao Chen and Jibo Wu. The prediction of Chongqing's GDP based on the LASSO method and chaotic whale group algorithmâback propagation neural networkâARIMA model. *Scientific Reports*, 13(1), 2023. Article ID 15002.
14. Jing Luo, Wentao Tang, Yao Chen, Xue Chen, and Huan Zhou. Dynamical analysis and synchronization control of flux-controlled memristive chaotic circuits and its fpga-based implementation. *Results in Physics*, 54, 2023. Article ID 107085.
15. Liang Yang and Qiang Lai. Construction and implementation of discrete memristive hyper-chaotic map with hidden attractors and self-excited attractors. *Integration*, 94, 2024. Article ID 102091.
16. Meitong Yu, Hongxiang Wang, Yuefeng Ji, and Yu Zhang. Ultra-wideband chaotic optical communication based on electro-optic differential feedback loop. *Optics Communications*, 545, 2023. Article ID 129729.
17. Siyang Liu, Xinlei An, Yue Wang, and Qianqian Shi. Design of a new multi-wing chaotic system and its application in color image encryption. *Optik*, 290, 2023. Article ID 171334.
18. Meitong Yu, Hongxiang Wang, and Yuefeng Ji. Secure WDM communication based on second order differential feedback electro-optical chaotic mutual injection synchronization. *Optics and Laser Technology*, 169, 2024. Article ID 110147.
19. Shaohui Yan, Binxian Gu, Ertong Wang, and Yu Ren. Finite-time synchronization of multi-scroll hyperchaotic system and its application in image encryption. *Mathematics and Computers in Simulation*, 206:391 â 409, 2023.
20. Sheng-Hao Jia, Yu-Xia Li, Qing-Yu Shi, and Xia Huang. Design and FPGA implementation of a memristor-based multi-scroll hyperchaotic system. *Chinese Physics B*, 31(7), 2022. Article ID 70505.

21. Wei Pan, Liping Chen, and Weidong Zhang. Design of a class of fractional-order hyperchaotic multidirectional multi-scroll attractors. *Mathematical Methods in the Applied Sciences*, 44(3):2416 â 2430, 2021.
22. Chunhua Wang, Hu Xia, and Ling Zhou. A memristive hyperchaotic multiscroll jerk system with controllable scroll numbers. *International Journal of Bifurcation and Chaos*, 27(6), 2017. Article ID 1750091.
23. Mei Zhang and Qingtao Han. Dynamic analysis of an autonomous chaotic system with cubic nonlinearity. *Optik*, 127(10):4315–4319, 2016.
24. Shuying Wang, Ling Hong, and Jun Jiang. An image encryption scheme using a chaotic neural network and a network with multistable hyperchaos. *Optik*, 268, 2022. Article ID 169758.
25. Ming-Hong Qin and Qiang Lai. Extreme multistability and amplitude modulation in memristive chaotic system and application to image encryption. *Optik*, 272, 2023. Article ID 170407.
26. Shaohui Yan, Binxian Gu, Yu Ren, Xi Sun, and Ertong Wang. Dynamical analysis of four-dimensional chaotic system and its application in image encryption. *Multimedia Tools and Applications*, 82(14):21509 â 21534, 2023.
27. Quan Xu, Liping Huang, Ning Wang, Han Bao, Huagan Wu, and Mo Chen. Initial-offset-boosted coexisting hyperchaos in a 2d memristive chialvo neuron map and its application in image encryption. *Nonlinear Dynamics*, 111(21):20447 â 20463, 2023.
28. Jie Zhang, Yan Guo, and Jinhao Guo. Design of memristor hyperchaotic circuit with burst oscillation and infinite attractor coexistence and its application. *Microelectronic Engineering*, 282, 2023. Article ID 112099.
29. Muhammet Emin Sahin, Géza Kolumbán, Serdar Ethem Hamamci, and Hasan Guler. Use of memristor-based chaotic circuits in AM-DCSK and FM-DCSK modulation. *Physica Scripta*, 98(10), 2023. Article ID 105228.
30. Yujuan Gu, Guodong Li, Xiangliang Xu, Xiaoming Song, and Si Wu. Multistable dynamics and attractors self-reproducing in a new hyperchaotic complex lü system. *Chaos*, 33(9), 2023. Article ID 093112.

Chapter 14
Secure Communication System Based on Multistability: Evaluation Using the SCAMPER Method for Innovation Projects

C. E. Rivera-Orozco, J. H. García-López, M. R. Ramírez-Jiménez, K. Pulido-Hernández, N. A. Gómez-Torres, L. Serrano-Zúñiga, M. T. Solorio-Núñez, and R. Jaimes-Reátegui

14.1 Introduction

In cryptography, a coding or encryption mechanism is usually used, with which an initial message, called plaintext, is taken, and this is encoded; is that, the message that itself cannot be read, called ciphertext, is changed. To read the ciphertext, it is necessary to decode this message; this process is called decoding. This should only be done by someone authorized. Otherwise, it is said that the encryption has been broken. Encoding and decoding work using a security key that is an instrument to hide and reveal the information that you want to protect, according to [1], p.373.

Communication using chaos has attracted much attention due to the inherent properties of chaotic systems, such as pseudorandom behavior and spread spectrum according to [2], p.963, [3], p.343,[4],p.384, [5], p.50, [6], p.153. Since chaos can be treated as deterministic motion, it can be easily decoded. In recent years, a growing interest in applications of chaos in communication has been motivated by the fact that chaotic systems can be synchronized, as [7], p.1403; this allows the transmission of information using a chaotic spread spectrum signal, as stated in the work of [8].

The main advantage of communication systems based on chaotic synchronization over conventional systems is that they allow the implementation of a coherent receiver, i.e., the received chaotic waveform contains all possible sampling func-

C. E. Rivera-Orozco (✉) · J. H. García-López · M. R. Ramírez-Jiménez · M. T. Solorio-Núñez · R. Jaimes-Reátegui
Optics, Complex Systems and Innovation Laboratory, Centro Universitario de los Lagos, Universidad de Guadalajara, Lagos de Moreno, Jalisco, Mexico
e-mail: carmen.rivera@academicos.udg.mx

K. Pulido-Hernández · N. A. Gómez-Torres · L. Serrano-Zúñiga
Departamento de Ciencias Básicas, Centro Universitario de la Ciénega, Universidad de Guadalajara, Ocotlan, Jalisco, Mexico

E. Campos-Cantón et al. (eds.), *Complex Systems and Their Applications*,
https://doi.org/10.1007/978-3-031-51224-7_14

tions, while in the case of incoherent reception, only one or more properties of the sampling function are estimated. In addition, chaotic modulation works better under conditions with multiple propagation paths, because the cross-correlation between chaotic time series segments is much lower than for periodic signals. Some early work on chaotic cryptography can be found in the references of [9], where the idea of encrypting an information flow based on one-dimensional chaotic maps is presented [10], p.3031, also published a paper on synchronization of chaotic systems, which was an excellent tool for [11], developments on communication security. Chaotic cryptography has also taken two different paths without affecting each other: chaotic digital encryption, which is described in the work of [12], p.927, [13], p.465, and the chaotic synchronization presented by the authors [14], p.230, [15] and [16]. The main advantage between these two ways is that in the first case, the cryptosystem needs a standard key, while in the second case, the key is the parameters of the system. Other advantages of the chaotic synchronization scheme are the simple analog implementation for communication security.

Chaotic cryptography can be implemented with very fast electronic and optical components. Although synchronization-based communication is widely used to encrypt information according to [17] and [18], there are still serious problems in the security of these systems. The main advantage of chaotic cryptography is the high sensitivity to the initial conditions used as keys, but these keys are inefficient for communication systems based on chaotic synchronization, because in this case, the system parameters are used as secret keys. However, according to [19], p.2, the parameters can be recovered by a brute force synchronization attack in which the hacker minimizes the synchronization error to adjust the parameters of a virtual system. Recent work has proposed using chaotic maps to improve the encryption of information prior to transmission. The main drawback of this method is that the synchronization error caused by the same message can be easily cryptanalyzed to recover the information. The great importance of communication systems and new computer technologies leads to the fact that the problem of security of the transmitted information increases. It is important to address these types of security and privacy issues as they impact our fundamental rights and our collective ability to trust secure communication systems. The use of encryption, authentication, and access control technologies for secure communications systems is an appropriate solution to prevent attempted espionage and theft during data transmission. One contribution to solving this problem is the use of signals, chaotic carriers generated by components operating in the nonlinear domain, such as multistable nonlinear oscillators, which behave chaotically.

14.2 Context

In the Laboratory of Optics, Complex Systems and Innovation of the Los Lagos University Center of the University of Guadalajara, an analysis was carried out with the filters of the SCAMPER method, which made it possible to show its individual and integral application in projects for communication systems and to propose new solutions to problems involving chaos theory. The SCAMPER method is applied to a research problem involving a secure communication system based on multistability and consisting of chaotic and multistable transmitter/receiver elements by using two public/private communication channels. Based on the obtained results, the SCAMPER method shows that its process leads to the procedure used to define the methodology and scientific results. **Appendix** provides an overview of how the problem of a communication system is treated in its mathematical analysis, moving from a single channel to two channels. Accordingly, the method can be used to develop research schemes from the point of view of innovation and creativity. With the implementation of a cryptographic algorithm based on a system and method based on multistable chaotic dynamical systems, security in the transmission of encrypted information over telecommunications can be strengthened, guaranteeing greater security in the transmission of information, robustness to noise, and other external disturbances. With this in mind, this work explored ideas for innovative solutions to ensure security in communication systems with immunity to cyberattacks.

14.3 Methodology

The method SCAMPER is used to develop new ideas or, as in the case of the Optics, Complex Systems and Innovation Laboratory, to propose improvements to existing systems or processes. It was introduced by Eberle [20] and consists of stimulating creativity by asking specific questions about a particular topic or problem. Seven filters are used to break down creative blocks and discover new ways of working through a series of targeted brainstorms with a starting point of key engineering words.

Figure 14.1 is the representation scheme: Nassi-Schneiderman N-S diagram aims to generalize the approach to perform each of the different filters used by the method SCAMPER. It can be deduced that each filter has a close relationship with a keyword and the essential approaches in and for the filters of the method.

The method SCAMPER came at just the right time to solve the research problem posed in the Laboratory of Optics, Complex Systems and Innovation with the intention of improving a communication system (problem) to propose improvements to existing systems or processes. We started our work as a team with SCAMPER and clearly defined the goal, which was to define and build a secure communication system based on chaotic and multistable transmitter/receiver elements using two communication channels, public/private communication.

Technology (communication system) was studied in general terms with the questions: What can we substitute?; see Fig. 14.2; what can we combine?; see Fig. 14.3; what can we adapt?; see Fig. 14.4; what can we modify?; see Fig. 14.5; what can we propose?; see Fig. 14.6; what can we eliminate?; see Fig. 14.7; and what can we reorder?; see Fig. 14.8. The previous questions served to stimulate

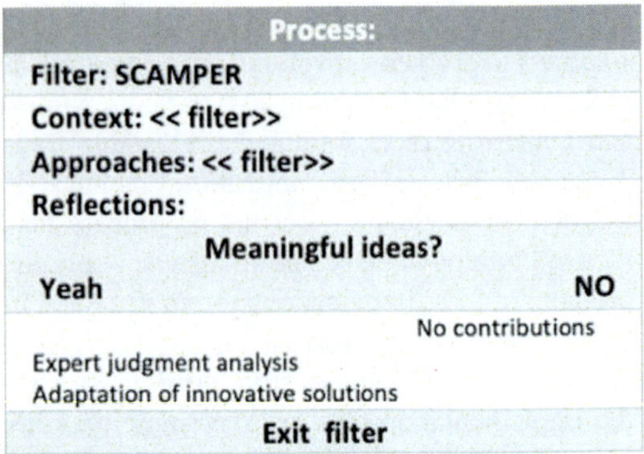

Fig. 14.1 Diagram N-S: generalization for addressing filters with SCAMPER

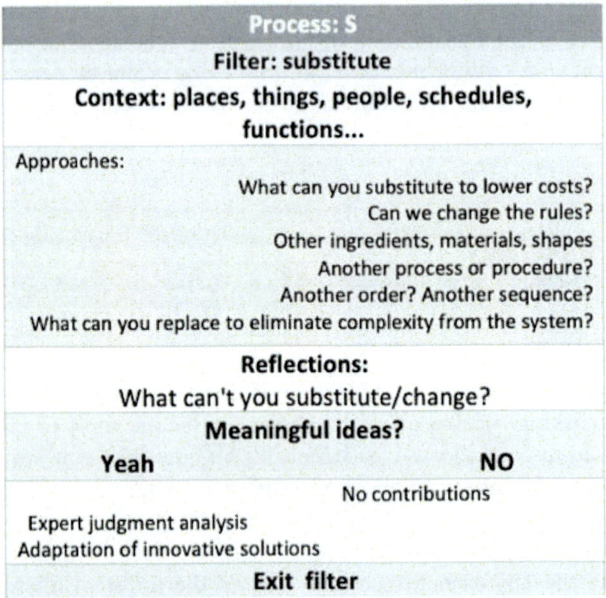

Fig. 14.2 Diagram N-S 1: substitute filter with SCAMPER

creativity by going through each letter of the method SCAMPER. Then the exercise was repeated reflectively to analyze in more detail the ideas that emerged. In a later stage, the main ideas were discussed based on the presented filters of the method SCAMPER, which allowed the experts to reflect on the proposals to solve the research problem and explore alternatives. The N-S diagrams are presented in Figs. 14.2, 14.3, 14.4, 14.5, 14.6, 14.7 and 14.8 to illustrate the application of each letter of the method SCAMPER.

In the new secure communication system based on multistability and consisting of transmitter/receiver based on multistable chaotic oscillators, the use of communication channels is rearranged to define what is transmitted over a public or

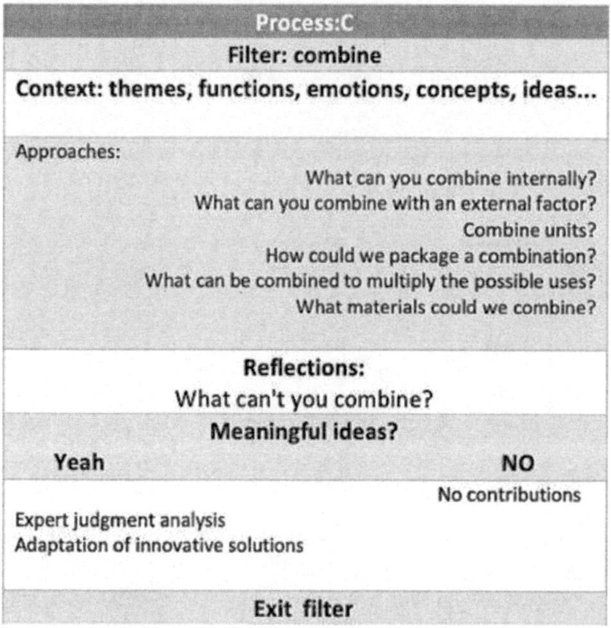

Fig. 14.3 Diagram N-S 2: combine filter with SCAMPER

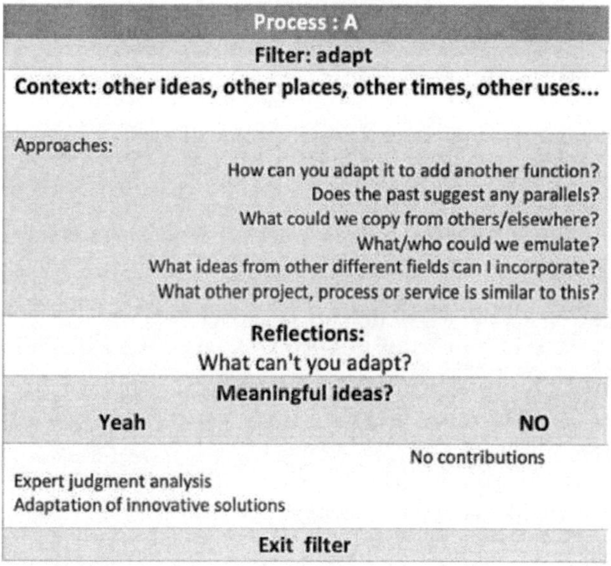

Fig. 14.4 Diagram N-S 3: adapt filter with SCAMPER

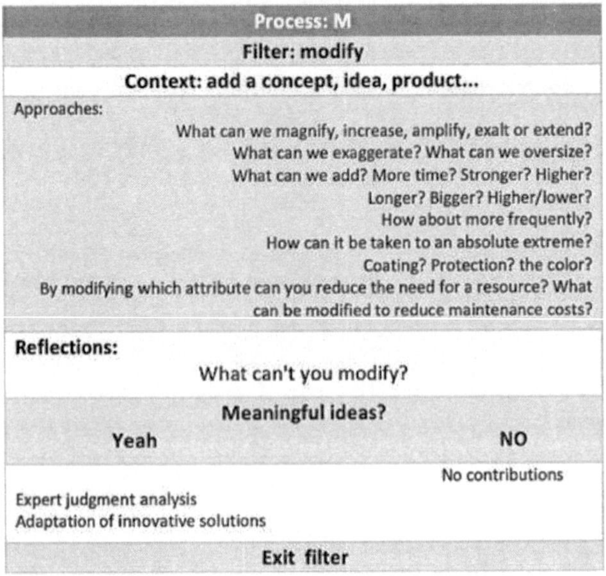

Fig. 14.5 Diagram N-S 4: modify filter with SCAMPER

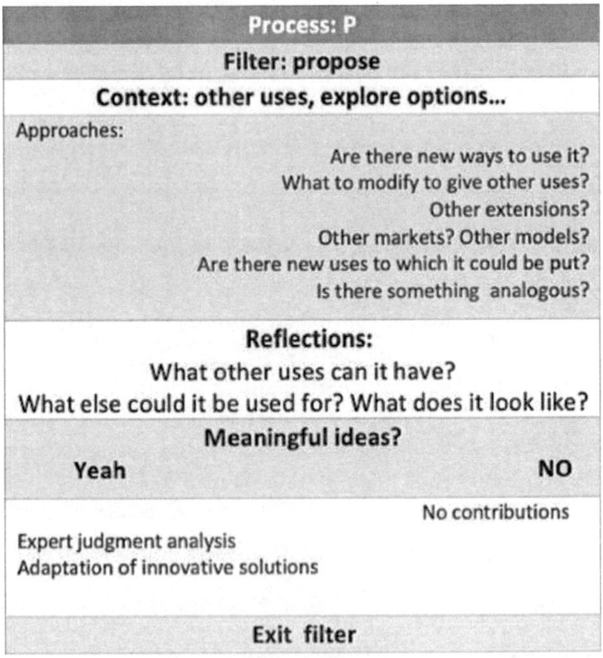

Fig. 14.6 Diagram N-S 5: propose filter with SCAMPER

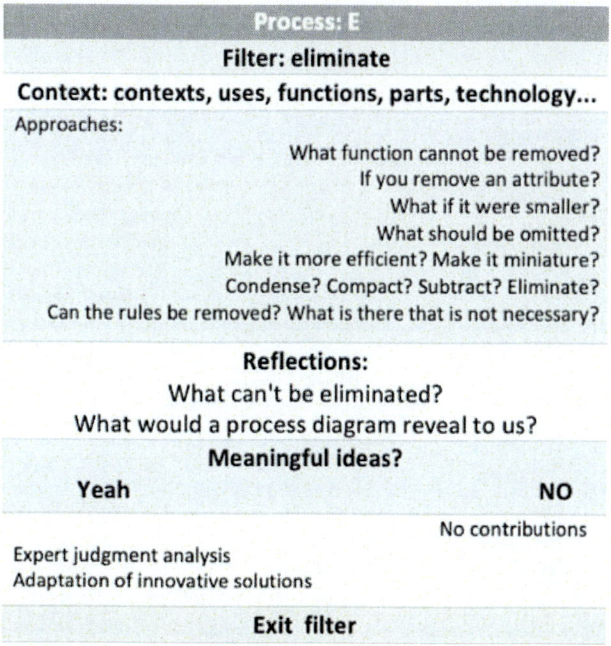

Fig. 14.7 Diagram N-S 6: eliminate filter with SCAMPER

private channel. The private channel allows synchronization of the oscillators of the transmitter and receiver. Identical chaotic switches in the transmitter and receiver discretely change the initial conditions of the oscillators at intervals shorter than the synchronization time. They act as secret keys that allow the chaotic attractors in the transmitter and receiver oscillators to be changed simultaneously. The information is transmitted over the public channel, with a very large number of information signal packets appended in a time-shifted manner to a very large number of multistable chaotic masking signals within the same time series.

14.4 Conclusions

The paper describes in detail how the SCAMPER method was used in analyzing a proposal with the intention of generating new ideas for invention, encouraging creative thinking and exploration of different approaches. In this particular case, a secure multistability-based communication system was used, consisting of a chaotic and multistable transmitter/receiver using a public and a private communication channel. Security was achieved by exploiting the properties of chaotic multistable systems [21].

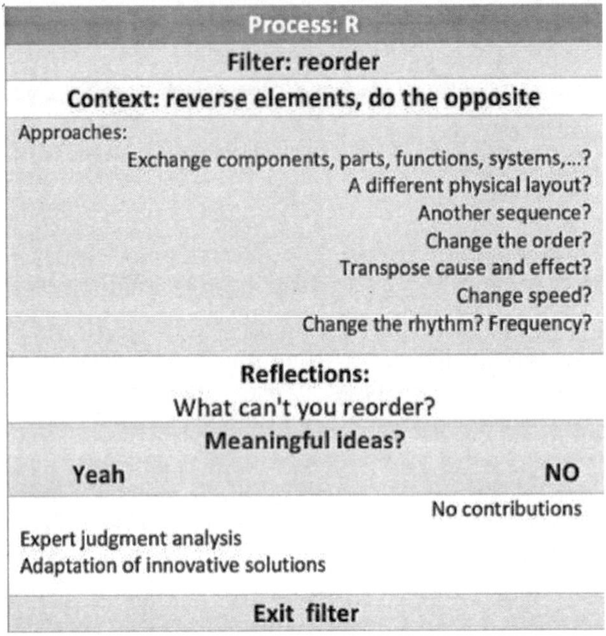

Fig. 14.8 Diagram N-S 7: reorder filter with SCAMPER

Each letter in SCAMPER represents a series of questions with a different approach to idea generation. Using these questions as a guide, the SCAMPER method helps apply strategies to stimulate creative minds in developing alternatives in idea generation by stimulating creative thinking in a short period of time and helping to solve problems quickly by going into depth. The method helps in finding innovative solutions to problems and in finding ideas that differentiate products or services from others, and it is also useful in improving and varying products or services you already have. It can be used when teams are stuck at any stage of project or product development; when you have a stubborn problem that seems impossible to solve; when you want to innovate a product or service or look for alternatives; or when you need to look at a problem from different perspectives.

Acknowledgments To the University of Guadalajara and to the collaborators of the UDG-CA-1038: Optics, Complex Systems and Innovation and UDG-CA-730: Teaching-Learning Networks and Technologies (UDG-CA-730).

Appendix

The concept of extreme multistability has been observed in two different categories of nonlinear systems. These categories include dynamical systems affected by external forces, where a dissipative system is affected by a conservative system [22], and high-dimensional systems with a certain type of nonlinearity [23, 24].

In the study of [21], the main focus is on the latter category of multistable systems, where extreme multistability occurs due to nonlinear interactions.

Burkin and Kuznetsova have explored various techniques to generate extremely multistable systems that contain an infinite network of hidden attractors [25]. In addition, Chakraborty and Poria have conducted research on synchronization of coupled extremely multistable systems [26].

Many chaotic communication systems are typically implemented in a single-channel master-slave configuration. In this configuration, both the master and slave oscillators are identical chaotic systems, as shown in Fig. 14.9.

In a chaotic masking method, as described in [21], an information message labeled $m_T(t)$ is encoded into one of the master oscillator's state variables, namely $y_T(t)$, which serves as a chaotic carrier. The resulting signal $s(t) = m_T(t)+y_T(t)$ is then transmitted to the receiver and serves two main purposes: (i) synchronizing the transmitter oscillator with the receiver oscillator and (ii) recovering the information at the receiver.

However, it is important to note that the chaotic signal $y_R(t)$ generated by the receiver cannot achieve complete synchronization with the transmitted signal $s(t)$. This is because the transmitted signal contains the message signal, resulting in a synchronization error $e(t) = m_R(t) - m_T(t)$ in the retrieved signal $m_R(t) = s(t) - y_R(t)$ at the receiver. This synchronization error can have a negative effect on the quality of the communication.

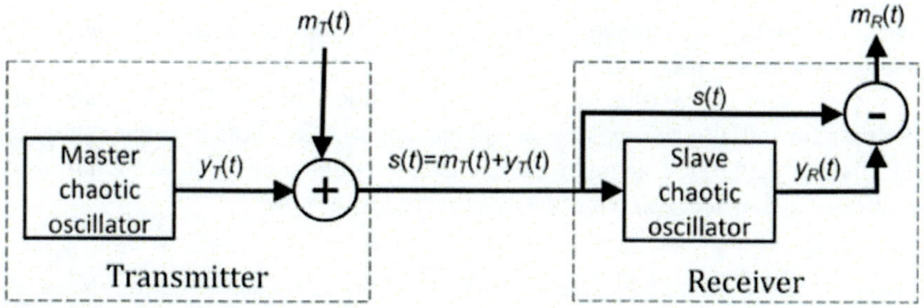

Fig. 14.9 Simple chaotic-masking communication scheme in a single-channel master-slave configuration

Chaotic System Demonstrating Extreme Multistability

Let us consider a six-dimensional dynamical system

$$\begin{aligned}\dot{x} &= -y - z,\\ \dot{y} &= x + av,\\ \dot{z} &= b - cz + uw,\\ \dot{u} &= x - y - z - u,\\ \dot{v} &= u + av,\\ \dot{w} &= b + w(u - c),\end{aligned} \tag{14.1}$$

where a dot means the derivative with respect to time. For positive parameters $a > 0$, $b > 0$, and $c > 0$, the system in Eq. (14.1) has a single hyperbolic fixed point at $[c - b, 1, 1, c - b, (b - c)/a, 1]$. Using the parameter values $a = 0.2$, $b = 0.2$, and $c = 5.7$, the eigenvalues $\lambda_1 = -1.0$, $\lambda_2 = 0$, $\lambda_3 = -5.7$, $\lambda_4 = 0$, $\lambda_5 = -1.4i$, and $\lambda_6 = -1.4i$, i.e., the equilibrium point is a spiral saddle [27]. Originally, the system Eq. (14.1) was introduced as two coupled three-dimensional Rössler systems x, y, z and u, v, w with special nonlinear coupling [28].

Under the parameter settings studied, the system described in Eq. (14.1) exhibits an intriguing phenomenon known as extreme multistability, as discussed in reference [27].

This behavior is visually represented in Fig. 14.10, which shows a bifurcation diagram illustrating the local maxima of the variable x with respect to the initial condition u. This diagram represents a Feigenbaum tree of period doublings leading to chaos. It is noteworthy here that even small variations in the initial conditions can steer the system into different attractors.

In the analysis, examining the region where $u(0)$ falls within the interval $[-7, -3]$. Within this range, the system exhibits the coexistence of an infinite number of chaotic attractors.

Figure 14.11 shows illustrative examples of time series for Eq. (14.1) and phase portraits for initial conditions $u_L = -7$ and $u_H = -3$. These examples show that the chaotic regimes are asymptotically stable, although they do not exhibit transient chaos, since transient behavior is only observed for $t < 40$.

The Basic Idea

In the proposed communication system, both the transmitter T and the receiver R contain identical six-dimensional oscillators given as

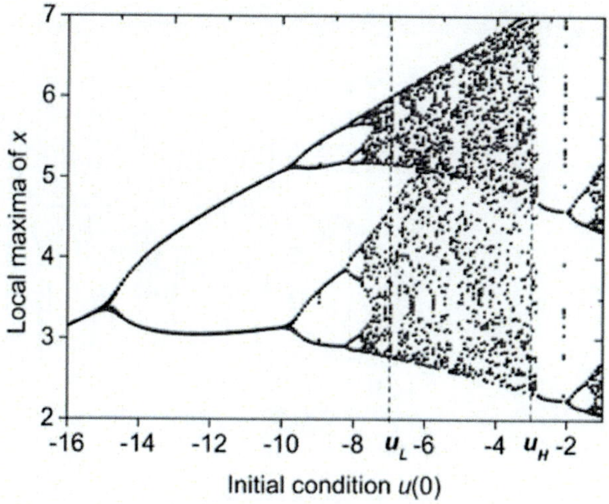

Fig. 14.10 Bifurcation diagram of local maxima of x versus initial condition $u_1(0)$ representing a cascade of period doubling bifurcations, for $a = 0.2$, $b = 0.2$, and $c = 5.7$ in Eq. (14.1). $u_L = -7$ and $u_H = -3$ are the minimum and maximum values of initial conditions used in our communication system

$$
\begin{aligned}
\dot{x_T} &= -y_T - z_T,\\
\dot{y_T} &= x_T + av_T,\\
\dot{z_T} &= b - cz_T + u_T w_T,\\
\dot{u_T} &= x_T - y_T - z_T - u_T,\\
\dot{v_T} &= u_T + av_T,\\
\dot{w_T} &= b + w_T(u_T - c),\\
\dot{x_R} &= -y_R - z_R + k(x_T - x_R),\\
\dot{y_R} &= x_R + av_R,\\
\dot{z_R} &= b - cz_R + u_R w_R,\\
\dot{u_R} &= x_R - y_R - z_R - u_R,\\
\dot{v_R} &= u_R + av_R,\\
\dot{w_R} &= b + w_R(u_R - c),
\end{aligned}
\tag{14.2}
$$

where the subindices T and R refer to the systems in the transmitter and receiver, respectively, and k is the coupling strength.

The basic concept of the proposed communication system is to encode information by embedding it into coexisting chaotic attractors. These attractors are switched using a dynamic key generated by a discrete chaotic system. The number

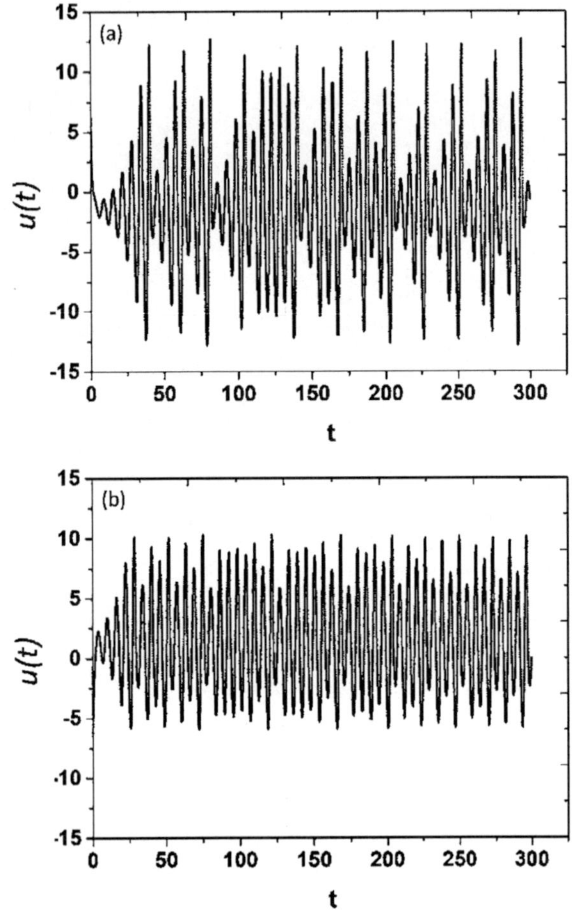

Fig. 14.11 Two coexisting chaotic attractors for initial conditions (**a**) $u_L = -7$ (upper), and (**b**) $u_H = -3$ (lower)

of these coexisting attractors is infinite, and each of them is characterized by chaotic behavior, which ensures a high degree of information obfuscation.

With a sufficiently strong coupling, denoted by k, the chaotic system of the receiver synchronizes with that of the transmitter, even if it starts from different initial conditions. Nevertheless, any change in the initial conditions of the transmitter system leads to a switch to a different coexisting attractor, temporarily breaking the synchronization. To restore synchronization, a certain synchronization time is required.

During this transition period, the information cannot be retrieved by the receiver and therefore is not transmitted. Once synchronization is restored, the information encoded in a different chaotic attractor is sent to the receiver. However, the interval at which the information is transmitted should not be too long to prevent the synchronization from becoming ineffective. Therefore, periodic switching between chaotic attractors provides robust communication security, as potential intruders do not have the necessary time to synchronize their virtual receiver systems with the transmitter.

The proposed communication system includes two channels, a private channel and a public channel, as shown in Fig. 14.11. The private channel is used for synchronization between receiver-transmitter synchronization, while the public channel is used for information transmission. This two-channel communication not only increases security but also improves the quality of communication by using different state variables for synchronization and information transmission.

Specifically, x_T and x_R are used for synchronization, while u_T and u_R are used for encoding and decoding information, respectively. Switching between coexisting attractors is achieved by changing the initial condition of the variable u_T within the range $u(0)\epsilon[-7, -3]$, where the attractors exhibit chaotic behavior, as shown in Fig. 14.10.

The selection of a particular attractor is achieved using a chaotic map. In this case, the logistic map is explored, although other chaotic maps can be used. Consequently, this communication system is a hybrid system that combines both continuous-time and discrete-time chaotic systems.

The communication system in Fig. 14.12 operates with two channels, consisting of a private and a public channel. The private channel is intended for the synchro-

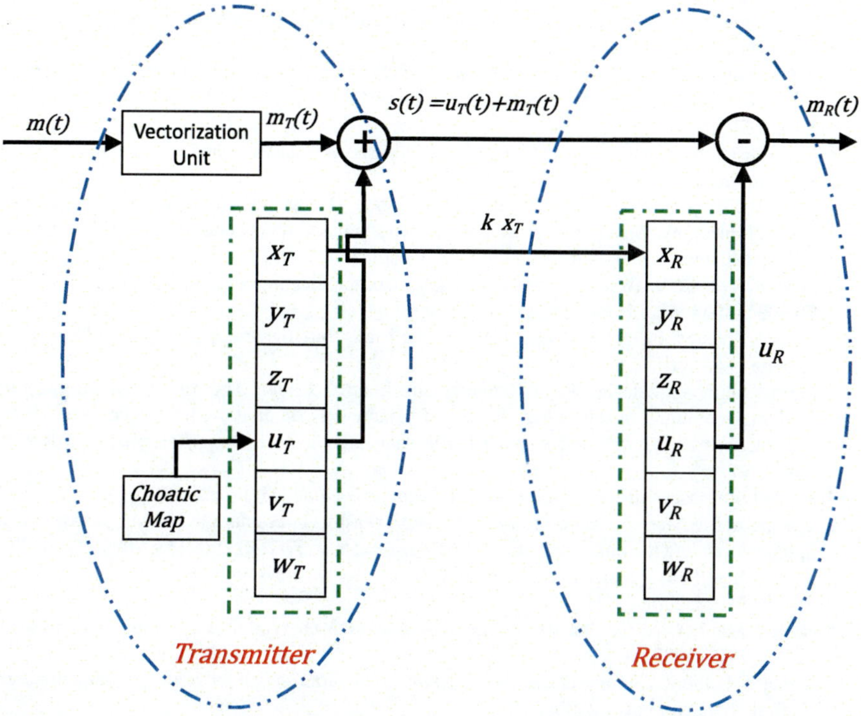

Fig. 14.12 Two-channel communication scheme containing private and public channels

nization process between receiver and the transmitter, which is facilitated by the variable x_T and the coupling coefficient k. The public channel, on the other hand, is used exclusively for the transmission of information.

In this system, $m(t)$ represents the input message, $m_T(t)$ corresponds to the information packets, $s(t)$ denotes the encrypted signal, $m_R(t)$ is the decoded signal at the receiver containing the synchronization error, and $m_{out}(t)$ represents the recovered message.

References

1. E. Alvarez, A. Fernández, P. García, J. Jiménez, and A. Marcano. New approach to chaotic encryption. *Physics Letters A*, 263(4–6):373–375, December 1999.
2. Valerio Annovazzi-Lodi, Silvano Donati, and Alessandro Scire. Synchronization of chaotic injected-laser systems and its application to optical cryptography. *IEEE journal of quantum electronics*, 32(6):953–959, 1996.
3. Apostolos Argyris, Dimitris Syvridis, Laurent Larger, Valerio Annovazzi-Lodi, Pere Colet, Ingo Fischer, Jordi Garcia-Ojalvo, Claudio R Mirasso, Luis Pesquera, and K Alan Shore. Chaos-based communications at high bit rates using commercial fibre-optic links. *Nature*, 438(7066):343–346, 2005.
4. Peter Ashwin. Synchronization from chaos. *Nature*, 422(6930):384–385, 2003.
5. MS Baptista. Cryptography with chaos. *Physics letters A*, 240(1–2):50–54, 1998.
6. MS Baptista, EE Macau, and Celso Grebogi. Integrated chaos-based communication. *Acta Astronautica*, 54(3):153–157, 2004.
7. Yuan-Zhao Yin. Experimental demonstration of chaotic synchronization in the modified chua's oscillators. *International Journal of Bifurcation & Chaos in Applied Sciences & Engineering*, 7(6), 1997.
8. Jiri Fridrich. Symmetric ciphers based on two-dimensional chaotic maps. *International Journal of Bifurcation and chaos*, 8(06):1259–1284, 1998.
9. JH García-López, R Jaimes-Reategui, R Chiu-Zarate, D Lopez-Mancilla, R Ramirez-Jimenez, and AN Pisarchik. Secure computer communication based on chaotic rössler oscillators. *The Open Electrical & Electronic Engineering Journal*, 2(1), 2008.
10. Scott Hayes, Celso Grebogi, and Edward Ott. Communicating with chaos. *Phys. Rev. Lett.*, 70:3031–3034, May 1993.
11. Ljupco Kocarev, Zbigniew Galias, and Shiguo Lian. *Intelligent computing based on chaos*, volume 184. Springer, 2009.
12. Géza Kolumbán, Michael Peter Kennedy, and Leon O Chua. The role of synchronization in digital communications using chaos. i. fundamentals of digital communications. *IEEE Transactions on circuits and systems I: Fundamental theory and applications*, 44(10):927–936, 1997.
13. Panos Louridas. *Real-world algorithms: a beginner's guide*. MIT Press, 2017.
14. Cleverson MP Marinho, Elbert EN Macau, and Takashi Yoneyama. Chaos over chaos: A new approach for satellite communication. *Acta Astronautica*, 57(2–8):230–238, 2005.
15. C Masoller. Noise-induced resonance in delayed feedback systems. *Physical review letters*, 88(3):034102, 2002.
16. Robert Matthews. On the derivation of a âchaoticâ encryption algorithm. *Cryptologia*, 13(1):29–42, 1989.
17. Robert C Hilborn. *Chaos and nonlinear dynamics: an introduction for scientists and engineers*. Oxford university press, 2000.

18. Jean-Luc Schwartz, Nicolas Grimault, Jean-Michel Hupé, Brian CJ Moore, and Daniel Pressnitzer. Multistability in perception: binding sensory modalities, an overview. *Philosophical Transactions of the Royal Society B: Biological Sciences*, 367(1591):896–905, 2012.
19. Dimitar Solev, Predrag Janjic, and Ljupco Kocarev. Introduction to chaos. In *Chaos-Based Cryptography: Theory, Algorithms and Applications*, pages 1–25. Springer, 2011.
20. B Eberle. Creative games and activities for imagination development, 2008.
21. Alexander N Pisarchik, Rider Jaimes-Reátegui, C Rodríguez-Flores, JH García-López, Guillermo Huerta-Cuéllar, and F Javier Martín-Pasquín. Secure chaotic communication based on extreme multistability. *Journal of the Franklin Institute*, 358(4):2561–2575, 2021.
22. Ying-Cheng Lai and Celso Grebogi. Complexity in hamiltonian-driven dissipative chaotic dynamical systems. *Physical Review E*, 54(5):4667, 1996.
23. Hongyan Sun, Stephen K Scott, and Kenneth Showalter. Uncertain destination dynamics. *Physical Review E*, 60(4):3876, 1999.
24. Calistus N Ngonghala, Ulrike Feudel, and Kenneth Showalter. Extreme multistability in a chemical model system. *Physical Review E*, 83(5):056206, 2011.
25. IM Burkin and OI Kuznetsova. On some methods for generating extremely multistable systems. In *Journal of Physics: Conference Series*, volume 1368, page 042050. IOP Publishing, 2019.
26. Priyanka Chakraborty and Swarup Poria. Extreme multistable synchronisation in coupled dynamical systems. *Pramana*, 93:1–13, 2019.
27. Albert CJ Luo. *Regularity and complexity in dynamical systems*. Springer, 2012.
28. CR Hens, R Banerjee, U Feudel, and SK Dana. How to obtain extreme multistability in coupled dynamical systems. *Physical Review E*, 85(3):035202, 2012.

Chapter 15
An Image Compression and Encryption Approach with Convolutional Layers, Two-Dimensional Sparse Recovery, and Chaotic Dynamics

Pooyan Rezaeipour-Lasaki, Aboozar Ghaffari, Fahimeh Nazarimehr, and Sajad Jafari

15.1 Introduction

With the rapid development of multimedia communication and internet technology, more and more multimedia data are transmitted through the network. As an important form of multimedia data, images are widely used in daily life [1]. Images are transmitted through different networks; therefore, securing these images is an essential topic [2]. Image encryption is a necessary approach to secure and convert data to a noise-like and meaningless form, and its understanding is difficult [3, 4]. The cryptosystems are mainly divided into two categories, which are symmetric and asymmetric encryptions [5]. In symmetric one, the encryption key is the same as the decryption key, while asymmetric encryption requires that the encryption key be different from the decryption key [6]. In chaotic systems-based image encryption systems, secret keys are usually used as the initial values or parameters of the chaotic systems, and pseudorandom sequences generated by the iterations of chaotic systems are employed to transform plain images into noise-like cipher images via

P. Rezaeipour-Lasaki · A. Ghaffari
Department of Electrical Engineering, Iran University of Science and Technology, Tehran, Iran
e-mail: pooyan_rezaeipour@elec.iust.ac.ir; aboozar_ghaffari@iust.ac.ir

F. Nazarimehr
Department of Biomedical Engineering, Amirkabir University of Technology (Tehran Polytechnic), Tehran, Iran
e-mail: f_nazarimehr@aut.ac.ir

S. Jafari (✉)
Department of Biomedical Engineering, Amirkabir University of Technology (Tehran Polytechnic), Tehran, Iran

Health Technology Research Institute, Amirkabir University of Technology (Tehran Polytechnic), Tehran, Iran
e-mail: sajadjafari@aut.ac.ir

E. Campos-Cantón et al. (eds.), *Complex Systems and Their Applications*,
https://doi.org/10.1007/978-3-031-51224-7_15

scrambling operations [7, 8]. Various operations and architectures were proposed to encrypt data, such as DNA coding [9–11], Substitution box(S-box) [12, 13], Neural Network [14, 15] and XOR operation [16–18].

Due to the limited bandwidth and storage capacity in the transmission network, it is necessary to measure and compress the original image. Various techniques are used for compression and encryption at different stages of the approach, such as compression-then-encryption (CTE) [19, 20], encryption-then-compression (ETC) [21, 22], and simultaneous compression encryption (SCE) [23, 24]. Compressed sensing, also known as sparse representation, is a powerful method for encrypting and compressing at the same time [9]. A linear measurement matrix multiplication compresses the image. In order to increase the security level, these measurement matrices were generated by a chaotic system with the initial values. In the decryption process, the original image is recovered by sparse recovery and solving an optimization problem [25]. In the sparse recovery problem, we cannot recover the original image directly or in a well-defined form and should disjoint the optimization problem [26]. For decoupling the joint optimization problem, we can use the two frequently employed approaches, which are the alternating directions method of multipliers (ADMM) [27] and half-quadratic splitting (HQS) [28].

It has been shown that the scrambling of the pixels of the matrix can enhance the compression and reconstruction performance [29]. For satisfying the scrambling and permutating the coefficients randomly, usually, two approaches of zigzag scan [30, 31] and chaotic confusion [3] were proposed. An approach for reducing computational complexity, reducing required memory, and also increasing the security level involved sparse decomposition of a two-dimensional signal using two random measurement (sensing) matrices in two directions [32]. Therefore, certain encryption methods have suggested using two measurement matrices to achieve better performance, as shown in Ref. [33]. The construction of the measurement matrix is one of the most significant parts of compressive sensing, which relates to the ratio of signal compression and the accuracy of signal reconstruction [34]. Chaotic systems (CSs) are utilized in CS-based encryption methods to generate the measurement matrices in the different approaches, such as orthogonal random matrix using the concept of singular value decomposition [3], Hadamard transform [35, 36] and Toeplitz matrix [37].

In this research, a novel image encryption approach that is based on CS and 2D sparse recovery is presented. This approach consists of some important components. The first step is convolutional layer-based encryption (CLE), which includes chaotic scrambling, convolutional layers, and downsampling. Then, the encrypted image is compressed by two random orthogonal sensing matrices. It quantized and mapped in the interval [0, 255]. In order to increase security level, maximize image entropy, and reduce correlation of neighboring pixels, the quantized matrix is scrambled one more time and encrypted by the simple XOR operation. All of the components use chaotic time series to improve the performance of the cryptosystem. For decryption, the sparse recovery problem comes into play. This process includes convolutional layer-based decryption and sparse decomposition, which we call convolutional layers-based sparse decomposition (CLSD). To solve the CLSD problem, we propose an iterative solver based on the HQS method. HQS decouples the optimization problem into two iterative subproblems. The first step is equivalent to the denoising module, while the second is a projection onto the feasible subspace.

Experimental results demonstrate that the proposed approach represents satisfactory reconstruction performance and is robust versus different attacks. Section 15.2 investigates sparse representation and chaotic flow. In Sect. 15.3, the main idea and the proposed encryption and decryption methods are stated explicitly. Section 15.4 indicates the experimental outcomes along with security evaluation. Section 15.5 concludes the research.

15.2 Preliminaries

Here, the sparse representation and the applied chaotic flow are summarized.

15.2.1 Sparse Representation

Signal decomposition is a crucial aspect of signal processing, involving the representation vector $\mathbf{x} \in R^n$ as a linear combination of the basic functions $\mathbf{d}_i \in R^n, 1 \leq i \leq m$. In other words, $\mathbf{x}$ can be expressed as $\mathbf{x} = \mathbf{d}_1 s_1 + \cdots + \mathbf{d}_m s_m = \mathbf{Ds}$, where $\mathbf{D} = [\mathbf{d}_1, \ldots, \mathbf{d}_m]$, and $\mathbf{s} = [s_1, .., s_m]^T$. The sparsity assumption has been increasingly utilized in various applications for representing signals and images. The assumption is that the signal is k-sparse, indicating that it contains a maximum of k nonzero entries in a learned dictionary or a transform domain like discrete cosine transform (DCT) or wavelet.

Compressed sensing aims to reconstruct an unknown sparse signal $\mathbf{s} \in R^m$ from a collection of underdetermined measurements $\mathbf{x} = \mathbf{Ds} \in R^n (m > n)$, with $\mathbf{D} \in R^{n \times m}$ representing the measurement matrix. The problem of sparse recovery can be expressed as follows:

$$Min \; \|\mathbf{s}\|_0 \quad s.t. \quad \mathbf{x} = \mathbf{Ds}, \tag{15.1}$$

the notation $\|\mathbf{s}\|_0$ represents the ℓ_0 norm of $\mathbf{s}$, which corresponds to the count of nonzero elements in $\mathbf{s}$. If we define $spark(\mathbf{D})$ as the lowest number of columns in $\mathbf{D}$ that are dependent, there is only one (unique) solution $\mathbf{s}$ to this problem if $\|\mathbf{s}\|_0 < spark(\mathbf{D})/2$. The condition of uniqueness is a crucial aspect of this problem. Finding the sparsest solution of (1) typically involves a combinatorial search and is generally considered NP-hard. In recent years, various techniques have been developed, including matching pursuit (MP) [38], orthogonal matching pursuit (OMP) [39], basis pursuit (BP) [40], and smoothed ℓ_0 (SL0) [41].

In certain applications, applying compressive sensing to signals in two dimensions (2D) is necessary. To compress a sparse matrix $\mathbf{S} \in R^{m_1 \times m_2}$, two measurement matrices $\mathbf{D}_1 \in R^{n_1 \times m_1}$ and $\mathbf{D}_2 \in R^{n_2 \times m_2}$ were employed, resulting in the form $\mathbf{X} = \mathbf{D}_1\mathbf{S}\mathbf{D}_2^T$. The process of sparse recovery using two measurement matrices involves solving the following optimization problem:

$$Min \ \|\mathbf{S}\|_0 \quad s.t. \quad \mathbf{X} = \mathbf{D}_1\mathbf{S}\mathbf{D}_2^T. \tag{15.2}$$

This approach has two advantages: reduced required memory and computational complexity. Several studies have demonstrated the uniqueness of this 2D problem [32, 42]. In reference [32], this sparse recovery problem's uniqueness was studied, specifically focusing on the spark definition. The necessary conditions (but not sufficient) regarding the distribution of nonzero entries of $\mathbf{S}$ are (1) $\|\mathbf{S}\|_0 < spark(\mathbf{D}_1) \times spark(\mathbf{D}_2)/\ 4$, (2) ($\ell_0$ norm of each column) $< spark(\mathbf{D}_1)/2$ and (3) (ℓ_0 norm of each row) $< spark(\mathbf{D}_2)/2$. These conditions significantly influence the compression performance in image encryption.

Several algorithms have been introduced to solve 2D compressed sensing, including 2D orthogonal matching pursuits (2D-OMP) [43], 2D projected gradient (2DPG) algorithm [44], and 2D smoothed ℓ_0 (2D-SL0) [32]. This chapter presents an encryption-compression method using 2D measurement matrices. The corresponding image recovery problem is also solved with HQS.

15.2.2 Chaotic Flow

Here, a chaotic flow [45] is employed for the purpose of image encryption. The suitability of chaotic dynamics for encryption applications originates from their sensitivity to initial conditions. The following equations describe the system with the parameters $a = b = 0.7$:

$$\begin{aligned} x' &= y, \\ y' &= z, \\ z' &= -x + ax^2 - y^2 + bxy + xz. \end{aligned} \tag{15.3}$$

Figure 15.1 illustrates the chaotic behavior of the system when initialized with the conditions $[-1.36, 8.62, 3.91]$.

15.3 The Encryption and Decryption Approach

Here, we present a new convolutional layer (CL)-compression method using filter banks, two-dimensional sparse representation, and the XOR operator. The basis of the method is three steps. At first, input is encrypted via CL blocks and chaotic

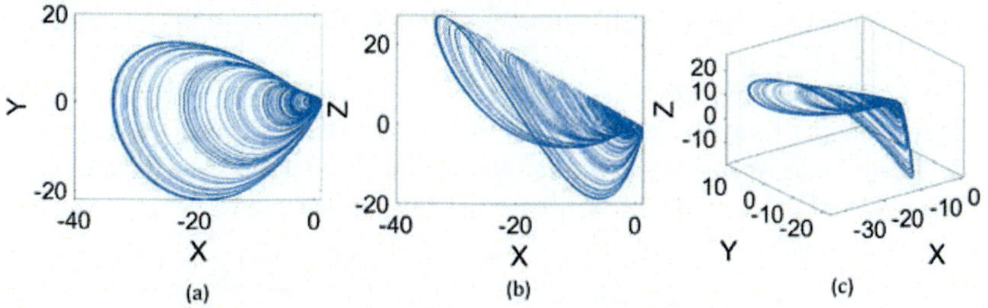

Fig. 15.1 Chaotic dynamics with initial values $[x_{01}, y_{01}, z_{01}] = [-1.36, 8.62, 3.91]$; **(a)** 2D projection in $X{-}Y$ plane; **(b)** 2D projection in $X{-}Z$ plane; **(c)** 3D dynamic

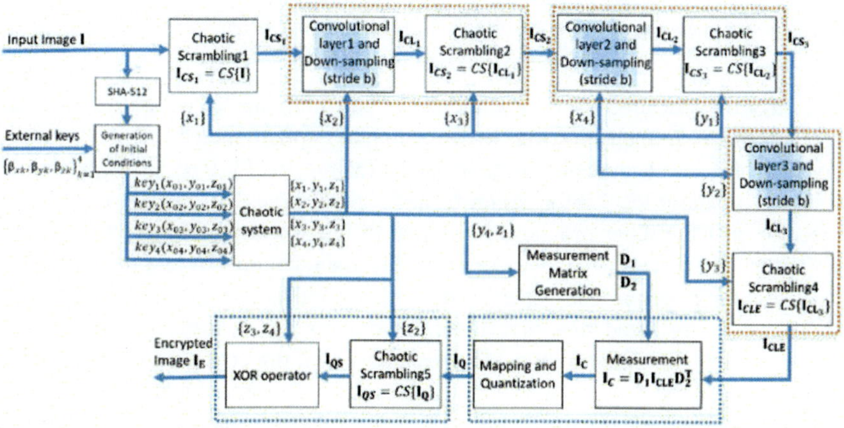

Fig. 15.2 The encryption method

scrambling. Then, the image is compressed via two matrices. At last, an encryption operator based on XOR is used. The proposed method is shown in Fig. 15.2.

15.3.1 Chaotic Shuffling

Applying chaotic shuffling to $\mathbf{X} \in R^{n \times m}$ is a crucial stage for a cryptography process. This procedure is employed to enhance the degree of safety for encryption and reduce the correlation between neighboring elements within $\mathbf{X}$. The process of chaotic shuffling can be outlined as follows:

1. Chaos generation: creating the chaotic signals $Y = \{y_i\}$ by utilizing the flow with a set of initial values.
2. Transform: changing the sequence $Y = \{y_i\}$ into an integer sequence $Y^d = \{y_i^d\}$ where $y_i^d = mod(y_i \times 10^{15}, 2^{12})$.
3. Sort: ordering through a series Y^d

$$[Y_s^d, index_s^Y] = sort(Y^d),$$

the sequencing operator $[.,.] = sort(.)$ is applied to sort the elements, Y_s^d represents the sorted sequence, while $index_s^Y$ denotes the index sequence of Y_s^d, which consists of a series of integer numbers with random perturbations.

4. Scramble: shuffling the matrix $\mathbf{X}$ based on the signal $index_s^Y$.

This approach is employed five times, each time with various initial values. This operation and its inverse are defined as the functions of $\mathbf{X}^s = CS(\mathbf{X}, Y)$ and $\mathbf{X} = CS^{-1}(\mathbf{X}^s, Y)$, respectively.

15.3.2 Convolutional Layer

The suggested method uses three blocks of convolutional layers before the compression stage. In the following, we are going to explain one CL block. This layer consists of a set of mask filters (filter bank) $\{\mathbf{M}_i \in R^{b\times b}\}, i = 1, \ldots, b^2$ that are utilized to extract local information from the input. These filters are orthonormal, meaning that the inner product of any two different masks, $\mathbf{M}_i$ and $\mathbf{M}_j$, is zero. Using b^2 filters leads to an increase in the data volume by a factor of b^2. A downsampling operation is implemented with a stride of b samples in two directions to reduce the required memory. The filter outputs are concatenated to form the $\mathbf{I}_{CL} \in R^{n_x \times n_y}$ matrix, which is structured based on the convolutional layers. The n_x/b and n_y/b ratios must be integer numbers. If not, we can add additional columns and rows of zero values to the original image (which is named zero padding) to achieve integer ratios. In the simulations, we have considered three different values for the parameter b: 1, 8, and 16. After the CL block, chaotic scrambling is utilized to disperse the local information obtained from the convolutional layers across all domains of the matrix. The entire process of the proposed procedure for one CL block can be summarized as follows:

1. Filter Construction
   - Chaos Generation: creating the chaotic sequence $Y = \{y_i\}$ with length of b^4.
   - Transform: changing the signal $Y = \{y_i\}$ into $Y^* = \{y_i^*\}$: $y_i^* = mod(y_i \times 2^{15}, 2^{10})/2^9 - 1$.
   - Random Matrix Generation: reshaping the signal Y^* to construct $\mathbf{Y}^M \in R^{b^2 \times b^2}$.
   - Orthogonal Random Matrix Generation: computing the singular value decomposition $\mathbf{Y}^M = \mathbf{U}\mathbf{\Sigma}\mathbf{V}^T$ where $\mathbf{\Sigma} \in R^{b^2 \times b^2}$ is a diagonal matrix containing singular values. The columns of matrices $\mathbf{U} \in R^{b^2 \times b^2}$ and $\mathbf{V} \in R^{b^2 \times b^2}$ are singular vectors, and both of them are orthogonal matrices.
   - Filter Generation: reshaping each row of $\mathbf{U}$ to construct the mask filters $\{\mathbf{M}_i \in R^{b\times b}\}$.

2. Convolution: The procedure involves convolving the filters with the input matrix **I**, downsampling the output of each filter using a stride of b samples in two directions, and constructing the CL matrix $\mathbf{I}_{CL}$ by concatenating all the filter outputs.

This operation is defined as the function of $\mathbf{I}_{CL} = CL(\mathbf{I}, Y)$.

15.3.3 Chaotic Measurement

This matrix plays a vital role in compressed sensing. In this approach, the encrypted figure $\mathbf{I}_{CLE} \in R^{n_x \times n_y}$ is measured and compressed by two measurement matrices $\mathbf{D}_1 \in R^{m_x \times n_x}$ and $\mathbf{D}_2 \in R^{m_y \times n_y}$. In this case, we generate the measurement matrices by utilizing a unitary matrix that is constructed using a random matrix derived from the chaotic flow. Measurement matrices $\mathbf{D}_1$ and $\mathbf{D}_2$ are generated using two different chaotic sequences. Note that the process of generating the two matrices is the same. The process of generating matrix $\mathbf{D}_1$ can be broken down into five steps, which are outlined below:

1. Chaotic Sequence Generation: creating the chaotic sequence $Y = \{y_i\}$ with length of n_x^2.
2. Transform: changing the signal $Y = \{y_i\}$ into a signal $Y^* = \{y_i^*\}$: $y_i^* = mod(y_i \times 2^{15}, 2^{10})/2^9 - 1$.
3. Random Matrix Generation: the sequence Y^* is reshaped to construct the matrix $\mathbf{Y}^M \in R^{b^2 \times b^2}$.
4. Orthogonal Random Matrix Generation: computing the singular value decomposition: $\mathbf{Y}^M = \mathbf{U}\boldsymbol{\Sigma}\mathbf{V}^T$.
5. Slicing: generating the matrix $\mathbf{D}_1$, with the first m_x rows of matrix **U**.

15.3.4 The Proposed Encryption Process with Convolutional Layers

The encryption method (Fig. 15.2) comprises three blocks. They are explained as the following.

15.3.4.1 Generating Initial Values

The proposed encryption process involves the generation of 12 chaotic sequences using 4 different initial conditions. To enhance the security level, the initial conditions for generating the chaotic sequences are determined by the SHA-512 hash function, which considers the original image's characteristics in the key generation. This operation increases the method's resistance against the chosen-plaintext attack. The following steps generate the keys.

1. Divide the hash value: Reshaping 512 bits of the SHA value to 64 blocks with length 8 and corresponding decimal numbers of all blocks s_k, $k = 1, \ldots, 64$.
2. Key computation: Generating the initial conditions (x_{0k}, y_{0k}, z_{0k}), $k = 1, \ldots, 4$, by the XOR operation:

$$x_{0k} = \frac{(s_{15(k-1)+1} \oplus s_{15(k-1)+2} \oplus \ldots \oplus s_{15(k-1)+5} \oplus s_{60+k}) - 128}{256} + \beta_{xk},$$

$$y_{0k} = \frac{(s_{15(k-1)+6} \oplus s_{15(k-1)+7} \oplus \ldots \oplus s_{15(k-1)+10}) - 128}{256} + \beta_{yk}, \quad (15.4)$$

$$z_{0k} = \frac{(s_{15(k-1)+11} \oplus s_{15(k-1)+12} \oplus \ldots \oplus s_{15(k-1)+15}) - 128}{256} + \beta_{zk}.$$

where β_{xk}, β_{yk}, and β_{zk} are the external keys.

15.3.4.2 Convolutional Layer-Based Encryption (CLE)

This block improves the security level for potential attacks and enhances image reconstruction during decryption. According to this block, the convolutional layer sparse decomposition (CLSD) problem is proposed to restore the encrypted image. Experimental outcomes present that the CLSD method is better than other image encryption techniques. All steps of the CLE are as follows:

1. Chaotic scrambling:

$$\mathbf{I}_{CS1_1} = CS\{\mathbf{I}, X_1\}.$$

   This step enhances degree of safety by reducing the correlation between the elements.
2. Convolution layer encryption:

$$\mathbf{I}_{CL_1} = CL\{\mathbf{I}_{CS1_1}, X_2\}.$$

3. Chaotic scrambling block as the role of activation function in deep models:

$$\mathbf{I}_{CS_1} = CS\{\mathbf{I}_{CL_1}, X_3\}.$$

   Note that the CLE is implemented by sequentially repeating steps 2 and 3 for three times:

$$\mathbf{I}_{CL_2} = CL\{\mathbf{I}_{CS_1}, X_4\}, \qquad \mathbf{I}_{CS_2} = CS\{\mathbf{I}_{CL_2}, Y_1\}.$$

$$\mathbf{I}_{CL_3} = CL\{\mathbf{I}_{CS_2}, Y_2\}, \qquad \mathbf{I}_{CLE} = CS\{\mathbf{I}_{CL_3}, Y_3\}.$$

We found out empirically that this approach leads to better outcomes. The above steps are called as $\mathbf{I}_{CLE} = CLE\{\mathbf{I}\}$. This operation is reversible, and its inverse operation is defined as $\mathbf{I} = CLE^{-1}\{\mathbf{I}_{CLE}\}$.

15.3.4.3 Compression

The encrypted image $\mathbf{I}_{CLE}$ is compressed by using two chaotic measurement matrices $\mathbf{D}_1 \in R^{m_x \times n_x}$ and $\mathbf{D}_2 \in R^{m_y \times n_y}$:

$$\mathbf{I}_C = \mathbf{D}_1 \mathbf{I}_{CLE} \mathbf{D}_2^T$$

The compression ratio (CR) is assumed to be the same value in x and y directions, i.e., $\sqrt{CR} = m_x/n_x = m_y/n_y$. Because of the utilization of chaotic systems for generating the measurement matrices, image encryption's security level increases. We can uniquely recover the original image from the compressed and encrypted version according to the uniqueness theorem of the sparse recovery problem.

The CL and compression blocks act as a neural network that includes CL as a filter and an average pooling consecutively. It means that the compression block and measurement matrix play the role of averaging, which is a linear operator. It can be investigated as a future work.

15.3.4.4 Quantization and Mapping

This step maps and quantizes the compressed figure $\mathbf{I}_C$ within the range [0, 255]:

$$\mathbf{I}_Q = \left[255 \frac{\mathbf{I}_C - min(\mathbf{I}_C)}{max(\mathbf{I}_C) - min(\mathbf{I}_C)} \right],$$

$[x]$ rounds the value of x to the nearest integer number. Each pixel in $\mathbf{I}_Q$ can be expressed using a single byte.

15.3.4.5 Final Encryption

To enhance the security level, we employ this block that utilizes chaotic scrambling and XOR at the end to generate a meaningless image. This block consists of the following two steps:

1. Scrambling the quantized and mapped image $\mathbf{I}_Q$:

$$\mathbf{I}_{QS} = CS\{\mathbf{I}_Q, Z_2\}.$$

This step enhances the security level by reducing the correlation between the elements of $\mathbf{I}_Q$.

2. The XOR operator is applied during the last step to achieve a uniform distribution for the encrypted image $\mathbf{I}_E$. The scrambled output $\mathbf{I}_{QS}$ is XORed with a random sequence of the chaotic signal. It can be further explained as follows:
   - Generating the chaotic sequences $Z_3 = \{z_{3i}\}$ and $Z_4 = \{z_{4i}\}$ with the length of $m_x \times m_y$.
   - The sequences Z_3 and Z_4 are transformed into integer values in the interval $[0, 255]$:

$$z_{3i}^* = mod(z_{3i} \times 10^{15}, 2^8), \qquad z_{4i}^* = mod(z_{4i} \times 10^{15}, 2^8).$$

   - $\mathbf{I}_{QS}$ and $Z_3^* = \{z_{3i}^*\}$ and $Z_4^* = \{z_{3i}^*\}$ are converted to the unsigned integer with 8 bits. The encrypted figure is acquired through the XOR operator:

$$\mathbf{I}_E = \mathbf{I}_{QS} \oplus [Z_3^* \oplus Z_4^*].$$

The proposed image encryption method utilizes the chaotic flow with 4 sets of initial values to generate 12 chaotic sequences.

15.3.5 *The Image Decryption Process Based on Convolutional Layers and Sparse Recovery*

Now, we will restore the original image by reversing all the steps of the encryption process. For image recovery, the receiver has to possess the key sources and the initial values of the chaotic systems. The following steps show the inverse process in detail:

1. Obtaining $\mathbf{I}_{QS}$ from $\mathbf{I}_E$ by the XOR operator: $\mathbf{I}_{QS} = \mathbf{I}_E \oplus [Z_3^* \oplus Z_4^*]$
2. Obtaining $\mathbf{I}_Q$ by chaotic scrambling's inverse operation $CS^{-1}\{.\}$: $\mathbf{I}_Q = CS^{-1}\{\mathbf{I}_{QS}, Z_2\}$
3. Obtaining the compressed and measured matrix $\mathbf{I}_C$ by the inverse mapping as follows:

$$\mathbf{I}_C = ((max(\mathbf{I}_C) - min(\mathbf{I}_C))/255)\mathbf{I}_Q + min(\mathbf{I}_C),$$

   $max(\mathbf{I}_C)$ and $min(\mathbf{I}_C)$ are also gotten from the sender.
4. The recovery of the original picture with the CLSD problem is outlined as:

$$CLSD: \quad argmin_{\mathbf{I}}\ Reg(\mathbf{I}) \quad s.t. \quad \mathbf{I}_c = \mathbf{D}_1 \mathbf{I}_{CLE} \mathbf{D}_2^T, \quad \mathbf{I}_{CLE} = CLE\{\mathbf{I}\}. \tag{15.5}$$

The $Reg(\mathbf{I})$ cost function is a regularization term such as total variation, smoothness, or sparseness in the wavelet domain. In this problem, the original image is restored through a sparse reconstruction algorithm. In this image recovery problem, the encrypted form of the original image is measured. For recovering the original image, directly applying traditional sparse recovery approaches is not feasible due to the unconventional constraint of the sparse recovery problem in the method. The proposed CLSD problem combines two steps of CLE and compression into a unified process. Therefore, to restore the original image, it is necessary to simultaneously perform the inverse of the steps mentioned above. We refer to this problem as convolutional layers-based sparse decomposition (CLSD). This problem is optimized with an optimization method named half-quadratic splitting (HQS), explained in the next subsection.

15.3.5.1 Convolutional Layers-Based Sparse Decomposition (CLSD)

In this section, a method is presented to solve the CLSD problem. The compressed matrix $\mathbf{I}_C$ is acquired from an underdetermined measurement following a forward process model $\mathbf{I}_C = \mathbf{D}_1 \mathbf{I}_{CLE} \mathbf{D}_2^T$. Since $\mathbf{D}_1$ and $\mathbf{D}_2$ are ill-posed matrices, a unique solution from solving the forward model is not guaranteed. It has been shown that the unique solution is obtainable by adding a non-smooth sparsity-promoting function (a regularization term) in which the sparse representation of the recovered signal is held. Therefore, to restore the original picture, we redefine the CLSD problem (15.5) in the following minimization manner:

$$CLSD: \quad \hat{\mathbf{I}} = argmin_{\mathbf{I}}\ J(\mathbf{I}), \qquad J(\mathbf{I}) = Reg(\mathbf{I}) + \delta_C(\mathbf{I}), \tag{15.6}$$

where $\delta_C(\mathbf{X})$ denotes the function for the feasible set $C = \{\mathbf{X} \in R^{n_x \times n_y} | \mathbf{I}_C = \mathbf{D}_1 \mathbf{I}_{CLE} \mathbf{D}_2^T,\ \mathbf{I}_{CLE} = CLE\{\mathbf{X}\}\}$, representing the subspace of measurement data, defined as

$$\delta_C(\mathbf{X}) = \begin{cases} 0, & \text{if } \mathbf{X} \in C, \\ +\infty, & \text{if } \mathbf{X} \notin C. \end{cases} \tag{15.7}$$

Here, the regularization term can be a non-convex function like ℓ_0-norm or any function that measures the sparseness in a transform like wavelet, DCT, or Fourier.

The plug-and-play half-quadratic splitting (PnP-HQS) method [26] is employed to solve the CLSD problem. Rather than directly solving the minimization problem in (15.6), we utilize the PnP approach, a highly effective technique in sparse signal recovery. The concept behind PnP is to separate the minimization function into two components: fidelity and a prior term. To disjoint the minimization process, a common approach is to employ a variable splitting technique. In this case, we opt to utilize HQS.

The HQS optimizes $J(\mathbf{I})$ by adding an auxiliary variable $\mathbf{Z} \in R^{n_x \times n_y}$ to the CLSD as:

$$\hat{\mathbf{I}}, \hat{\mathbf{Z}} = argmin_{\mathbf{I},\mathbf{Z}}\ J_{HQS}(\mathbf{I}, \mathbf{Z}) \quad s.t. \quad \mathbf{Z} = \mathbf{I}$$
$$J_{HQS}(\mathbf{I}, \mathbf{Z}) = Reg(\mathbf{I}) + \delta_C(\mathbf{Z}). \tag{15.8}$$

This optimization can be solved as

$$\hat{\mathbf{I}}, \hat{\mathbf{Z}} = argmin_{\mathbf{I},\mathbf{Z}}\ J_\alpha(\mathbf{I}, \mathbf{Z}),$$
$$J_\alpha(\mathbf{I}, \mathbf{Z}) = Reg(\mathbf{I}) + \delta_C(\mathbf{Z}) + \frac{1}{2\alpha}||\mathbf{I} - \mathbf{Z}||_F^2, \tag{15.9}$$

where $\alpha > 0$ is the penalty parameter and $||\mathbf{X}||_F = \sqrt{\Sigma x_{ij}^2}$ is the Frobenius norm. Reducing value of α imposes a penalty on the deviation from the constraint $\mathbf{Z} = \mathbf{I}$. In this optimization, a decreasing sequence α_t ($\alpha_t < \alpha_{t-1}$) is considered. The HQS method changes this problem into two disjoint blocks for each iteration t:

$$\mathbf{I}^t = argmin_{\mathbf{I}}\ Reg(\mathbf{I}) + \frac{1}{2\alpha_t}||\mathbf{I} - \mathbf{Z}^{t-1}||_F^2,$$
$$\mathbf{Z}^t = argmin_{\mathbf{Z}}\ \delta_C(\mathbf{Z}) + \frac{1}{2\alpha_t}||\mathbf{I}^t - \mathbf{Z}||_F^2. \tag{15.10}$$

Consequently, the minimization of Eq. (15.9) is converted into an iterative optimization process involving two steps for solving subproblems. The first step applies the prior condition, while the second step ensures that the reconstructed signal lies within the feasible subspace C. The first block of Eq. (15.10) can be interpreted as the denoising of $\mathbf{Z}^{t-1}$. So, we define $\mathbf{I}^t$ as

$$\mathbf{I}^t = Denoiser(\mathbf{Z}^{t-1}, \alpha_t), \tag{15.11}$$

where $Denoiser(., \alpha_t)$ represents an arbitrary denoising module with the parameter α_t that considers the noise level. The second term in Eq. (15.10) is a projection onto the feasible subspace C. It is defined as:

$$\mathbf{Z}^t = Proj_C(\mathbf{I}^t). \tag{15.12}$$

This Euclidean projection of $\mathbf{I}^t$ onto C can be obtained with some calculations in the following form:

$$\mathbf{I}_{CLE}^t = CLE\{\mathbf{I}^t\},$$
$$\mathbf{I}_{CLE}^{proj} = \mathbf{I}_{CLE}^t - \mathbf{D}_1^\dagger(\mathbf{D}_1\mathbf{I}_{CLE}^t\mathbf{D}_2^T - \mathbf{I}_C)(\mathbf{D}_2^\dagger)^T, \tag{15.13}$$
$$\mathbf{Z}^t = CLE^{-1}\{\mathbf{I}_{CLE}^{proj}\},$$

where $\mathbf{D}^\dagger = \mathbf{D}^T(\mathbf{D}\mathbf{D}^T)^{-1}$ is the pseudo inverse (Moore-Penrose inverse) of $\mathbf{D}$.

Denoiser is the main part of the PnP algorithm. The quality of the reconstruction improves as the denoiser becomes more effective. In this research, we use two kinds of denoisers. One of them is Weiner, and the other one is block-matching 3D (BM3D) [46]. Compared with each other, BM3D presents better results, but it has more computational load. In future work, we can utilize a deep model or a series of CNN models for the denoiser because of their preferable accuracy and speed.

At the last step of the decryption process, to reduce the quantization noise, which is in the projection step, we use another denoiser. The denoiser type is the same as the one used in the CLSD approach. We figured out that the final procedure leads to better results even if it causes insensible enhancements. The image reconstruction method is presented in Algorithm 1.

Algorithm 1 The image reconstruction with the convolutional layers based sparse decomposition

Input: $\mathbf{I}_C, \mathbf{D}_1, \mathbf{D}_2, \alpha_t$ (decreasing sequence), α_f

Initialization: $\mathbf{Z}^0$

1– Pseudo inverses: $\mathbf{I}^0_{CLE} = \mathbf{D}_1^{\dagger}\mathbf{I}_C(\mathbf{D}_2^{\dagger})^T$

2– Convolution layer based decryption: $\mathbf{Z}^0 = CLE^{-1}\{\mathbf{I}^0_{CLE}\}$

For $t = 1, \ldots, J$

Denoiser: $\mathbf{I}^t = Denoiser(\mathbf{Z}^{t-1}, \alpha_t)$

Projection:

(1) Convolution layer based encryption: $\mathbf{I}^t_{CLE} = CLE\{\mathbf{I}^t\}$

(2) Projection on the linear subspace: $\mathbf{I}^{proj}_{CLE} = \mathbf{I}^t_{CLE} - \mathbf{D}_1^{\dagger}(\mathbf{D}_1\mathbf{I}^t_{CLE}\mathbf{D}_2^T - \mathbf{I}_C)(\mathbf{D}_2^{\dagger})^T$

(3) Convolution layer based decryption: $\mathbf{Z}^t = CLE^{-1}\{\mathbf{I}^{proj}_{CLE}\}$

End For

Final Denoiser: $\hat{\mathbf{I}} = Denoiser(\mathbf{Z}^J, \alpha_f)$

Output: $\hat{\mathbf{I}}$

15.4 Experimental Outcomes

Here, various simulations are conducted to validate the performance of the suggested encryption method using different metrics. For performance evaluation, pictures of Cameraman, Lena, Fundus, and MRI, each with a dimension of 256 × 256, are utilized. During the experiments, the external keys are configured as $[\beta_{x1}, \beta_{y1}, \beta_{z1}, \beta_{x2}, \beta_{y2}, \beta_{z2}, \beta_{x3}, \beta_{y3}, \beta_{z3}, \beta_{x4}, \beta_{y4}, \beta_{z4}] = [-1.36, 8.62, 3.91, -1.67, 8.98, 3.67, -1.43, 8.75, 3.35, -1.85, 8.25, 3.47]$. The decrypted and reconstructed pictures are evaluated by using two metrics: the peak signal-to-noise ratio (PSNR) and the mean square deviation (MSE). These measures are:

$$MSE = \frac{1}{n_x n_y}\Sigma_{i,j}(\mathbf{I}(i, j) - \hat{\mathbf{I}}(i, j))^2, \tag{15.14}$$

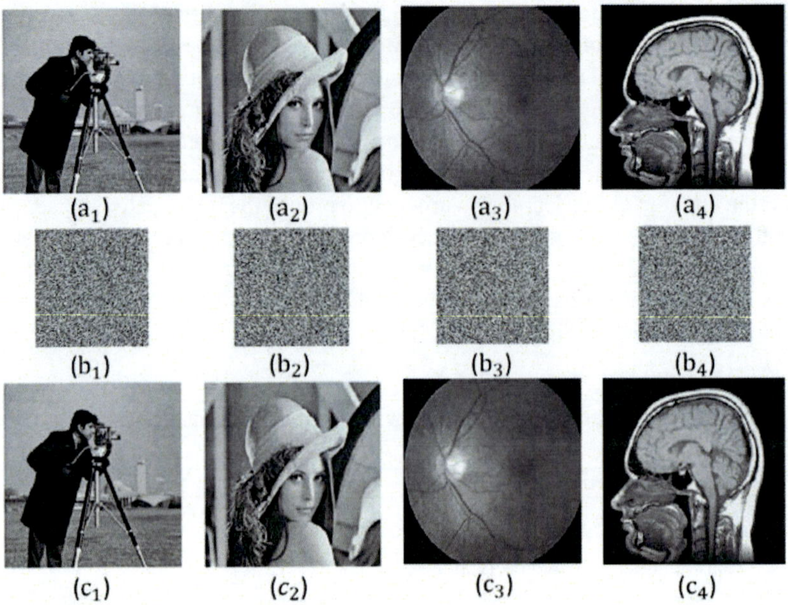

Fig. 15.3 Experimental results: ($\mathbf{a}_1$)–($\mathbf{a}_4$) original images; ($\mathbf{b}_1$)–($\mathbf{b}_4$) encrypted images; ($\mathbf{c}_1$)–($\mathbf{c}_4$) decrypted images

$$PSNR = 10\log\left(\frac{255^2}{MSE}\right), \tag{15.15}$$

$\mathbf{I} \in R^{n_x \times n_y}$ and $\hat{\mathbf{I}} \in R^{n_x \times n_y}$ represent the original and the decrypted images, respectively. Figure 15.3 displays the outcomes of the method for various test pictures.

15.4.1 Compression Performance

Here, the performance of the convolution-compression method is measured by the quality of the decrypted image at different compression ratios (CRs). The decrypted images for various compression ratios (CRs) are presented in Fig. 15.4. The visual quality of the decrypted images is satisfactory. This demonstrates the efficacy of the proposed CLSD approach in successfully recovering the compressed and encrypted picture. Figure 15.5 illustrates PSNR values for various CRs and test images. Table 15.1 shows the evaluation of various schemes [33, 47–50] with compressed sensing, comparing them with the method. It is evident that the proposed encryption approach outperforms other schemes [33, 47–50] at the same CR.

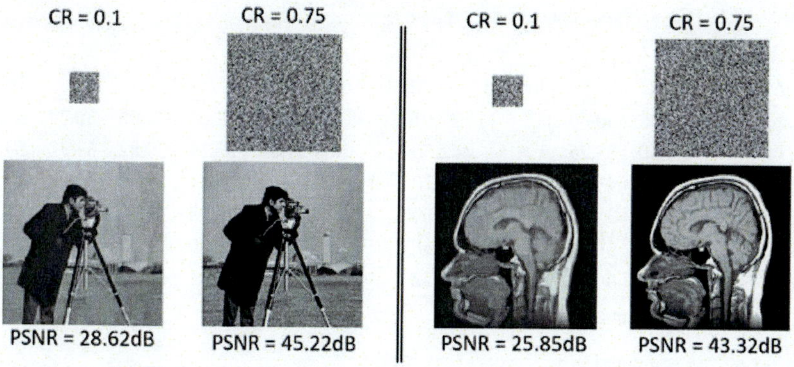

Fig. 15.4 The outcomes of various compression ratios: the first row and the second row show the encrypted and decrypted images

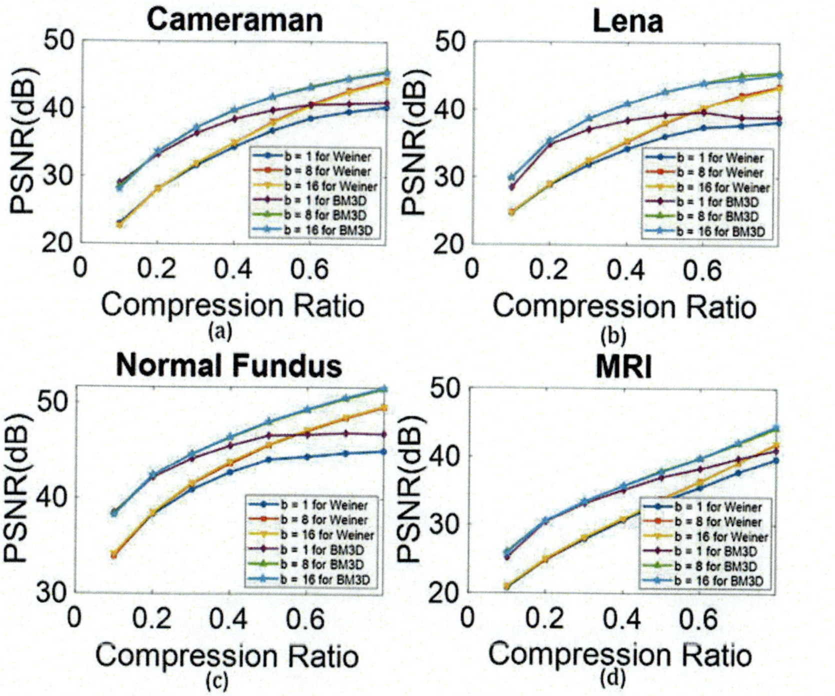

Fig. 15.5 PSNR of the decrypted pictures for different CRs based on two approaches of Weiner and BM3D for (**a**) Cameraman, (**b**) Lena, (**c**) Normal fundus, and (**d**) MRI

In order to study the feasibility of the encryption method, the computational complexity is investigated by measuring the computation time. Figure 15.6 shows the runtime of the algorithm. The runtime can vary depending on factors such

Table 15.1 PSNR(dB) of the encrypted pictures

Image 256*256	CR	Weiner method	BM3D method	[33]	[47]	[48]	[49]	[50]
Lena	0.25	30.69	37.61	31.72	28.67	26.06	30.77	–
	0.50	38.25	42.50	37.33	32.14	29.82	33.78	34.59
	0.75	43.34	45.70	41.63	–	29.56	32.28	–
Cameraman	0.25	30.27	35.89	30.99	27.95	25.23	–	–
	0.50	39.03	41.98	37.39	32.89	29.43	–	37.51
	0.75	44.02	45.22	42.52	–	28.93	–	–

– These results are not in the refs.

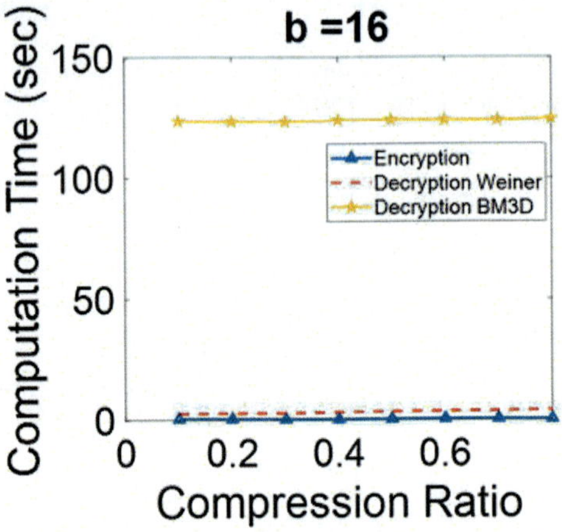

Fig. 15.6 The computation times for the different compression ratios

as programming skill and language used, making it an inappropriate measure for evaluating the previously mentioned computational complexity.

15.4.2 *Security Performance Analyses*

Here, the security of the method is analyzed by employing two essential measures: key sensitivity and keyspace analysis.

15.4.2.1 Key Sensitivity

The effectiveness of a cryptosystem is primarily determined by the keystream, which plays a crucial role. Enhancing security is achieved by considering key sensitivity as a significant aspect. The sensitivity can be observed from two perspectives. During the encryption, even a slight change in the key should lead to a completely different encrypted image. From another standpoint, during

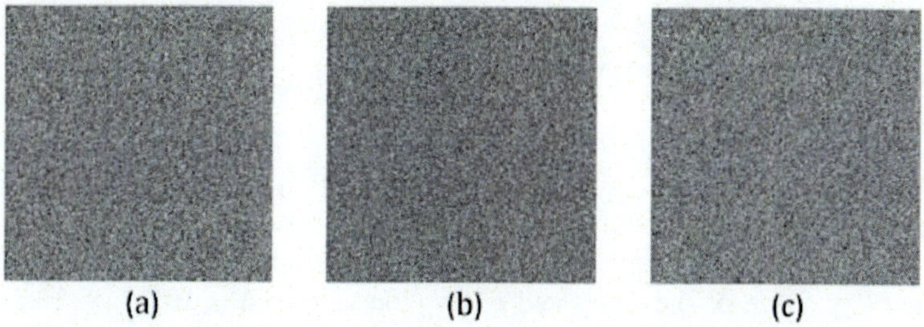

Fig. 15.7 Decrypted Lena image with wrong keys: (**a**) with $\beta_{x1} + 10^{-15}$; (**b**) with $\beta_{y1} + 10^{-15}$; (**c**) with $\beta_{z1} + 10^{-15}$

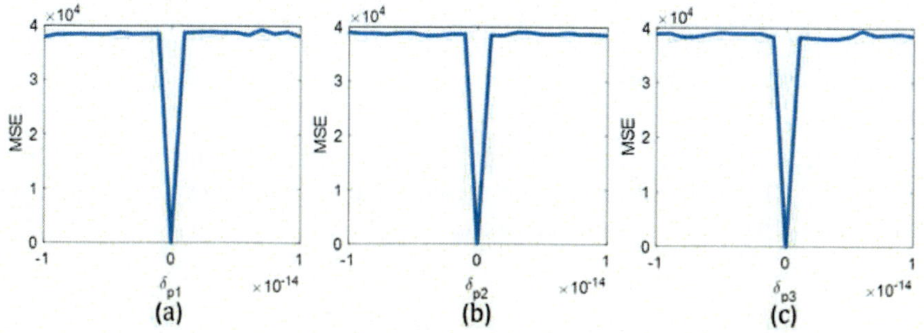

Fig. 15.8 MSE curves concerning the variation of the external keys: (**a**) $\beta_{x1} + \delta_{p1}$; (**b**) $\beta_{y1} + \delta_{p2}$; (**c**) $\beta_{z1} + \delta_{p3}$

the decryption, a slight change in the correct key will yield a picture which is entirely different. This method utilizes the sensitivity of the chaotic flow to initial values to establish a safe method concerning the keys, which are defined as $[\beta_{x1}, \beta_{y1}, \beta_{z1}, \beta_{x2}, \beta_{y2}, \beta_{z2}, \beta_{x3}, \beta_{y3}, \beta_{z3}, \beta_{x4}, \beta_{y4}, \beta_{z4}]$. Now, we illustrate the impact of slight changes on the decryption method. A small value of 10^{-15} is added to one of the keys. Figure 15.7 displays the decrypted Lena image with a CR value of 0.5 for three incorrect keystreams. Using incorrect keystreams has resulted in meaningless images that lack any visual resemblance to the original image. Figure 15.8 illustrates the MSEs, depicting the changes of three keys and its impact on the similarity between the original and decrypted pictures. The curves demonstrate that the minimum MSE value is achieved when using the correct key. The observation suggests that the method exhibits a high level of sensitivity to the external keys.

15.4.2.2 Key Space

The key space is a crucial aspect of a secure cryptosystem. This metric estimates the required complexity for attacking the cryptosystem through key search. The increased size of the key space makes the brute force attack impractical. Based on

Table 15.2 Key space of our method and others

Algorithm	Proposed method	[33]	[47]	[48]	[49]	[50]
Key space	2^{598}	2^{398}	2^{349}	2^{232}	2^{449}	2^{249}

the research [51], it is recommended that the key space should exceed a size of 2^{100}. The precision of a double-precision number is approximately 10^{-15}. Hence, the total key space of the suggested encryption approach is $10^{180} \approx 2^{598}$, which is considered sufficiently large. This key space prevents exhaustive searching from being feasible. Table 15.2 presents a comparison of the key spaces among various approaches [33, 47–50].

15.4.3 *Statistical Analysis*

Enhancing the level of randomness of the encrypted image leads to a high level of security. Here, the randomness of the encrypted picture is evaluated from the statistical perspective using three criteria.

15.4.3.1 Histogram

A histogram serves as a metric to evaluate the performance of an encryption approach. In a secure cryptosystem, an ideal histogram exhibits a uniform distribution, which enhances the resistance of the encryption process against statistical attacks. Figure 15.9 displays the histograms of the original pictures, CLE pictures, and final encrypted pictures for a CR of 0.5. The proposed approach ensures that different images at various stages of encryption exhibit similar histograms. This similarity in histograms presents the safety of the approach.

15.4.3.2 Entropy

Information entropy is a metric used to quantify the level of randomness in the encrypted image. A higher level of entropy indicates the security level of the encrypted pictures. The uniform distribution of a discrete variable maximizes the Shannon entropy. The maximum value is $\log_2(M)$, where M represents the number of states of the discrete variable. In our method, the number of gray levels denoted as M is 256. Therefore, the maximum entropy in this experiment is 8. Table 15.3 displays the entropy values of the original and the encrypted pictures at various compression ratios. The findings indicate that the entropy of the encrypted pictures is maximized and approximately equal to 8.

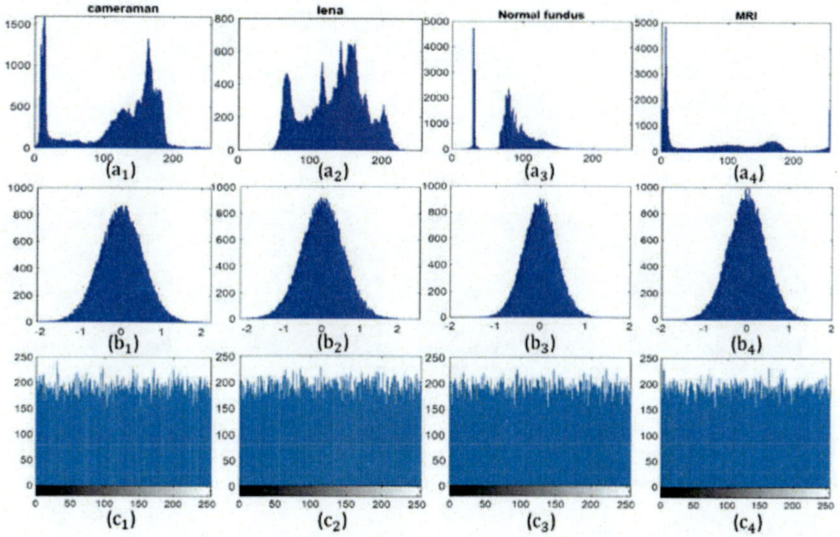

Fig. 15.9 Histogram of original, CLE, and encrypted pictures. **(a$_i$)** The original **(b$_i$)**. The corresponding CLE **(c$_i$)** The encrypted images

Table 15.3 Entropy of the pictures

		Encrypted image		
	Original image	CR = 0.25	CR = 0.5	CR = 0.75
Lena	7.6365	7.9898	7.9945	7.9960
Cameraman	7.0257	7.9883	7.9943	7.9967
Normal fundus	6.3953	7.9890	7.9946	7.9963
MRI	6.9868	7.9888	7.9951	7.9961

15.4.3.3 Correlation

The previously mentioned metrics consider the uniformity of the distribution but do not compute the correlation between pixels. In natural images, there is a high correlation between neighboring pixels in all directions. To ensure secure image encryption, it is desirable to minimize the correlation of pixels. The correlation between neighboring pixels in the horizontal direction is effectively reduced in the encrypted images, as demonstrated by the joint distribution depicted in Fig. 15.10 for both the original and encrypted pictures. The depicted graphs demonstrate that the encrypted image exhibits a nearly uniform distribution in two dimensions. The correlation coefficient (CC) is a metric used in here. For comparison purposes, Table 15.4 displays the CC values of various schemes [33, 47, 49] in three different directions. The findings indicate that the suggested encryption technique decreases the correlation among pixels, resulting in improved performance.

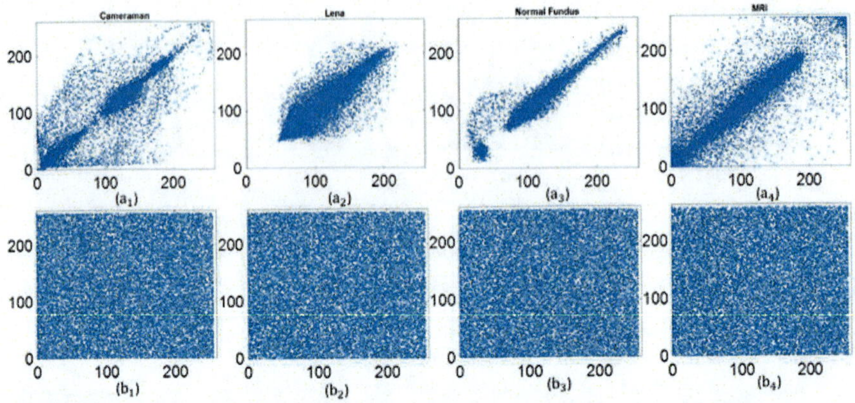

Fig. 15.10 The correlation distribution of two pixels in the horizontal direction for ($\mathbf{a}_i$) the original pictures. ($\mathbf{b}_i$) the encrypted pictures

Table 15.4 Correlation coefficients

Image	Lena			Cameraman		
Algorithm	Horizontal	Vertical	Diagonal	Horizontal	Vertical	Diagonal
Original image	0.9459	0.9721	0.9212	0.9556	0.9738	0.9340
Proposed method	0.0008	−0.0082	−0.0038	0.0044	0.0001	0.0057
[33]	0.0037	0.0078	0.0019	0.0037	−0.0037	0.0094
[47]	0.0088	0.0008	0.0022	0.0040	0.0088	0.0180
[49]	−0.0046	−0.0002	0.0005	–	–	–

– These results are not in the refs.

15.4.4 *Noise Attack*

During the transmission process, images are commonly subject to various noise models. The robustness of algorithm is studied when exposed to various noise attacks, including Gaussian noise (GN), speckle noise (SN), and salt and pepper noise (SPN). So, the Lena image is selected as the test, and the CR is set to 0.5. Figure 15.11 demonstrates the performance of the algorithm with various noise models at different intensities. Figure 15.12 displays the recovery efficiency of the encryption method across various noise intensities. Based on the results, we can conclude that GN and SPN attacks have the most and the least significant impact on the recovery performance, respectively. Also, there is a certain level of resistance to the SN attack. So, the proposed encryption method possesses a strong capability to successfully restore the original image even when subjected to different noise attacks.

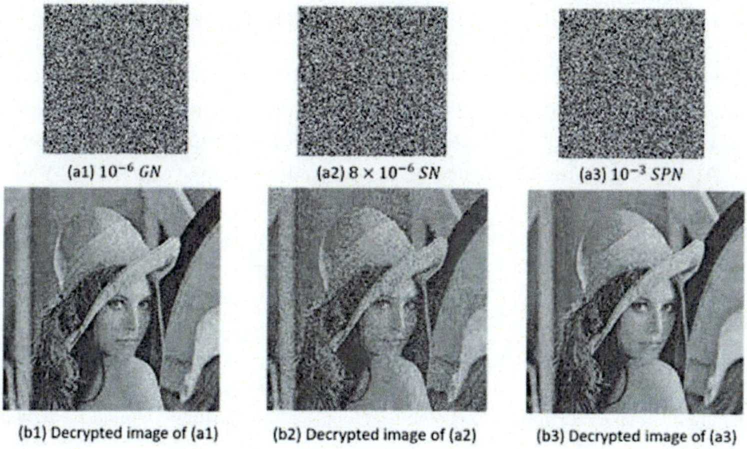

Fig. 15.11 Encrypted and decrypted pictures in the presence of GN, SN, and SPN noises

Fig. 15.12 PSNR between decrypted and original pictures with various noises

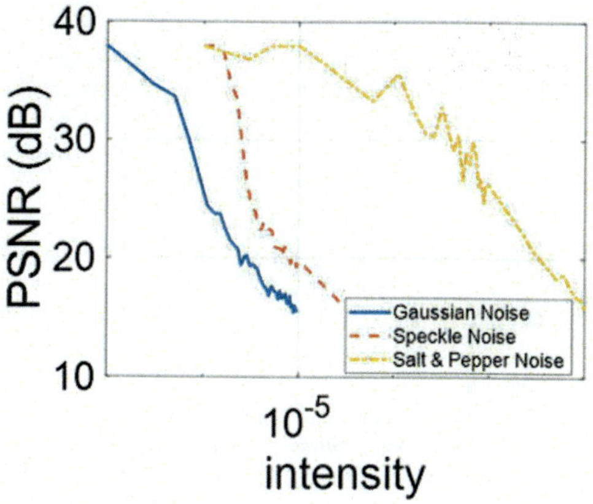

15.5 Conclusion

This research proposed a novel image encryption system using two-dimensional sparse recovery and chaotic flow. In the first block, the convolutional layers with downsampling and chaotic scrambling were used to encrypt the input image more securely. Next, the encrypted image was compressed via two orthogonal measurement matrices. After mapping and quantization, to reduce the correlation among neighboring pixels of the matrix, another chaotic scrambling followed by an XOR operator was used. The chaotic scrambling and the XOR operator were employed to increase the security level of the system. Statistical analysis showed that the encrypted image has a uniform distribution and achieves the maximum entropy. For the decryption process and image recovery, a convolutional layer sparse decomposition (CLSD) problem was presented. For solving the CLSD, HQS iterative solver was proposed. This technique makes it easier and splits the problem into two denoising and projection steps. The experimental results demonstrated that the proposed encryption system has a high-security level and outstanding performance in comparison to other methods. In future work, we want to add a deep model to the denoiser to enhance the performance by improving the accuracy and speed. Also, we can boost the performance of the projection section by changing and modifying its definition.

References

1. Xiaoling Huang, Youxia Dong, Guodong Ye, Wun-She Yap, and Bok-Min Goi. Visually meaningful image encryption algorithm based on digital signature. *Digital Communications and Networks*, 9(1):159–165, 2023.
2. Sara T Kamal, Khalid M Hosny, Taha M Elgindy, Mohamed M Darwish, and Mostafa M Fouda. A new image encryption algorithm for grey and color medical images. *IEEE Access*, 9:37855–37865, 2021.
3. Aboozar Ghaffari. Image compression-encryption method based on two-dimensional sparse recovery and chaotic system. *Scientific Reports*, 11(1):369, 2021.
4. Narendra K Pareek, Vinod Patidar, and Krishan K Sud. Image encryption using chaotic logistic map. *Image and vision computing*, 24(9):926–934, 2006.
5. Gustavus J Simmons. Symmetric and asymmetric encryption. *ACM Computing Surveys (CSUR)*, 11(4):305–330, 1979.
6. Yuling Luo, Xue Ouyang, Junxiu Liu, and Lvchen Cao. An image encryption method based on elliptic curve elgamal encryption and chaotic systems. *IEEE Access*, 7:38507–38522, 2019.
7. Yong Zhang. A chaotic system based image encryption scheme with identical encryption and decryption algorithm. *Chinese Journal of Electronics*, 26(5):1022–1031, 2017.
8. Claude E Shannon. Communication theory of secrecy systems. *The Bell system technical journal*, 28(4):656–715, 1949.
9. Aboozar Ghaffari. Image encryption-compression method via encryption based sparse decomposition. *Multimedia Tools and Applications*, pages 1–32, 2023.
10. Hend Ali Mohammed Ali Basha, Ashraf Shawky Seliem Mohra, Tamer Omar Mohamed Diab, and Wageda Ibrahim El Sobky. Efficient image encryption based on new substitution box using dna coding and bent function. *IEEE Access*, 10:66409–66429, 2022.
11. Adélaïde Nicole Kengnou Telem, Hilaire Bertrand Fotsin, and Jacques Kengne. Image encryption algorithm based on dynamic dna coding operations and 3d chaotic systems. *Multimedia Tools and Applications*, 80:19011–19041, 2021.
12. Abhishek Kumar and Sokat Tejani. S-box architecture. In *Futuristic Trends in Network and Communication Technologies: First International Conference, FTNCT 2018, Solan, India, February 9–10, 2018, Revised Selected Papers 1*, pages 17–27. Springer, 2019.
13. Wassim Alexan, Mohamed ElBeltagy, and Amr Aboshousha. Rgb image encryption through cellular automata, s-box and the lorenz system. *Symmetry*, 14(3):443, 2022.
14. Zhenlong Man, Jinqing Li, Xiaoqiang Di, Yaohui Sheng, and Zefei Liu. Double image encryption algorithm based on neural network and chaos. *Chaos, solitons & fractals*, 152:111318, 2021.
15. Eva Volna, Martin Kotyrba, Vaclav Kocian, and Michal Janosek. Cryptography based on neural network. In *ECMS*, pages 386–391, 2012.

16. Amitava Nag, Jyoti Prakash Singh, Srabani Khan, Saswati Ghosh, Sushanta Biswas, D Sarkar, and Partha Pratim Sarkar. Image encryption using affine transform and xor operation. In *2011 International conference on signal processing, communication, computing and networking technologies*, pages 309–312. IEEE, 2011.
17. Komal D Patel and Sonal Belani. Image encryption using different techniques: A review. *International Journal of Emerging Technology and Advanced Engineering*, 1(1):30–34, 2011.
18. Fawad Masood, Maha Driss, Wadii Boulila, Jawad Ahmad, Sadaqat Ur Rehman, Sana Ullah Jan, Abdullah Qayyum, and William J Buchanan. A lightweight chaos-based medical image encryption scheme using random shuffling and xor operations. *Wireless Personal Communications*, 127(2):1405–1432, 2022.
19. Xinyan Li, Di Xiao, Huajian Mou, Dianjun Lu, and Mei Peng. A compressive sensing based image encryption and compression algorithm with identity authentication and blind signcryption. *IEEE Access*, 8:211676–211690, 2020.
20. Yuanyuan Sun, Rudan Xu, Lina Chen, and Xiaopeng Hu. Image compression and encryption scheme using fractal dictionary and julia set. *IET Image Processing*, 9(3):173–183, 2015.
21. Bo Zhang, Di Xiao, and Yong Xiang. Robust coding of encrypted images via 2d compressed sensing. *IEEE Transactions on Multimedia*, 23:2656–2671, 2020.
22. T Suguna and R Shanmugalakshmi. Secure image communication through adaptive deer hunting optimization based vector quantization coding of perceptually encrypted images. *Wireless Personal Communications*, 116:2239–2260, 2021.
23. Zhihua Gan, Jianqiang Bi, Wenke Ding, and Xiuli Chai. Exploiting 2d compressed sensing and information entropy for secure color image compression and encryption. *Neural Computing and Applications*, 33:12845–12867, 2021.
24. Xiaoling Huang, Guodong Ye, Huajin Chai, and Ou Xie. Compression and encryption for remote sensing image using chaotic system. *Security and Communication Networks*, 8(18):3659–3666, 2015.
25. Mohamed-Jalal Fadili and J-L Starck. Monotone operator splitting for optimization problems in sparse recovery. In *2009 16th IEEE International conference on image processing (ICIP)*, pages 1461–1464. IEEE, 2009.
26. Arash Rasti-Meymandi, Aboozar Ghaffari, and Emad Fatemizadeh. Plug and play augmented hqs: Convergence analysis and its application in mri reconstruction. *Neurocomputing*, 518:1–14, 2023.
27. Stephen Boyd, Neal Parikh, Eric Chu, Borja Peleato, Jonathan Eckstein, et al. Distributed optimization and statistical learning via the alternating direction method of multipliers. *Foundations and Trends® in Machine learning*, 3(1):1–122, 2011.
28. Donald Geman and Chengda Yang. Nonlinear image recovery with half-quadratic regularization. *IEEE transactions on Image Processing*, 4(7):932–946, 1995.
29. Yushu Zhang, Jiantao Zhou, Fei Chen, Leo Yu Zhang, Kwok-Wo Wong, Xing He, and Di Xiao. Embedding cryptographic features in compressive sensing. *Neurocomputing*, 205:472–480, 2016.
30. Tinghuan Chen, Meng Zhang, Jianhui Wu, Chau Yuen, and You Tong. Image encryption and compression based on kronecker compressed sensing and elementary cellular automata scrambling. *Optics & Laser Technology*, 84:118–133, 2016.
31. Hao Fang, Sergiy A Vorobyov, Hai Jiang, and Omid Taheri. Permutation meets parallel compressed sensing: How to relax restricted isometry property for 2d sparse signals. *IEEE Transactions on Signal Processing*, 62(1):196–210, 2013.
32. Aboozar Ghaffari, Massoud Babaie-Zadeh, and Christian Jutten. Sparse decomposition of two dimensional signals. In *2009 IEEE international conference on acoustics, speech and signal processing*, pages 3157–3160. IEEE, 2009.
33. Balakrishnan Sriram, Aboozar Ghaffari, Karthikeyan Rajagopal, Sajad Jafari, and Esteban Tlelo-Cuautle. A chaotic map with trigonometric functions: Dynamical analysis and its application in image encryption based on sparse representation and convolutional filters. *Optik*, 273:170379, 2023.

34. Cai Zhuoran, Zhao Honglin, Jia Min, Wang Gang, and Shen Jingshi. An improved hadamard measurement matrix based on walsh code for compressive sensing. In *2013 9th International Conference on Information, Communications & Signal Processing*, pages 1–4. IEEE, 2013.
35. Liya Zhu, Huansheng Song, Xi Zhang, Maode Yan, Liang Zhang, and Tao Yan. A novel image encryption scheme based on nonuniform sampling in block compressive sensing. *IEEE Access*, 7:22161–22174, 2019.
36. Gesen Zhang, Shuhong Jiao, and Xiaoli Xu. Compressed sensing and reconstruction with semi-hadamard matrices. In *2010 2nd International Conference on Signal Processing Systems*, volume 1, pages V1–194. IEEE, 2010.
37. Ronald A DeVore. Deterministic constructions of compressed sensing matrices. *Journal of complexity*, 23(4–6):918–925, 2007.
38. Stéphane G Mallat and Zhifeng Zhang. Matching pursuits with time-frequency dictionaries. *IEEE Transactions on signal processing*, 41(12):3397–3415, 1993.
39. Guangqi Shao, Yanping Wu, A Yong, Xiao Liu, and Tiande Guo. Fingerprint compression based on sparse representation. *IEEE transactions on image processing*, 23(2):489–501, 2013.
40. Scott Shaobing Chen, David L Donoho, and Michael A Saunders. Atomic decomposition by basis pursuit. *SIAM review*, 43(1):129–159, 2001.
41. Hosein Mohimani, Massoud Babaie-Zadeh, and Christian Jutten. A fast approach for overcomplete sparse decomposition based on smoothed ℓ^0 norm. *IEEE Transactions on Signal Processing*, 57(1):289–301, 2008.
42. Armin Eftekhari, Massoud Babaie-Zadeh, and Hamid Abrishami Moghaddam. Two-dimensional random projection. *Signal processing*, 91(7):1589–1603, 2011.
43. Yong Fang, JiaJi Wu, and BorMin Huang. 2d sparse signal recovery via 2d orthogonal matching pursuit. *Science China Information Sciences*, 55:889–897, 2012.
44. Gao Chen, Defang Li, and Jiashu Zhang. Iterative gradient projection algorithm for two-dimensional compressive sensing sparse image reconstruction. *Signal Processing*, 104:15–26, 2014.
45. Zahra Faghani, Fahimeh Nazarimehr, Sajad Jafari, and Julien C Sprott. Simple chaotic systems with specific analytical solutions. *International Journal of Bifurcation and Chaos*, 29(09):1950116, 2019.
46. Kostadin Dabov, Alessandro Foi, Vladimir Katkovnik, and Karen Egiazarian. Image denoising by sparse 3-d transform-domain collaborative filtering. *IEEE Transactions on image processing*, 16(8):2080–2095, 2007.
47. Wenji Bao and Congxu Zhu. A secure and robust image encryption algorithm based on compressive sensing and dna coding. *Multimedia Tools and Applications*, 81(11):15977–15996, 2022.
48. Xiuli Chai, Xiaoyu Zheng, Zhihua Gan, Daojun Han, and Yiran Chen. An image encryption algorithm based on chaotic system and compressive sensing. *Signal Processing*, 148:124–144, 2018.
49. Haiying Hu, Yinghong Cao, Ji Xu, Chenguang Ma, and Huizhen Yan. An image compression and encryption algorithm based on the fractional-order simplest chaotic circuit. *IEEE Access*, 9:22141–22155, 2021.
50. Yuling Luo, Yuting Liang, Shunsheng Zhang, Junxiu Liu, and Fangxiao Wang. An image encryption scheme based on block compressed sensing and chenâs system. *Nonlinear Dynamics*, 111(7):6791–6811, 2023.
51. Gonzalo Alvarez and Shujun Li. Some basic cryptographic requirements for chaos-based cryptosystems. *International journal of bifurcation and chaos*, 16(08):2129–2151, 2006.

Chapter 16
Comparative Study of the Viscoelastic Behavior on PLA Filaments, a Fractional Calculus Approach

Jesús Gabino Puente-Córdova and Karla Louisse Segura-Méndez

16.1 Introduction

Traditional calculus, or integer-order calculus, is a mathematical tool used to describe a number of natural phenomena. The applications range from physics, chemistry, biology, medicine, and engineering. The classical description of such phenomena is carried out by developing a model, obtaining an integer-order differential equation. The ability of modeling to describe or predict experimental results depends on the variables involved in the study and the mathematical operators employed.

Due to the above and the inherent error of the experimental measurements, the theoretical descriptions present considerable deviations. It has been shown in various scientific works that the integer order of differential or integral operators is a factor to consider for these deviations, since the result gives rise to functions whose geometric shape is restricted, this being an impediment to the adjustment of the theoretical curves with the experimental data. To minimize this problem, many theoretical and experimental works have successfully used differential and integral operators of fractional order.

This branch of mathematics is known by the name of fractional calculus or, in another way, the method of integration and derivation of arbitrary order. Although today it is considered a novel and innovative tool, the genesis of the fractional calculus occurs in 1695, in the exchange of correspondence between Leibniz and L'Hopital. However, the first reported application is given in 1823 by Niels Abel solving the tautochrone problem.

J. G. Puente-Córdova (✉) · K. L. Segura-Méndez
Facultad de Ingeniería Mecánica y Eléctrica, Universidad Autónoma de Nuevo León, Avenida Universidad s/n Cd. Universitaria, San Nicolás de los Garza, Nuevo León, Mexico
e-mail: jesus.puentecr@uanl.edu.mx; karla.seguramz@uanl.edu.mx

E. Campos-Cantón et al. (eds.), *Complex Systems and Their Applications*,
https://doi.org/10.1007/978-3-031-51224-7_16

Nowadays, the use of fractional operators requires an appropriate definition. Illustrious mathematicians such as Liouville, Riemann, Fourier, Laplace, Abel, Weyl, among others, made important contributions to establishing a definition of fractional derivative. Currently, two definitions are the most widely used, the Riemann-Liouville fractional derivative and the Caputo fractional derivative. When solving fractional order differential equations, the Riemann–Liouville definition has the disadvantage of not traditionally corresponding to initial conditions in physical systems, whereas Caputo's definition does. Other definitions used in science and engineering are: Grünwald-Letnikov, Caputo-Fabrizio, Atangana-Baleanu, Weyl, and the conformable derivative.

Nevertheless, the definitions of fractional derivatives are limited by their physical and geometric interpretation, since, when these are defined by an integral, they are considered as nonlocal. This means that the integral contains partial or total information about the history of the function, which allows it to be an ideal tool for modeling complex systems with memory [1]. From an engineering point of view, the geometric interpretation of the traditional derivation process of a function at a point is the slope of the tangent line to the curve at that point, while its physical interpretation can be associated with a rate of change. Regarding an integral of integer order, its geometric interpretation is related to the area under the curve, and its physical interpretation can be associated with an accumulation [2].

However, today it is not clear how a fractional derivative should be interpreted physically and geometrically. In the literature, there are works that try to give an answer to this problem; for example, Podlubny in 2001 presented an interesting proposal in this regard [3].

Based on the above, this paper presents a comparative study of viscoelastic models of complex systems, using derivatives of arbitrary order. As a model material, PLA has been selected, which is a thermoplastic of biological origin, compostable and biodegradable, with good mechanical strength, great optical properties, and good processing capacity [4]. For these qualities, it has a wide variety of applications such as food packaging, biomedical industry, drug delivery, tissue engineering, biocomposites, additive manufacturing, and even in the automotive industry. The aim of this work is to demonstrate the usefulness of fractional calculus to describe the real behavior at molecular and macroscopic level of viscoelasticity in a complex biopolymer material. To achieve this, we used the fractional Maxwell model and Voigt-Kelvin model to fit experimental data from a stress relaxation test. This work is structured as follows: in Sect. 16.2, the fundamentals of modeling in viscoelastic materials are presented; in Sect. 16.3 experimental methodology; Sect. 16.4 results and discussion; and finally the conclusions are presented.

16.2 Fractional Calculus and Viscoelasticity

Materials such as polymers, plastics, rubbers, biological tissues, and some metal alloys used in science and engineering have a behavior called viscoelastic. This

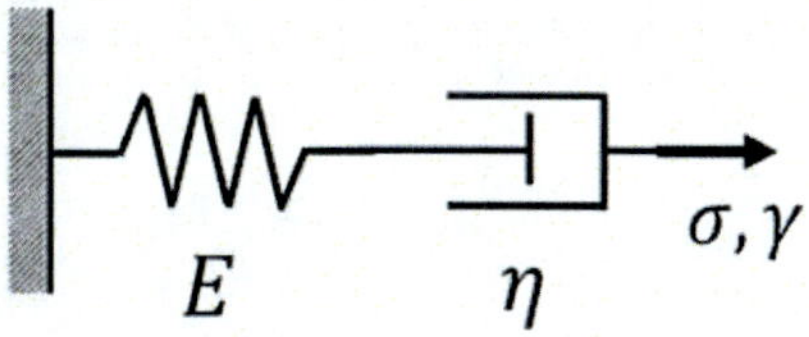

Fig. 16.1 Classical Maxwell model

behavior can be described by a linear combination of elastic and viscous properties. Subsequently, the modeling of this behavior can be carried out through the construction of mechanical arrangements consisting of springs and dashpots. Mathematical models such as Maxwell, Voigt-Kelvin, Zener, and Burgers represent viscoelastic response as a first approximation, and in particular, the mechanical stress relaxation and creep tests. They are generally described by constitutive equations (differential equations) that relate the dependence of stress with strain [5].

An interesting fact is the way to construct the mechanical arrangement. This basically depends on the scale time of the properties arising from the viscoelastic material. The Maxwell model, shown in Fig. 16.1, consists of the mechanical arrangement in series of an elastic spring (E) and a viscosity dashpot (η) whom each mechanical response corresponds to Hooke's law of ideal solids and Newton's law of pure viscous liquids, respectively. The spring is an element that represents the elastic component of the material (stored energy), while the dashpot represents the viscous component (dissipated energy). The ratio between viscosity (η) and elastic modulus (E) is called relaxation time (τ) or response time.

For the Maxwell model, one can obtain a linear differential equation of first order, Eq. 16.1. The operator D_t^nrepresents a derivative respect to time t of integer-order n. σ is the stress, and γ is the strain. This model only considers two parameters: E and τ.

$$ED_t^1\gamma = \frac{1}{\tau}D_t^0\sigma + D_t^1\sigma. \tag{16.1}$$

Various mathematical techniques are employed to solve differential equations. In this work, the Laplace transform is used. To solve Eq. 16.1, two cases can be considered: stress relaxation and creep experiments. Creep tests consist in applying a constant step stress over the time, while strain is monitored. Stress relaxation test consists in applying a constant step strain in a viscoelastic material, while the stress is monitored in the time. Considering the hereditary integrals and according to the linear viscoelastic theory can obtain two fundamental expressions, the compliance function $J(t)$ and the relaxation function $E(t)$, for these, respectively, tests [6]. In this sense, these functions are obtained for the classical Maxwell model. Thus, the differential Equation 16.1 is transformed to Laplace domain:

$$Es\gamma(s) = \frac{\sigma(s)}{\tau} + s\sigma(s). \tag{16.2}$$

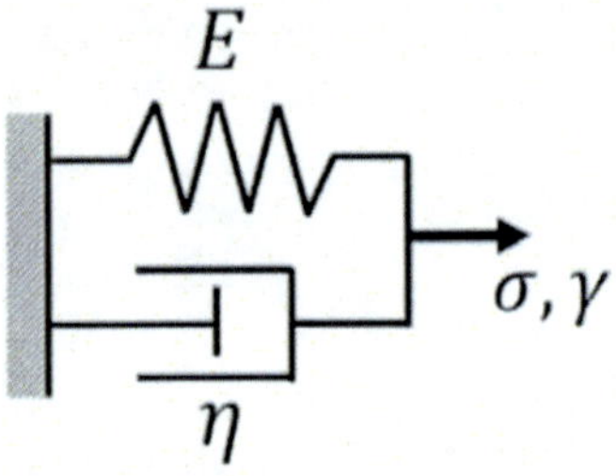

Fig. 16.2 Classical Voigt-Kelvin model

Rearranging the previous equation to satisfy the relationship of $J(s)$:

$$J(s) = \frac{\gamma(s)}{s\sigma(s)} = \frac{s + \frac{1}{\tau}}{Es^2}. \tag{16.3}$$

Using the inverse Laplace transform, the solution of differential Equation 16.1 is obtained for the compliance function $J(t)$:

$$J(t) = \frac{1}{E} + \frac{t}{\tau E}. \tag{16.4}$$

To obtain the relaxation function $E(t)$, a similar process is performed using the Laplace transform. Equation 16.2 is arranged to satisfy the relationship of $E(s)$:

$$E(s) = \frac{\sigma(s)}{s\gamma(s)} = \frac{E}{s + \frac{1}{\tau}}. \tag{16.5}$$

Then, using the inverse Laplace transform, we obtain the relaxation function $E(t)$:

$$E(t) = E \cdot exp\left(-\frac{t}{\tau}\right). \tag{16.6}$$

The solutions are obtained from the differential equation of the classical Maxwell model producing particular functions with a characteristic linear and ideal response. From the academic point of view, this model is used to teach the phenomenon of relaxation in polymers. However, this behavior is far from experimental results observed in materials, due principally to the use of integer-order derivatives.

Regarding the Voigt-Kelvin model, this consists in the mechanical arrangement in parallel of a spring with a dashpot, Fig. 16.2. In this model, it is considered that the elastic and viscous components manifest themselves on the same time scale.

From this model, we obtain a linear differential equation of first order, Eq. 16.7.

$$\frac{1}{E}D_t^0\sigma = D_t^0\gamma + \tau D_t^1\gamma. \tag{16.7}$$

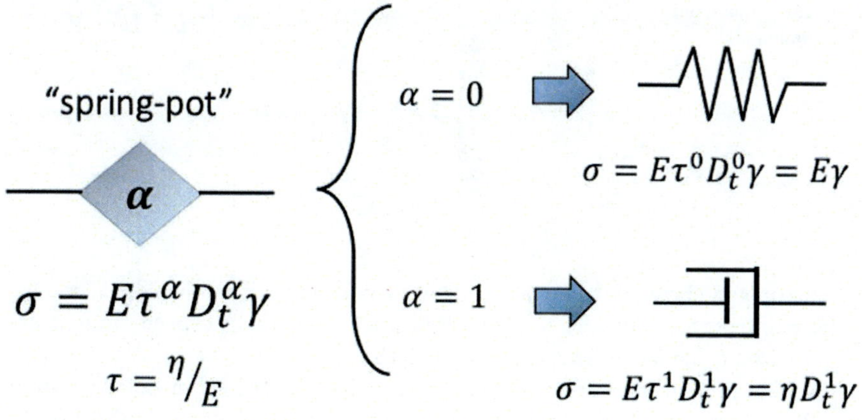

Fig. 16.3 Spring-pot element

Equation 16.7 is solved using the Laplace transform method to obtain the compliance function:

$$J(t) = \frac{1}{E}\left[1 - exp\left(-\frac{t}{\tau}\right)\right]. \tag{16.8}$$

And Eq. 16.7 is also solved to obtain the relaxation function:

$$E(t) = E + \eta\delta(t). \tag{16.9}$$

Here, $\delta(t)$ is the Dirac delta function. This model is very useful as a first approximation to represent a creep test; it only considers two physical parameters.

Since the viscoelastic behavior is intermediary between a solid ideal elastic and a viscous ideal liquid, it is possible to obtain a constitutive element called spring-pot or Scott-Blair element, Fig. 16.3. Its constitutive equation makes use of a fractional derivative, whose order varies between 0 and 1. When the order is equal to 0, Hooke's Law is obtained, when the value is equal to 1, Newton's Law is obtained.

In this work, we use the Caputo fractional derivative, Eq. 16.10

$$D_t^{\alpha}\gamma = \frac{1}{\Gamma(1-\alpha)}\int_0^t \frac{\gamma'(y)}{(t-y)^{\alpha}}dy. \tag{16.10}$$

where α is the fractional order with $0 < \alpha < 1$ and $\Gamma(\cdot)$ represents the gamma function. It has been shown in the literature that the constitutive equation of the spring-pot can be interpreted physically through hierarchical arrangements of an infinite number of springs and dashpots, i.e., this corresponds to a fractal arrangement [7–9]. A single spring-pot does not have the ability to represent the real viscoelastic behavior of polymeric materials. Therefore, an alternative is to replace

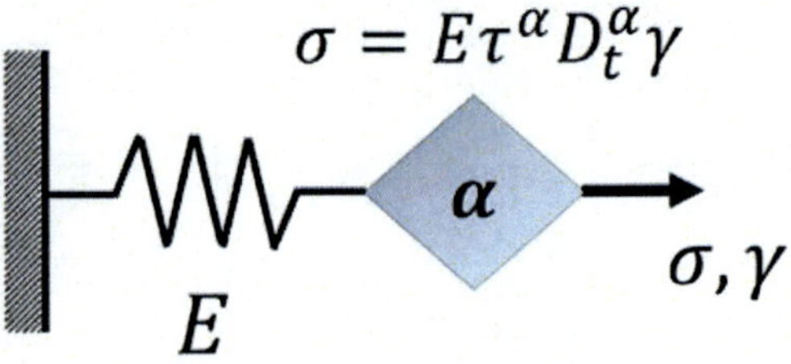

Fig. 16.4 Fractional Maxwell model

the dashpot in classical rheological models, to obtain fractional rheological models, whose differential equations must result in solutions with the capacity to represent viscoelastic behavior at the experimental level.

Moreover, if we use this constitutive element in a rheological model to substitute the dashpot, we can describe the viscoelastic behavior of a material. In this case, we have the fractional Maxwell model (FMM) in Fig. 16.4, were the dashpot is replaced by a spring-pot.

For this viscoelastic model, we obtain the fractional differential Equation 16.11.

$$ED_t^{\alpha}\gamma = D_t^{\alpha}\sigma + \tau^{-\alpha}\sigma. \tag{16.11}$$

In order to achieve the objective of this work, Eq. 16.11 is solved only for the relaxation function:

$$E(t) = E \cdot M_{\alpha}\left[\left(-\frac{t}{\tau}\right)^{\alpha}\right]. \tag{16.12}$$

Here, $M_{\alpha}(\cdot)$ corresponds to the Mittag-Leffler function, which is defined in the following way:

$$M_{\alpha}(z) = \sum_{n=0}^{\infty} \frac{z^n}{\Gamma(\alpha n + 1)}. \tag{16.13}$$

where z is a complex variable, $\Gamma(\cdot)$ is the gamma function, and $\alpha \geq 0$.

In the fractional Voigt-Kelvin model (FVKM) (Fig. 16.5), the viscous element is replaced by the spring-pot. Considering the constitutive equations, we obtain the fractional differential equation for FVKM:

$$\sigma = E\gamma + E\tau^{\alpha} D_t^{\alpha}\gamma. \tag{16.14}$$

According to Eq. 16.14, we use the Laplace transform to obtain the relaxation function $E(t)$, resulting in 16.15:

$$E(t) = E \cdot \left[1 + \left(\frac{\left(\frac{t}{\tau}\right)^{-\alpha}}{\Gamma(\alpha - 1)}\right)\right]. \tag{16.15}$$

Fig. 16.5 Fractional Voigt-Kelvin model

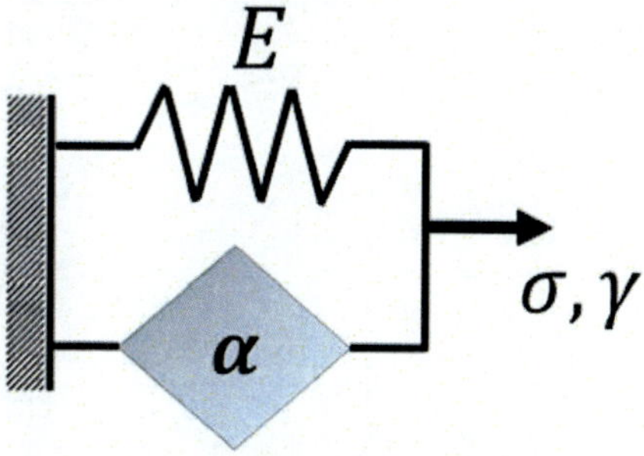

$$\sigma = E\tau^{\alpha} D_t^{\alpha} \gamma$$

As a summary, the following Table 16.1 presents the viscoelastic models used in this work, including the KWW (Kohlrausch-Williams-Watts) model that was obtained from reference [10]:

16.2.1 Testing the Response of the Fractional Models

In order to test the capability of the fractional models, we use in a heuristic way certain parameters that correspond to elastic modulus, relaxation time, and fractional order.

Theoretical Curves of Mathematical Models

In Fig. 16.6, we present the theoretical results of the Classical Maxwell model. The shape of the curve is typical for a common phenomenon related to energy dissipation, described by an exponential function. Furthermore, time relaxation varies in time, and the acquired curve maintains the same shape. Thus, the integer-order derivative does not have the capacity to describe experimental results for complex viscoelastic materials.

For the fractional Maxwell model in Fig. 16.7, we observed that by varying the fractional order, the shape of the curve changes. When the order is near to zero, an elastic response is obtained. And when the order is near to one, a viscous response is obtained. The intermediate values produce a response that is similar to the experimental data. This behavior is due to the Mittag-Leffler function, which generalizes an exponential function [11]. When $\alpha = 1$, we recover the solution for the classical Maxwell model.

For the results of FKVM in Fig. 16.8, the curves present a dependency on the fractional order, as increase the order the dissipation of energy is faster. However, the response is classical for power-law phenomena. When $a = 1$, we recover the solution for the classical VK model.

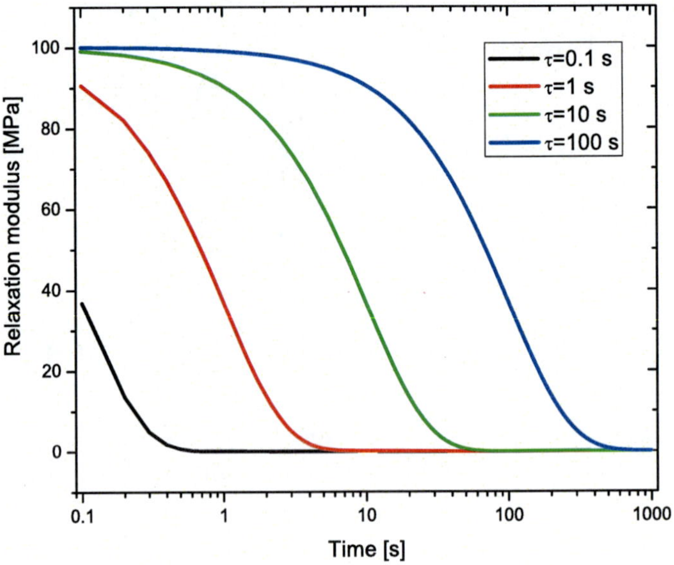

Fig. 16.6 Theoretical curve of the Classical Maxwell model

Table 16.1 Viscoelastic models

Mathematical model	Differential equation	Stress relaxation equation
Classical Maxwell	$E\tau D_t^1\gamma = \sigma + D_t^1\sigma\tau E$	$E\,(t) = E \cdot exp\left(-\frac{t}{\tau}\right)$
Fractional Maxwell	$ED_t^1\gamma = \frac{1}{\tau^\alpha}\sigma + D_t^\alpha\sigma$	$E\,(t) = E \cdot M_\alpha\left[\left(-\frac{t}{\tau}\right)^\alpha\right]$
Fractional Voigt-Kelvin	$\sigma = E\gamma + E\tau^\alpha D_t^1\gamma$	$E\,(t) = E \cdot \left[1 + \left(\frac{\left(\frac{t}{\tau}\right)^{-\alpha}}{\Gamma(1-\alpha)}\right)\right]$
KWW	$E\tau^{\beta-1}t^{1-\beta}D_t^1\gamma = \tau^\beta t^{1-\beta}D_t^1\sigma + D_t^0\sigma$	$E\,(t) = E \cdot exp^{-}\left(\frac{t}{\tau}\right)^\beta$

Regarding to KWW model, Fig. 16.9 presents the response of the theoretical curves calculated from differential equation in Table 16.1. It can be seen that by varying the value of the β, shape of the curve changes. As the value of β decreases, the curve tends to fall more steeply. When $\beta = 1$, the classic solution of the Maxwell model is obtained, where derivatives of fractional order have not been considered. This behavior can be justified by the fact that when beta increases, the viscosity represented by the dashpot imposes more strongly on the global response of the model, which translates into a lower rate of energy dissipation when beta increases, said in other words, the relaxation phenomenon is slower.

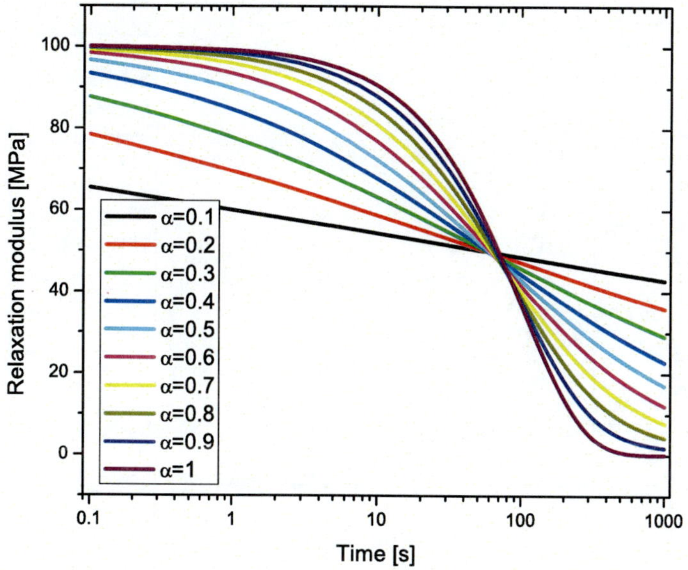

Fig. 16.7 Theoretical curve of the Fractional Maxwell model

16.3 Materials and Methods

In this section, the experimental methodology for the realization of this work is briefly detailed. Firstly, the materials used for the different tests were measured and identified according to each test. PLA or polylactic acid is the subject material due to its properties and type of polymer representative of biodegradable polymers. The PLA specimens were filaments of 70 cm in length and 1.75 mm in diameter. These were tested in the Shimadzu AGS-X universal testing machine that has a capacity of 10 kN of load. The test performed was a stress relaxation test.

16.3.1 Stress Relaxation Test

To elaborate this test, the specimens used in total were three per percentage of strain along the test. This is due to the time interval for each test programmed. Furthermore, depending on the polymer properties, the relaxation parameters change, since each one has different elastic modulus in addition to their viscoelastic behavior.

The specifications for this PLA stress relaxation test were three tests with different strain percentages of 5%, 10%, and 20%. The duration of each step strain was 1000 s.

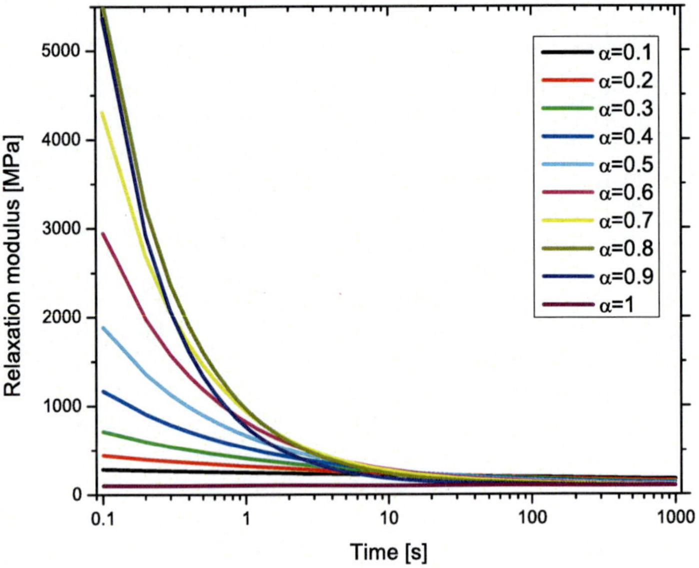

Fig. 16.8 Theoretical curve of the Fractional Voigt-Kelvin model

16.4 Results and Discussions

The viscoelastic behavior of polymeric materials is essential for the design and development of new products. In this sense, Figs. 16.10, 16.11, and 16.12 present the experimental results of the stress relaxation tests. To get a broader picture, measurements were made at three different strain levels (5%, 10%, and 20%). In general, the observed behavior is nonlinear and time dependent. At short times, an apparently constant relaxation modulus is observed, which can be associated with the elastic component of PLA filaments. As time elapses, the relaxation modulus begins to decrease following a power law.

When the polymeric material is subjected to an external agent (constant deformation), the material tends to return to its initial state; thus, a relaxation phenomenon is manifested. Phenomenologically, constitutive models and equations have been proposed in the literature to explain and describe what is observed at the experimental level. However, such efforts have been made using integer-order differential operators. An alternative to obtain a better representation of the viscoelastic behavior is to use generalized models, that is, multi-element models composed of a large number of springs and dashpots. Nevertheless, this results in a greater number of parameters, which end in lacking physical meaning. For this reason, in this work, derivative operators have been used under the framework of fractional calculus. A comparison is presented between the different responses offered by the classical Maxwell (two parameters), fractional Maxwell (three parameters), fractional Voigt-Kelvin (three parameters), and KWW model (three parameters).

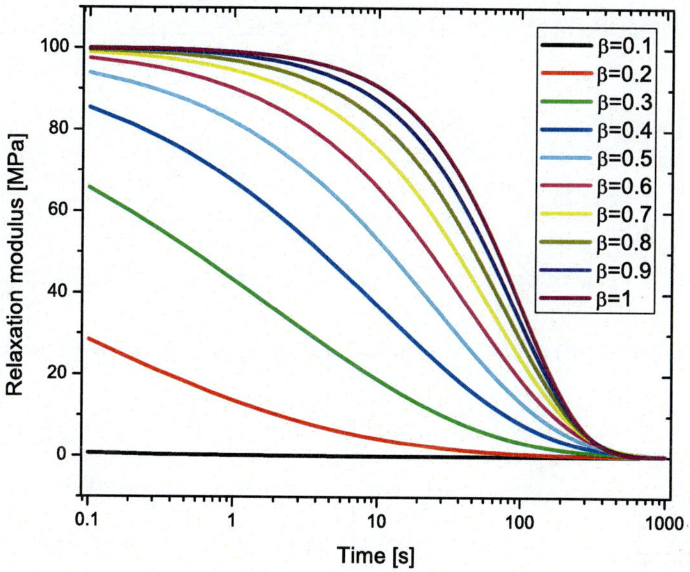

Fig. 16.9 Theoretical curve of the KWW model

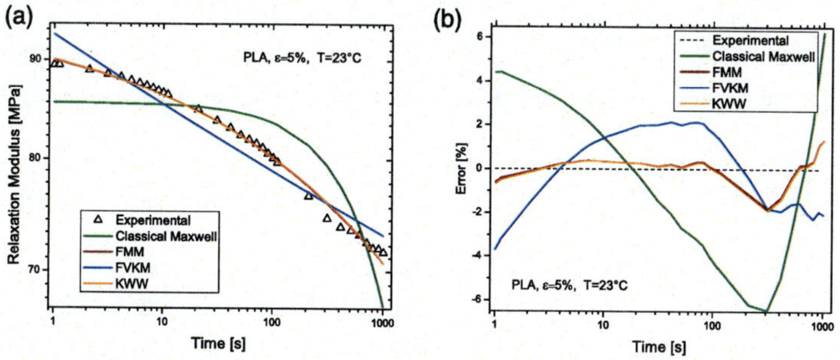

Fig. 16.10 Plots results for 5% (**a**) mathematical modeling (**b**) comparison of percentage errors

Referring to the results in Figs. 16.10, 16.11, and 16.12, it can be confirmed that Maxwell's classic model does not offer a close response to the relaxation phenomenon. The FMM is the one that presents the lowest total percentage error (Table 16.2) for the stress relaxation test. Likewise, in the individual steps, it was the one with the lowest error, which gives us to understand that, compared to the other models, it is the appropriate model for predicting the viscoelastic behavior of PLA. The FVKM presents an error 5.4 times greater than the FMM, resulting from a significant deviation between the prediction and the experimental data.

Finally, the KWW suggests a good prediction, but with a slightly larger error than the FMM. In this way, the classical Maxwell model is completely discarded due to a total error of 37%. Considering the above, it is demonstrated that a classical rheological model does not represent the real viscoelastic behavior of polymeric

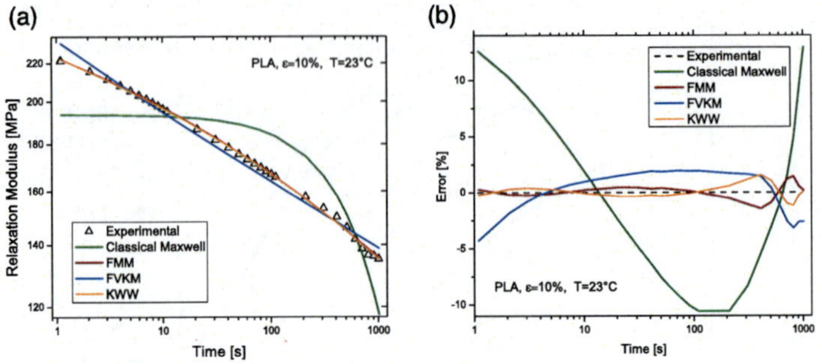

Fig. 16.11 Plots results for 10% (**a**) mathematical modeling (**b**) comparison of percentage errors

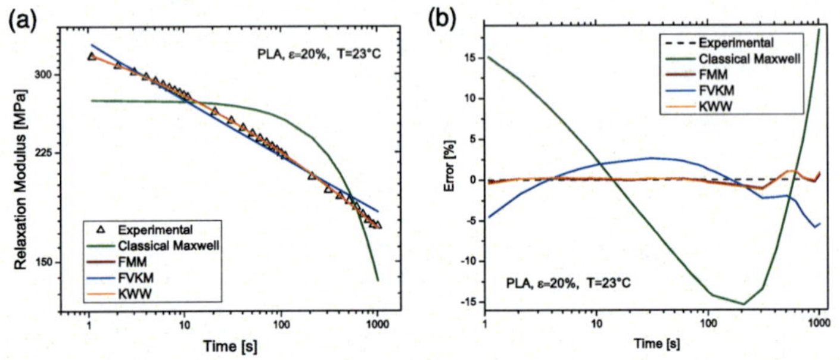

Fig. 16.12 Plots results for 20% (**a**) Mathematical modeling (**b**) Comparison of percentage errors

Table 16.2 Error percentage results of modeling

Error%	Classical	FMM	FVKM	KWW
5%	2.6	0.46	1.57	0.48
10%	12.45	0.84	3.59	0.82
20%	22.19	0.79	6.30	1
Total error	37.23	2.102	11.460	2.304
Average	12.41	0.7	3.82	0.77

materials, mainly because it uses differential operators of integer order. The use of nonlocal operators, that is, derivatives of non-integer or arbitrary order, adds degrees of freedom for the analysis of complex systems, such as viscoelastic materials.

Table 16.3 presents the parameters obtained from the adjustment of the experimental data, for each mathematical model. It is important to mention that a model is considered appropriate if a minimum error is obtained, for example, in engineering and science, it is recommended not to exceed an error of 5%, although on some occasions values greater than this are accepted. In addition, it is also important to consider the number of parameters involved in rheological models. The classical Maxwell model does not have the capacity to represent the viscoelastic behavior of polymers. However, as mentioned above, if N Maxwell elements are added in parallel, a better representation can be obtained, although this leads to a large

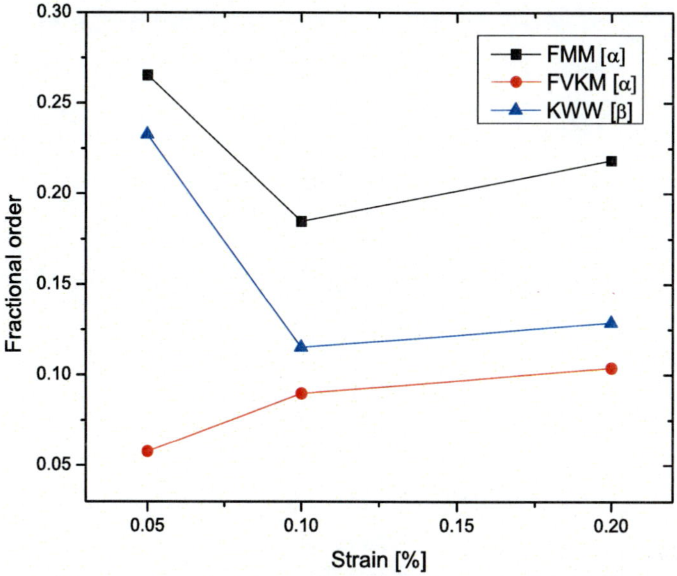

Fig. 16.13 Plot result of the comparison from the fractional orders α and β

number of parameters without physical interpretation. Regarding the results of the parameters E (elastic component) and τ (response time), it is observed that E increases as the deformation increases, except for the FVKM model. For τ, this decreases as the strain level increases, and vice versa for the FVKM.

Regarding the fractional order, for the FMM and FVKM models, the fractional order only takes values between 0 and 1. Figure 16.13 shows the results of the fractional order depending on the level of deformation, which also includes the beta exponent of the KWW model. It is observed that the values are closer to 0, which denotes that the behavior is for a viscoelastic solid. However, there is no trend or direct relationship between both parameters. In the literature, the origin of why different values of the fractional order are obtained is not recognized. The same definition of fractional derivative was used, in this case the Caputo derivative. Then, it is presumed that the values obtained for alpha are due to the arrangement of the constituent elements of the model (series vs. parallel), which leads us to the fact that the energy dissipation manifests itself on different time scales, therefore, resulting in values different. Furthermore, for the FMM, there is a solution that involves the Mittag-Leffler ML function, which generalizes to the exponential function. This ML function has two very interesting asymptotic behaviors: at short times, it behaves like a stretched exponential function (KWW function), and at long times, it behaves like a power law. The FVKM solution results in a power function, so under this assumption, it is possible to explain the difference between both models.

Table 16.3 Result parameters from each model

	Classical Maxwell			Fractional Maxwell			Fractional Voigt-Kelvin			KWW		
	5%	10%	20%	5%	10%	20%	5%	10%	20%	5%	10%	20%
E (MPa)	85.56	193.7	272.6	94.87	294.85	428.84	33.02	29.51	30.53	95.72	336.702	508.33
τ (s)	4175	2030	1514	923,724	7434.97	2733.85	54,950	4.27E+09	9.24E+09	171,020	2218.2	493.82
α	–	–	–	0.2654	0.1847	0.1777	0.05769	0.0862	0.1037	–	–	–
β	–	–	–	–	–	–	–	–	–	0.2326	0.115	0.1286

16.5 Conclusions

Fractional calculus is a powerful mathematical tool employed for the description of dynamic and complex systems, particularly with memory and dissipative properties. Viscoelastic materials present a complex behavior due to the intermediary elasticity and viscosity properties. In this work, we applied the fractional calculus to the modeling of the viscoelastic behavior of PLA filaments. The fractional Maxwell model and the fractional Voigt-Kelvin model were considered to achieve an understanding of the stress relaxation test and their corresponding molecular mobility.

References

1. Jesús G Puente-Córdova, Flor Y Rentería-Baltiérrez, José M Diabb-Zavala, Nasser Mohamed-Noriega, Mario A Bello-Gómez, and Juan F Luna-Martínez. Thermomechanical characterization and modeling of niti shape memory alloy coil spring. *Materials*, 16(10):3673, 2023.
2. Manuel Guía-Calderón, J Juan Rosales-García, Rafael Guzmán-Cabrera, Adrián González-Parada, and J Antonio Álvarez-Jaime. The differential and integral fractional calculus and its applications. *Acta universitaria*, 25(2):20–27, 2015.
3. Igor Podlubny, Ivo Petráš, and Tomas Skovranek. Fitting of experimental data using mittag-leffler function. 05 2012.
4. P McKeown and MD Jones. The chemical recycling of pla: A review, sustainable chemistry 1 (2020) 1–22.
5. Huanying Xu and Xiaoyun Jiang. Creep constitutive models for viscoelastic materials based on fractional derivatives. *Computers & Mathematics with Applications*, 73(6):1377–1384, 2017.
6. Tran Huu Nam, Iva Petríková, and Bohdana Marvalová. Experimental and numerical research of stress relaxation behavior of magnetorheological elastomer. *Polymer Testing*, 93:106886, 2021.
7. Flor Yanhira Rentería-Baltiérrez, Martín Edgar Reyes-Melo, Jesús Gabino Puente-Córdova, and Beatriz López-Walle. Correlation between the mechanical and dielectric responses in polymer films by a fractional calculus approach. *Journal of Applied Polymer Science*, 138(7):49853, 2021.
8. Edgar Reyes-Melo, Marco Garza, Virgilio González, Guerrero Salazar Carlos Alberto, Juan Martinez Vega, and Ubaldo Ortiz Mendez. Application of fractional calculus to the modeling of the complex magnetic susceptibility for systems containing nanometer-sized magnetic particles. 01 2009.
9. FY Rentería-Baltiérrez, ME Reyes-Melo, JG Puente-Córdova, and B López-Walle. Application of fractional calculus in the mechanical and dielectric correlation model of hybrid polymer films with different average molecular weight matrices. *Polymer Bulletin*, 80(6):6327–6347, 2023.
10. Jesús Gabino Puente-Córdova. La derivada conformable y sus aplicaciones en ingeniería. *Ingenierías*, 23(88):20–31, 2020.
11. Francesco Mainardi. Why the mittag-leffler function can be considered the queen function of the fractional calculus? *Entropy*, 22(12):1359, 2020.

Chapter 17
A New 4-D Highly Chaotic Two-Scroll System with a Hyperbola of Equilibrium Points and Its Circuit Simulation

Sundarapandian Vaidyanathan, Fareh Hannachi, and Aceng Sambas

17.1 Introduction

Chaos theory is a field of study in mathematics and physics that explores the behavior of particular nonlinear dynamical systems [1]. Chaotic systems exhibit a fascinating phenomenon called chaos under specific conditions. Chaos is characterized by its sensitivity to initial conditions, meaning that even tiny changes in the starting state can lead to widely different outcomes [2]. Chaos theory has many applications in science and engineering [3]. Chaotic systems are applicable in many areas such as weather systems [4], biological models [5, 6], ecological models [7, 8], stochastic models [9], mechanical systems [10], cellular neural networks [11, 12], fuzzy control [13, 14], image encryption [15, 16], etc.

In the chaos literature, there is good interest in finding and analysing highly chaotic systems or systems with large Lyapunov exponent due to their engineering applications such as secure communications and cryptography [17–19]. In this

S. Vaidyanathan (✉)
Centre for Control Systems, Vel Tech University, Chennai, India

Centre of Excellence for Research, Value Innovation & Entrepreneurship (CERVIE), UCSI University, Kuala Lumpur, Malaysia
e-mail: sundar@veltech.edu.in

F. Hannachi
Larbi Tebessi University - Tebessi, Tebessa, Algeria
e-mail: Fareh.hannachi@univ-tebessa.dz

A. Sambas
Faculty of Informatics and Computing, Universiti Sultan Zainal Abidin, Terengganu, Malaysia

Department of Mechanical Engineering, Universitas Muhammadiyah Tasikmalaya, Jawa Barat, Indonesia
e-mail: acengsambas@unisza.edu.my; acengs@umtas.ac.id

E. Campos-Cantón et al. (eds.), *Complex Systems and Their Applications*,
https://doi.org/10.1007/978-3-031-51224-7_17

work, we propose a novel 4D highly chaotic system with the largest Lyapunov exponent equal to 18.3581. The mathematical model of the new 4D highly chaotic system is obtained by modifying the dynamics of the Sprott-C chaotic system [20]. Due to its high complexity, the proposed highly chaotic system will have many applications in secure communication systems.

Coexisting attractors in chaotic systems [21, 22] refer to the coexistence of chaotic attractors that can exist simultaneously for the same set of parameter values but different values of initial states in a multistable chaotic system. In this work, we show the coexistence of chaotic attractors for different initial states and exhibit multistability of the new system. Offset boosting control [23, 24] for a chaotic system has important applications due to its broadband property and polarity control. In this work, we derive new results for offset boosting of the new chaotic system with variable boosting.

Since the introduction of the Chua circuit [25, 26], the study of chaotic circuits has gained considerable interest. Electronic circuits for chaotic systems are commonly utilized to generate chaotic signals and demonstrate the physical existence of chaotic systems [27, 28]. In this work, we design an electronic circuit for the new highly chaotic two-scroll system using MultiSim 14.2, which is useful for practical applications of the proposed chaotic system.

17.2 A New 4-D Highly Chaotic System

First, we state the Sprott-C chaotic system (1994, [20]) as follows:

$$\begin{cases} \dot{y}_1 = y_2 - y_1, \\ \dot{y}_2 = y_1 y_3, \\ \dot{y}_3 = 1 - y_2^2. \end{cases} \tag{17.1}$$

The Lyapunov exponents of the Sprott-C dynamical system (17.1) can be obtained for the initial state $Y_0 = (0.05, 0.05, 0.05)$ as follows:

$$\tau_1 = 0.1633, \ \tau_2 = 0, \ \tau_3 = -1.1633. \tag{17.2}$$

It is easy to observe that the Lyapunov exponents of the Sprott-C system (17.1) have the signs $(+, 0, -)$ with their sum equal to $-1 < 0$ and $\tau_1 = 0.1633 > 0$.

This shows that the Sprott-C dynamical system is a dissipative chaotic system with the maximal Lyapunov exponent (MLE) equal to $\tau_1 = 0.1633$.

The Kaplan-Yorke dimension for the Sprott-C dynamical system (17.1) can be evaluated as follows:

$$D_{KL} = 2 + \frac{\tau_1 + \tau_2}{|\tau_3|} = 2.1404. \tag{17.3}$$

The Sprott-C system (17.1) has two equilibrium points given by $E_1 = (1, 1, 0)$ and $E_2 = (-1, -1, 0)$.

In our work, we first obtain a new 3-D chaotic system by modifying the Sprott-C system (17.1) as follows:

$$\begin{cases} \dot{y}_1 = a(y_2 - y_1), \\ \dot{y}_2 = y_1 y_3, \\ \dot{y}_3 = c - y_1 y_2 \end{cases} \tag{17.4}$$

Basically, we have replaced the nonlinearity $y_2{}^2$ with $y_1 y_2$ in the third ODE of Sprott-C system (17.1).

Next, we take the parameter values as $a = 2.6$ and $c = 26$.

Basically, we have replaced the nonlinearity $y_2{}^2$ in the third differential equation of the Sprott-C system (17.1) with $y_1 y_2$ and taken new values for the parameters.

The Lyapunov exponents of the new 3-D system (17.4) can be obtained for the initial state $Y_0 = (0.05, 0.05, 0.05)$ as follows:

$$\tau_1 = 0.6308, \ \tau_2 = 0, \tau_3 = -3.2308. \tag{17.5}$$

It is easy to observe that the Lyapunov exponents of the new 3-D system (17.4) have the signs $(+, 0, -)$ with their sum equal to$-2.6 < 0$ and $\tau_1 = 0.6308 > 0$.

This shows that the new 3-D dynamical system (17.4) is a dissipative chaotic system with the maximal Lyapunov exponent (MLE) equal to $\tau_1 = 0.6308$.

The Kaplan-Yorke dimension for the new dynamical system (17.4) can be evaluated as follows:

$$D_{KL} = 2 + \frac{\tau_1 + \tau_2}{|\tau_3|} = 2.1952. \tag{17.6}$$

These calculations show that the new 3-D chaotic system (17.4) have the values of MLE and Kaplan-Yorke dimension greater than those of the Sprott-C chaotic system (17.1).

The new chaotic system (17.4) has two equilibrium points $E_1 = \left(\sqrt{c}, \sqrt{c}, 0\right)$ and $E_2 = \left(-\sqrt{c}, -\sqrt{c}, 0\right)$.

A 3-D phase portrait of the two-scroll chaotic attractor for the new 3-D chaotic system (17.4) for the initial state $Y_0 = (0.05, 0.05, 0.05)$ and $(a, c) = (2.6, 26)$ is plotted as in Fig. 17.1.

In this research work, we obtain a new 4-D two-scroll chaotic system by introducing an additional state variable and a few additional terms in the new chaotic system (17.4).

Fig. 17.1 3-D plot of the two-scroll attractor for the new 3-D chaotic system (17.4) for $(a, c) = (2.6, 26)$ and $Y_0 = (0.05, 0.05, 0.05)$

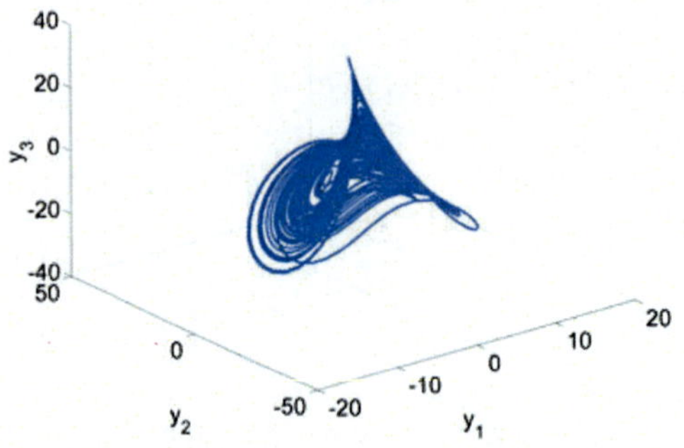

Thus, we consider the following new 4-D dynamical system described by

$$\begin{cases} \dot{y}_1 = a(y_2 - y_1) + by_2y_3 - y_4, \\ \dot{y}_2 = y_1y_3, \\ \dot{y}_3 = c - y_1y_2, \\ \dot{y}_4 = dy_3. \end{cases} \tag{17.7}$$

The system (17.7) has three quadratic nonlinearities and four parameters.

We denote the state of the 4-D system (17.7) as $Y = (y_1, y_2, y_3, y_4)$.

In this work, we show that the 4-D system (17.7) has a chaotic attractor when we set

$$a = 100, \ b = 0.1, \ c = 9000, \ d = 0.8. \tag{17.8}$$

This can be easily established as follows. When we take the initial state as $Y_0 = (0.05, 0.05, 0.05, 0.5)$ and the system parameters as indicated in (17.8), then the Lyapunov exponents of the 4-D system (17.7) can be calculated in MATLAB as follows:

$$\tau_1 = 18.3581, \tau_2 = 0, \tau_3 = 0, \tau_4 = -118.35687. \tag{17.9}$$

The Lyapunov exponents for the 4-D system (17.7) have the signs (+,0,0,−) and the total of the Lyapunov exponents in (17.9) equals $-99, 99877 < 0$. This calculation shows that the new 4-D system (17.7) is a dissipative chaotic system.

We note that the new 4-D chaotic system (17.7) has a large value of maximal Lyapunov exponent (MLE), viz. $\tau_1 = 18.3581$ which shows the high complexity of the system. Hence, the chaotic system (17.7) can be used for engineering applications like secure communications requiring chaotic systems with high complexity.

The balance points of the new 4-D chaotic system (17.7) are got by solving the equations

$$a(y_2 - y_1) + by_2y_3 - y_4 = 0, \tag{17.10a}$$

$$y_1y_3 = 0, \tag{17.10b}$$

$$c - y_1y_2 = 0, \tag{17.10c}$$

$$dy_3 = 0. \tag{17.10d}$$

From (17.10d), $y_3 = 0$. Since $y_3 = 0$, Eq. (17.10b) is satisfied immediately by all values of y_1.

When $y_3 = 0$, Eq. (17.10a) simplifies to

$$a(y_2 - y_1) - y_4 = 0, \tag{17.11}$$

which is a hyperplane in $\mathbf{R}^4$.

Thus, we see that the new 4-D chaotic system (17.7) has equilibrium points given by the set

$$S = \left\{ y \in \mathbf{R}^4 \mid a(y_2 - y_1) - y_4 = 0 \text{ and } y_1y_2 = c \right\}. \tag{17.12}$$

This shows that the 4-D chaotic system (17.7) has a hyperbola of equilibrium points in R^4. Since the 4-D chaotic system (17.7) has an infinite number of equilibrium points, it has hidden chaotic attractors.

For plotting the signal plots of the 4-D chaotic system (17.7), we take the initial state as $Y_0 = (0.05, 0.05, 0.05, 0.05)$ and the system parameters as in (17.8).

The 4-D chaotic system (17.7) depicts a two-scroll chaotic attractor as shown in Fig. 17.2

Also, the Kaplan-Yorke dimension of the highly chaotic system (17.7) is calculated as follows:

$$D_{KL} = 3 + \frac{\tau_1 + \tau_2 + \tau_3}{|\tau_4|} = 3.01970. \tag{17.13}$$

17.3 Dynamic Analysis of the New Four-Scroll Hyperchaotic System

In this section, we investigate numerically the dynamical behaviors of the new highly chaotic system (17.7) using the Lyapunov exponents spectrum (L_1, L_2, L_3, L_4) and bifurcation diagrams.

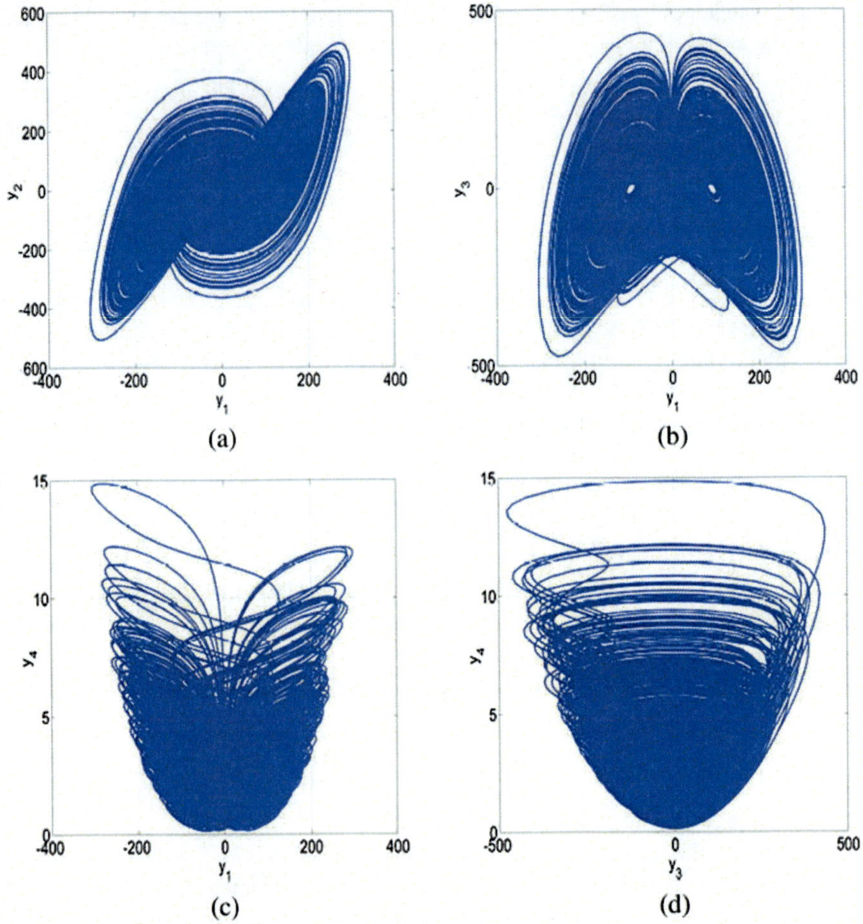

Fig. 17.2 MATLAB signal plots of the two-scroll highly chaotic system (17.7) for $(a, b, c, d) = (100, 0.1, 9000, 0.8)$ and $Y(0) = (0.05, 0.05, 0.05, 0.5)$ in various coordinate planes in $\mathbf{R}^4$. **(a)** (y1, y2) plane, **(b)** (y1, y3) plane, **(c)** (y1, y4) plane and **(d)** (y3, y4) plane

17.3.1 Varying the Parameter *a*

Figure 17.3 shows the Lyapunov exponents spectrum and the bifurcation diagram of the new highly chaotic system (17.7) with respect to parameter a.

We fix the values of the parameters b, c and d as $(b, c, d) = (0.1, 9000, 0.8)$. We vary a in the range [95, 105].

When $a \in [95, 105]$, the system (17.7) has only one positive Lyapunov exponent ($L_1 > 0$) and two zero Lyapunov exponents ($L_{2,3} = 0$) and one negative Lyapunov exponent ($L_4 < 0$) . Thus, the system (17.7) is chaotic for $a \in [95, 105]$.

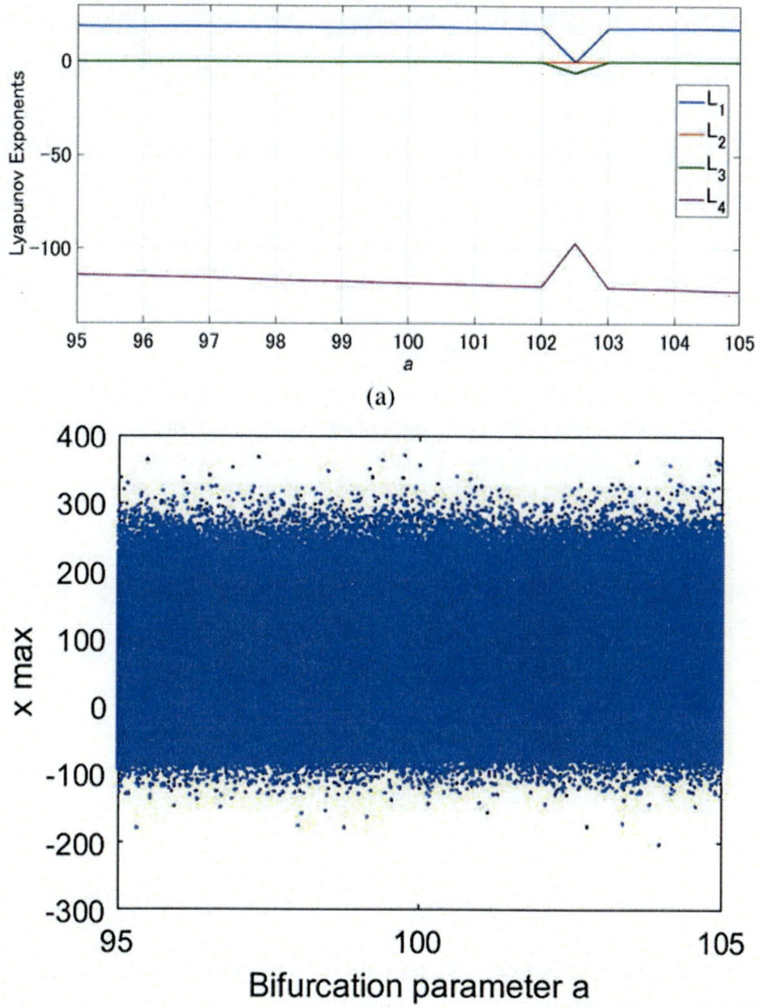

(b)

Fig. 17.3 Lyapunov exponents spectrum and bifurcation diagram of the 4-D system (17.7) for $a \in [95, 105]$

The values of the Lyapunov exponents are obtained for $a = 95$ as follows:

$$L_1 = 18.94, L_2 = 0, L_3 = 0, L_4 = -113.9. \tag{17.14}$$

The values of the Lyapunov exponents are obtained for $a = 98$ as follows:

$$L_1 = 18.57, L_2 = 0, L_3 = 0, L_4 = -116.6. \tag{17.15}$$

The values of the Lyapunov exponents are obtained for $a = 102.5$ as follows:

$$L_1 = 0.06913, L_2 = 0, L_3 = 0, L_4 = -96.81. \tag{17.16}$$

The values of the Lyapunov exponents are obtained for $a = 105$ as follows:

$$L_1 = 17.42, L_2 = 0, L_3 = 0, L_4 = -122.4. \tag{17.17}$$

17.3.2 *Varying the Parameter b*

Figure 17.4 shows the Lyapunov exponents spectrum and the bifurcation diagram of the new highly chaotic system (17.7) with respect to parameter b.

We fix the values of the parameters a, c and d as $(a, c, d) = (100, 9000, 0.8)$. We vary b in the range [0, 1].

When $b \in [0, 1)$, we note that the system (17.7) has only one positive Lyapunov exponent ($L_1 > 0$) and two zero Lyapunov exponents ($L_{2,3} = 0$) and one negative Lyapunov exponent ($L_4 < 0$) . Thus, the system (17.7) is chaotic for $b \in [0, 1)$.

The values of the Lyapunov exponents are obtained for $b = 0$ as follows:

$$L_1 = 19.04, L_2 = 0, L_3 = 0, L_4 = -119. \tag{17.18}$$

The values of the Lyapunov exponents are obtained for $b = 0.05$ as follows:

$$L_1 = 18.64, L_2 = 0, L_3 = 0, L_4 = -118.5. \tag{17.19}$$

The values of the Lyapunov exponents are obtained for $b = 0.25$ as follows:

$$L_1 = 16.77, L_2 = 0, L_3 = 0, L_4 = -116.6. \tag{17.20}$$

The values of the Lyapunov exponents are obtained for $b = 0.5$ as follows:

$$L_1 = 14.51, L_2 = 0, L_3 = 0, L_4 = -114.5. \tag{17.21}$$

The values of the Lyapunov exponents are obtained for $b = 1$ as follows:

$$L_1 = 0, L_2 = 0, L_3 = 0, L_4 = -99.9. \tag{17.22}$$

Since the Lyapunov exponents for $b = 1$ have the signs $(0, 0, 0, -)$, we deduce that the system (17.7) is periodic (existence of a limit cycle) in this case.

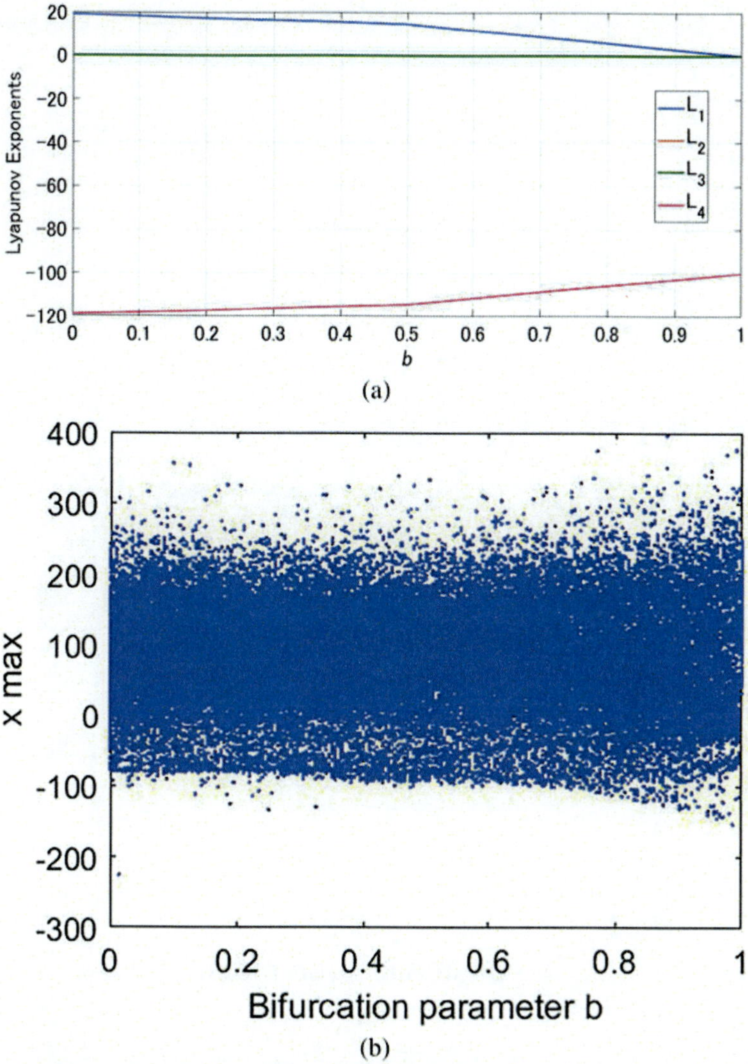

Fig. 17.4 Lyapunov exponents spectrum and bifurcation diagram of the 4-D system (17.7) for $b \in [0, 1]$

17.3.3 Varying the Parameter c

Figure 17.5 shows the Lyapunov exponents spectrum and the bifurcation diagram of the new highly chaotic system (17.7) with respect to parameter c.

We fix the values of the parameters a, b and d as $(a, b, d) = (100, 0.1, 0.8)$. We vary c in the range [8995, 9005].

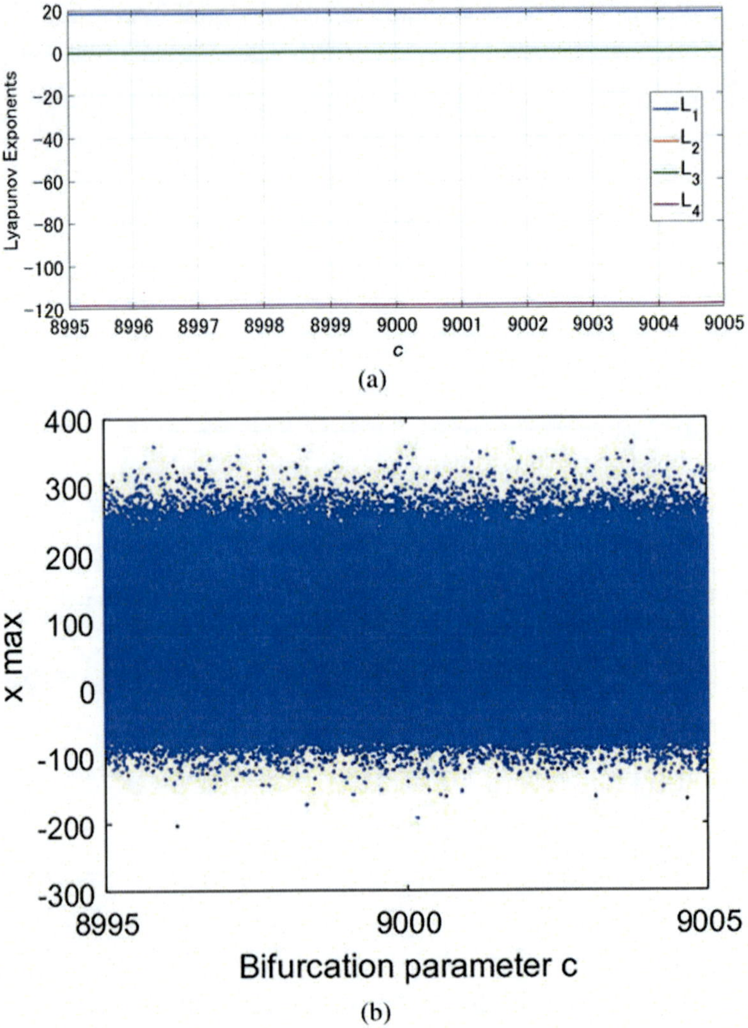

Fig. 17.5 Lyapunov exponents spectrum and bifurcation diagram of the 4-D system (17.7) for $c \in [8995, 9005]$

When $c \in [8995, 9005]$, we note that the system (17.7) has only one positive Lyapunov exponent ($L_1 > 0$) and two zero Lyapunov exponents ($L_{2,3} = 0$) and one negative Lyapunov exponent ($L_4 < 0$) . Thus, the system (17.7) is chaotic for $c \in [8995, 9005]$.

The values of the Lyapunov exponents are obtained for $c = 8995$ as follows:

$$L_1 = 18.40, L_2 = 0, L_3 = 0, L_4 = -118.40. \tag{17.23}$$

The values of the Lyapunov exponents are obtained for $c = 8998$ as follows:

$$L_1 = 18.36, L_2 = 0, L_3 = 0, L_4 = -118.40. \tag{17.24}$$

The values of the Lyapunov exponents are obtained for $c = 9002$ as follows:

$$L_1 = 18.37, L_2 = 0, L_3 = 0, L_4 = -118.40. \tag{17.25}$$

The values of the Lyapunov exponents are obtained for $c = 9005$ as follows:

$$L_1 = 18.30, L_2 = 0, L_3 = 0, L_4 = -118.30. \tag{17.26}$$

17.3.4 *Varying the Parameter d*

Figure 17.6 shows the Lyapunov exponents spectrum and the bifurcation diagram of the new highly chaotic system (17.7) with respect to parameter d.

We fix the values of the parameters a, b and c as $(a, b, c) = (100, 0.1, 9000)$. We vary d in the range $[0, 1]$.

When the parameter d varies in the range $[0, 1]$, we note that the system (17.7) has only one positive Lyapunov exponent ($L_1 > 0$) and two zero Lyapunov exponents ($L_{2,3} = 0$) and one negative Lyapunov exponent ($L_4 < 0$). Thus, the system (17.7) is chaotic for $d \in [0, 1)$.

The values of the Lyapunov exponents are obtained for $d = 0.5$ as follows:

$$L_1 = 18.35, L_2 = 0, L_3 = 0, L_4 = -118.30. \tag{17.27}$$

The values of the Lyapunov exponents are obtained for $d = 0.75$ as follows:

$$L_1 = 18.34, L_2 = 0, L_3 = 0, L_4 = -118.30. \tag{17.28}$$

The values of the Lyapunov exponents are obtained for $d = 1$ as follows:

$$L_1 = 18.37, L_2 = 0, L_3 = 0, L_4 = -118.40. \tag{17.29}$$

17.4 Multistability in the New 4D Four-Scroll Hyperchaotic System

In order to study the coexistence attractors and other characteristics of the system better, it is necessary to give some disturbance to the initial conditions under the condition of keeping the system parameters constant.

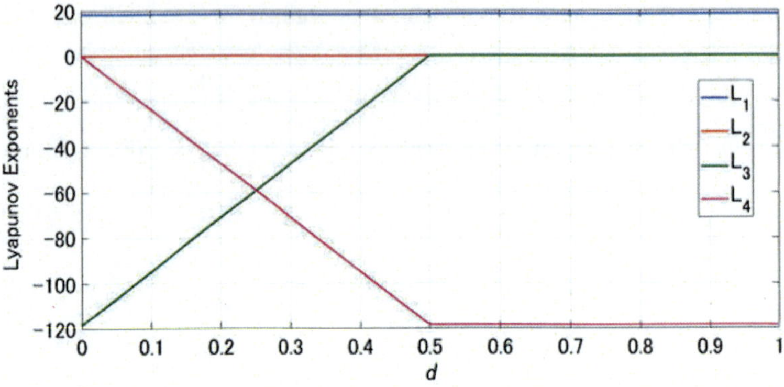

(a)

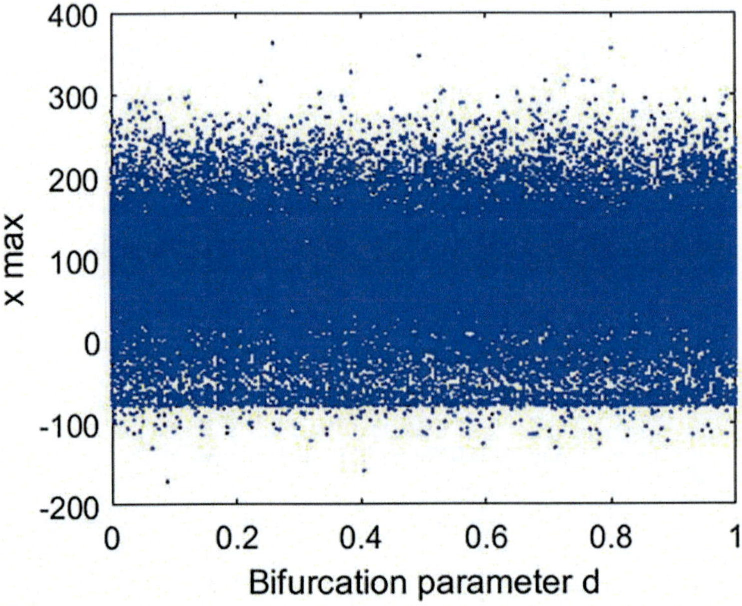

(b)

Fig. 17.6 Lyapunov exponents spectrum and bifurcation diagram of the 4-D system (17.7) for $d \in [0, 1]$

Figure 17.7 depicts the coexistence of three chaotic attractors with different initial states and same parameter values. Explicitly, we fix the parameter values as in the chaotic case, viz. $a = 160$, $b = 0.1$, $c = 9000$, and $d = 0.8$. We consider three trajectories, where the blue color trajectory corresponds to the initial state $X_0 = (0.05, 0.05, 0.05, 0.5)$, the red color trajectory corresponds to the initial state $Y_0 = (-0.05, -0.05, 0.05, 0.5)$, and the green color trajectory corresponds to $Z_0 = (-0.05, 0.05, -0.5, -0.5)$.

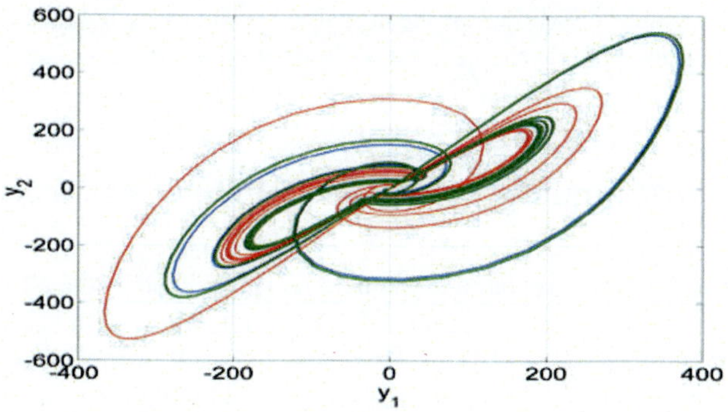

(a)

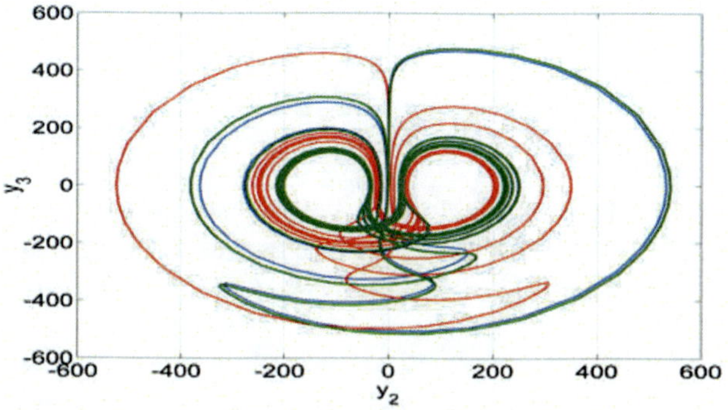

(b)

Fig. 17.7 Coexistence of three chaotic attractors for the two-scroll system (17.7) for the parameter values $(a, b, c, d) = (160, 0.1, 9000, 0.8)$ and the initial states $X_0 = (0.05, 0.05, 0.05, 0.5)$ (blue), $Y_0 = (-0.05, -0.05, 0.05, 0.5)$ (red) and $Z_0 = (-0.05, 0.05, -0.5, -0.5)$ (green)

17.5 Offset Boosting Control of the New Highly Chaotic System

In this section, we will discuss the offset boosting control of the new two-scroll highly chaotic system. When a variable appears only once in a nonlinear system, adding a constant m to it will produce an offset. Obviously, the state variable w only emerges in the first equation of the two-scroll chaotic system (17.7). Thus, we can obtain an offset-boosted system from the chaotic system (17.7) by replacing w with $w + m$ in the first equation of the system (17.7) as follows:

$$\begin{cases} \dot{x} = a(y - x) + bxz - (w + m), \\ \dot{y} = xz, \\ \dot{z} = c - xy, \\ \dot{w} = dz. \end{cases} \tag{17.30}$$

Figures 17.8 and 17.9 show that the signal w is effectively boosted from a bipolar signal to a unipolar signal when the offset-boosting parameter m is changed.

17.6 Circuit Simulation of the New Highly Chaotic System

In this section, the new 4-D chaotic system (17.7) is realized by the NI Multisim 14.2 platform. The electronic circuit design of the 4-D chaotic system (17.7) is shown in Fig. 17.10 in which TLO84ACN is selected as OPAMP. Applying the Kirchhoff's laws, the circuit presented in Fig. 17.10 is described by the following equations:

$$\begin{cases} \dot{Y}_1 = \frac{1}{R_1C_1}Y_2 - \frac{1}{R_2C_1}Y_1 + \frac{1}{10R_3C_1}Y_2Y_3 - \frac{1}{R_4C_1}Y_4, \\ \dot{Y}_2 = \frac{1}{10R_2C_2}Y_1Y_3, \\ \dot{Y}_3 = -\frac{1}{R_{10}C_3}V_1 - \frac{1}{10R_{11}C_3}Y_1Y_2, \\ \dot{Y}_4 = \frac{1}{R_{12}C_4}Y_3. \end{cases} \tag{17.31}$$

Here Y_1, Y_2, Y_3, Y_4 correspond to the voltages on the integrators $U1C$, $U3C$, $U5C$, and $U6C$, respectively. The values of components in the circuit are selected as follows: $R_1 = R_2 = R_{11} = 10$ kΩ, $R_3 = R_4 = 1000$ kΩ, $R_{12} = 1250$ kΩ, $R_i = 100$ kΩ, $i = 5, .., 10$, $C_1 = C_2 = C_3 = C_4 = 1$ nF.

MultiSim outputs of the circuit (17.31) are presented in Fig. 17.11. These results are consistent with the MATLAB simulation results for the new two-scroll highly chaotic system (17.7) depicted in Fig. 17.2.

17.7 Conclusions

The main contribution of this work is the modeling of a new 4-D highly chaotic two-scroll system with a hyperbola of equilibrium points. We derived the new highly chaotic system with a high value of maximal Lyapunov exponent ($L_1 = 18.3581$) by modifying the dynamics of the Sprott-C chaotic system (1994). We derived the dynamic properties of the new two-scroll chaotic system such as phase portraits, Lyapunov exponents, Kaplan-Yorke dimension, and equilibrium points. Since the new chaotic system has infinitely many number of equilibrium points, it has hidden attractors. Next, we carried out a bifurcation analysis of the new highly chaotic system using bifurcation diagrams and Lyapunov exponents. We also investigated offset-boosting control of the new highly chaotic system. Next, we derived results for multistability and coexisting chaotic attractors in the newly introduced 4-D chaotic two-scroll system. Finally, we designed an electronic circuit for the new highly chaotic two-scroll system using MultiSim 14.2, which is useful for practical applications of the proposed chaotic system.

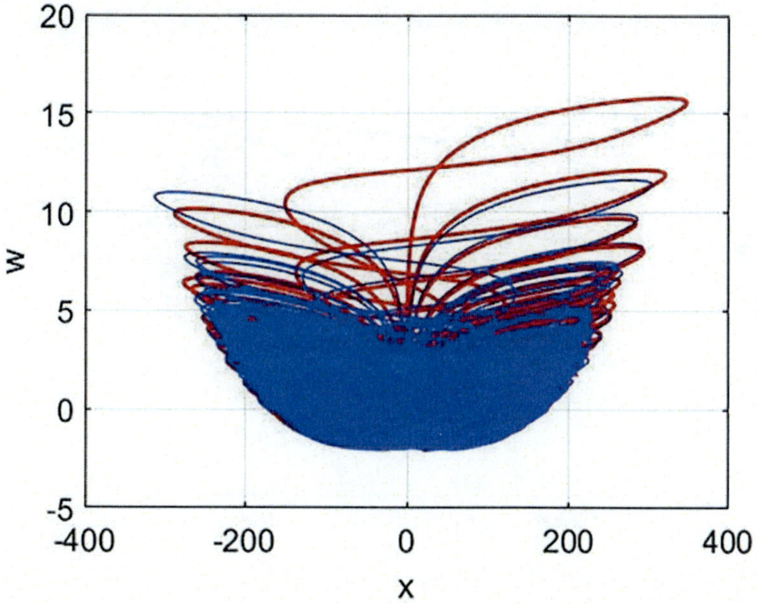

(a)

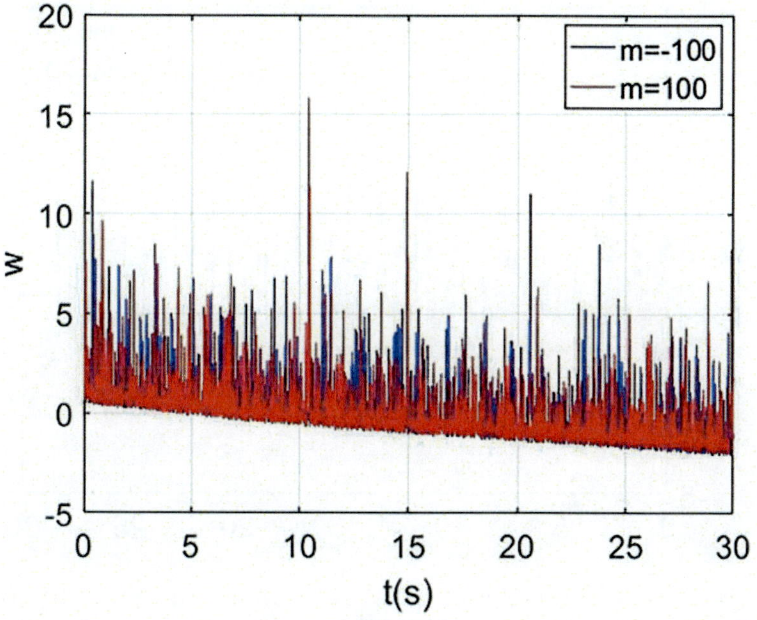

(b)

Fig. 17.8 Offset boosting plots of the system (17.30) for the initial state (0.05, 0.05, 0.05, 0.5) and different values of the offset boosting parameter, viz.: $m = -100$ (blue), $m = 100$ (red) : (**a**) (x, w) phase portraits and (**b**) w signal

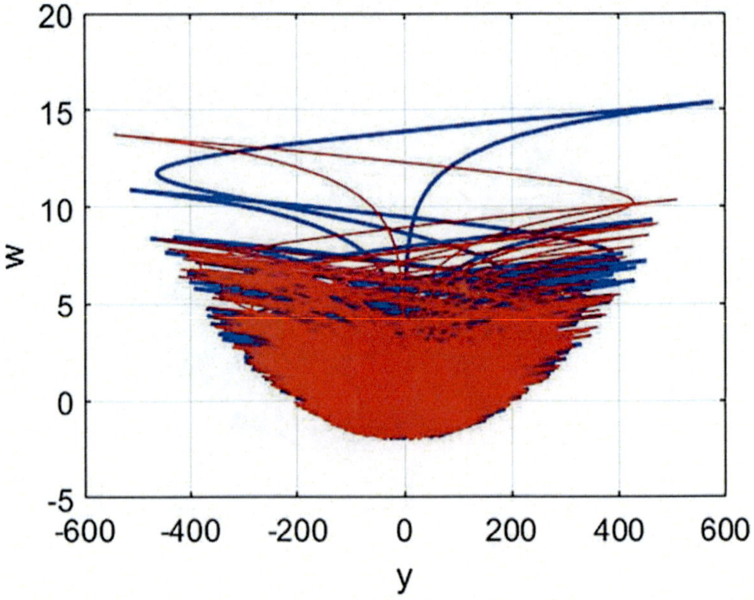

(a)

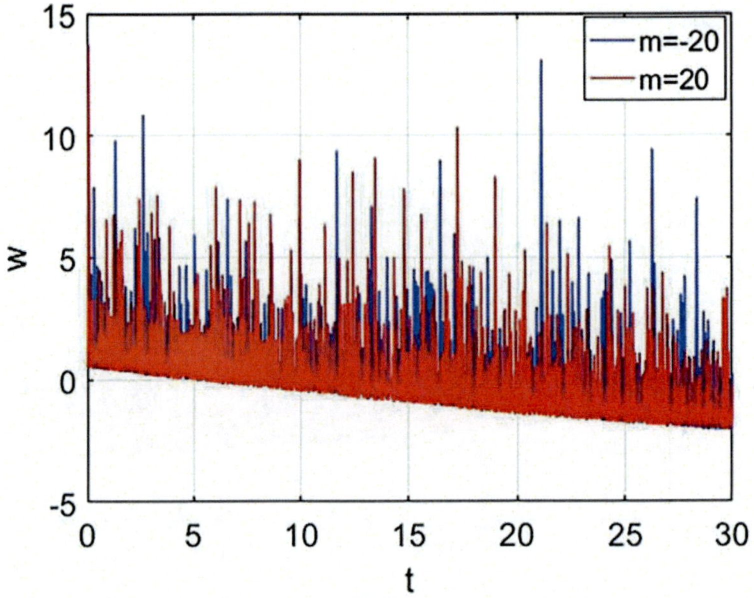

(b)

Fig. 17.9 Offset boosting plots of the system (17.30) for the initial state (0.05, 0.05, 0.05, 0.5) and different values of the offset boosting parameter, viz.: $m = -20$ (blue), $m = 20$ (red) : (**a**) (y, w) phase portraits and (**b**) w signal

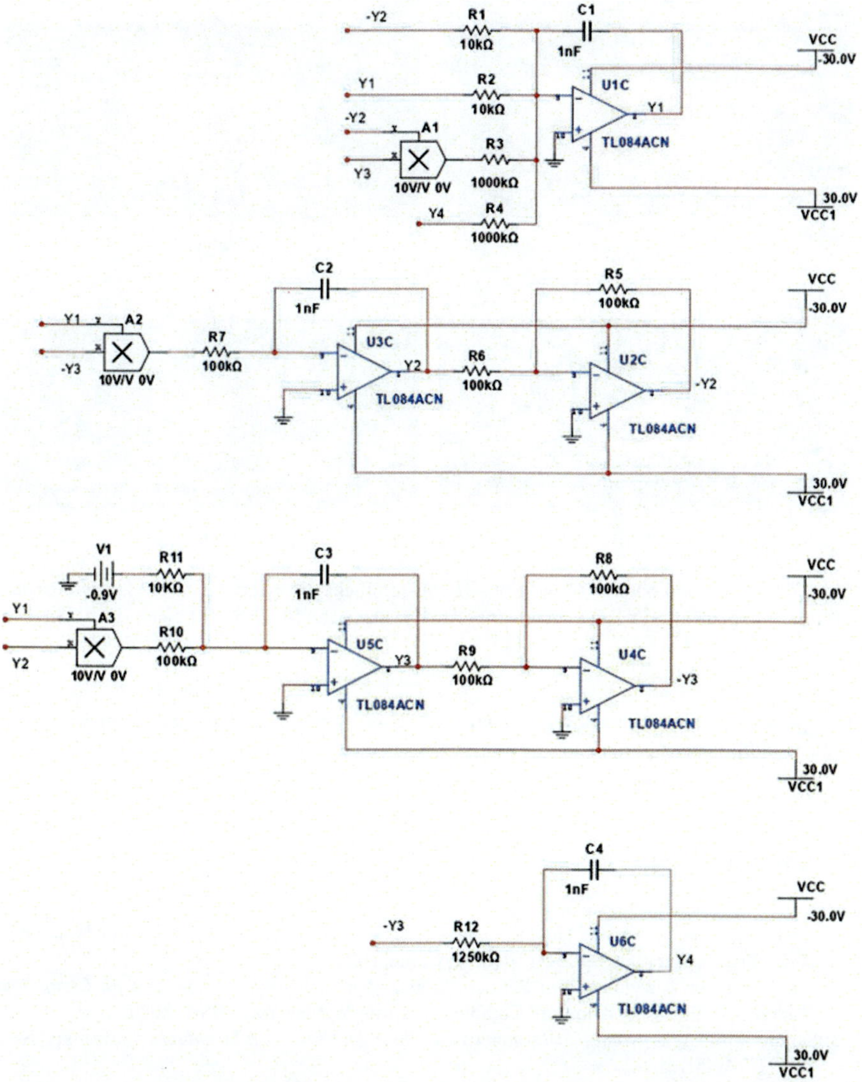

Fig. 17.10 Circuit design of the new 4-D two-scroll highly chaotic system (17.7). Note: High resolution figure is made available online

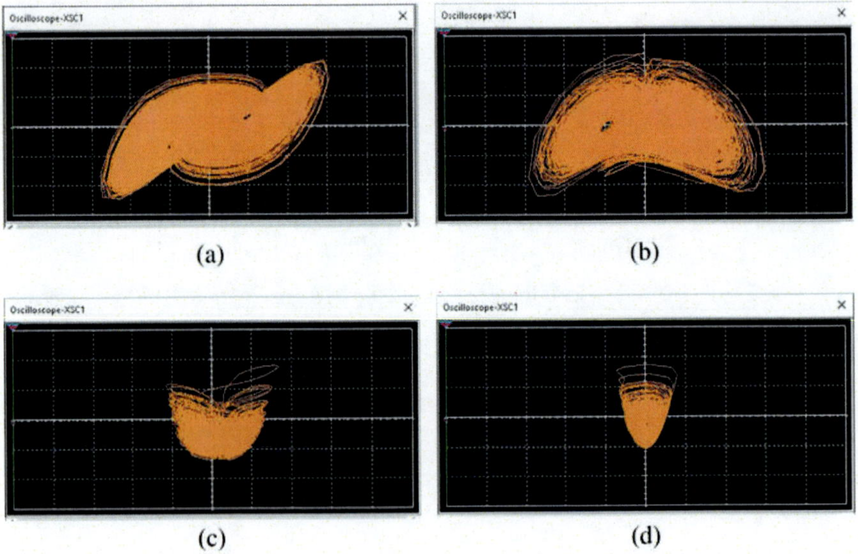

Fig. 17.11 MultiSim outputs of the new 4D two-scroll chaotic system (17.31) via oscilloscope. (**a**) (Y1, Y2) plane, (**b**) (Y1, Y3) plane, (**c**) (Y1, Y4) plane and (**d**) (Y3, Y4) plane

Competing Interests The authors have no conflicts of interest to declare that are relevant to the content of this chapter.

References

1. Kehui Sun, Shaobe He, and Huihai Wang. *Solution and Characteristic Analysis of Fractional-Order Chaotic Systems*. Springer, Beijing, China, 2023.
2. Marat Akhmet, Mechmet Onur Fen, and Ejaily Milad Alejaily. *Dynamics with Chaos and Fractals*. Nonlinear Systems and Complexity. Springer, New York, USA, 2020.
3. Sandro Wimberger. *Nonlinear Dynamics and Quantim Chaos: An Introduction*. Springer, New York, USA, 2016.
4. Jingzhuo Wang, Jing Chen, Hongqi Li, Haile Xue, and Zhizhen Xu. The roles of chaos seeding and multiple perturbations in convection-permitting ensemble forecasting over southern china. *Weather and Forecasting*, 38(9):1519–1537, 2023.
5. Feng Li. Incorporating fractional operators into interaction dynamics of a chaotic biological model. *Results in Physics*, 54, 2023. Article ID 107052.
6. Lorin Weiner and Janice L. Brissette. Finding meaning in chaos: a selection signature for functional interactions and its use in molecular biology. *FEBS Journal*, 290(16):3914–3927, 2023.
7. Sajan, Kapil Kumar Choudhary, and Balram Dubey. A non-autonomous approach to study the impact of environmental toxins on nutrient-plankton system. *Applied Mathematics and Computation*, 458, 2023. Article ID 128236.
8. Sivasamy Ramasamy, David Banjerdpongchai, and Poogyeon Park. A review of chaos control strategies for tri-trophic food chain ecological systems. *Engineering Journal*, 27(9):39–54, 2023.

9. Roland Pulch and Adrian Sandu. Rosenbrock-W methods for stochastic Galerkin systems. *Journal of Computational and Applied Mathematics*, 438, 2024. Article ID 115527.
10. Tianxiao Cao and Yuda Hu. Dual-mode coupling resonance and dynamic stability of axially moving ferromagnetic thin plate strips in alternating magnetic field. *Applied Mathematics and Computation*, 464, 2024. Article ID 128408.
11. Ibrahim Fennouh, Said Sadoudi, Djamal Teguig, and Camel Tanougast. Fpga implementation of orthogonal hyperchaotic sequences generator based on the cnn: application in multi-user chaotic communications. *Physica Scripta*, 98(10), 2023. Article ID 105231.
12. Qun Chen, Bo Li, Wei Yin, Xiaowei Jiang, and Xiangyong Chen. Bifurcation, chaos and fixed-time synchronization of memristor cellular neural networks. *Chaos, Solitons and Fractals*, 171, 2023. Article ID 113440.
13. Saeed Amiri, Seyed Mohsen Seyed Moosavi, Mehdi Forouzanfar, and Ebrahim Aghajari. Fuzzy composite nonlinear feedback sliding mode control for synchronization of the chaotic rikitake system subject to actuator saturation. *Iranian Journal of Science and Technology - Transactions of Electrical Engineering*, 47(4):1491–1508, 2023.
14. Van Nam Giap. Text message secure communication based on fractional-order chaotic systems with Takagi-Sugeno fuzzy disturbance observer and sliding mode control. *International Journal of Dynamics and Control*, 11(6):3109–3123, 2023.
15. Heping Wen and Yiting Lin. Cryptanalysis of an image encryption algorithm using quantum chaotic map and DNA coding. *Expert Systems with Applications*, 237, 2024. Article ID 121514.
16. Abdurrahim Toktas, UÄur Erkan, Suo Gao, and Chanil Pak. A robust bit-level image encryption based on Bessel map. *Applied Mathematics and Computation*, 462, 2024. Article ID 128340.
17. J.D. Annan and J.C. Hargreaves. Efficient parameter estimation for a highly chaotic system. *Tellus A: Dynamic Meteorology and Oceanography*, 56(5):520–526, 2004.
18. S. Vaidyanathan. A 3-D novel highly chaotic system with four quadratic nonlinearities, its adaptive control and anti-synchronization with unknown parameters. *Journal of Engineering Science and Technology Review*, 8(2):106 – 115, 2015.
19. Mohammad Ebrahim Aghili, Khosro Khandani, and Majid Parvizian. A highly chaotic fractional-order system with a four-wing attractor and its synchronization. *International Journal of Dynamics and Control*, 10:1199–1207, 2022.
20. J. C. Sprott. Some simple chaotic flows. *Physical Review E*, 50:647–650, 1994.
21. Haitao Qiu, Xuemei Xu, Zhaohui Jiang, Kehui Sun, and Can Cao. Dynamical behaviors, circuit design, and synchronization of a novel symmetric chaotic system with coexisting attractors. *Scientific Reports*, 13(1), 2023. Article ID 1893.
22. Liang Yang and Qiang Lai. Construction and implementation of discrete memristive hyperchaotic map with hidden attractors and self-excited attractors. *Integration*, 94, 2024. Article ID 102091.
23. Rameshbabu Ramar. Design of a new chaotic system with sine function: Dynamic analysis and offset boosting control. *Chaos Theory and Applications*, 5(2):118–126, 2023.
24. Nune Pratyusha and Santanu Mandal. Design and implementation of a novel circuit-based memristive non-autonomous hyperchaotic system with conservative and offset boosting for applications to image encryption. *Circuits, Systems, and Signal Processing*, 42(7):3812–3834, 2023.
25. Ronilson Rocha and Rene Orlando Medrano-T. Stability analysis of the Chua's circuit with generic odd nonlinearity. *Chaos, Solitons and Fractals*, 176, 2023. Article ID 114112.
26. Zhengkang Li and Xingbo Liu. Singular perturbation analysis in a coupled Chua's circuit with diffusion. *Chaos*, 33(10), 2023. Article ID 0152679.
27. Ndolane Sene. On the modeling and numerical discretizations of a chaotic system via fractional operators with and without singular kernels. *Mathematical Sciences*, 17(4):517–537, 2023.
28. Qiang Lai and Zhijie Chen. Dynamical analysis and finite-time synchronization of grid-scroll memristive chaotic system without equilibrium. *Chaos, Solitons and Fractals*, 176, 2023. Article ID 114118.

Index

E. Campos-Cantón et al. (eds.), *Complex Systems and Their Applications*,
https://doi.org/10.1007/978-3-031-51224-7